Joining of Advanced and Specialty Materials VI

Proceedings from Materials Solutions 2003 on
Joining of Advanced and Specialty Materials
13–15 October 2003
Pittsburgh, Pennsylvania

Editors
T.J. Lienert
V. Acoff
J.E. Indacochea
J.N. DuPont

Sponsored by

The Materials
Information Society

ASM International®
Materials Park, Ohio 44073
www.asminternational.org

First printing, May 2004

ISBN: 0-87170-801-9
SAN: 204-7586

ASM International®
Materials Park, OH 44073-0002
www.asminternational.org

Printed in the United States of America

Cover image courtesy W. Zhang, C.-H. Kim, and T. DebRoy, The Pennsylvania State University, University Park, Pennsylvania

Multiple copy reprints of individual articles are available from Technical Department, ASM International.

Preface

Development of new advanced and specialty materials continues at a fast pace. However, the application of many of these materials is limited by the ability to join them to similar and dissimilar materials. Research aimed at efficiently joining these materials must push forward at an equally fast rate to realize their ultimate potential.

The Sixth International Conference on Joining of Advanced and Specialty Materials held during Materials Solutions 2003 in Pittsburgh, Pennsylvania (October 13–15, 2003) marked our most successful symposium to date based on the number of presentations and sessions. The sessions covered a wide variety of topics including friction stir welding, solid-state processing, brazing and soldering, practical aspects of joining, TLP/interfacial reactions, NDT of weldments, fusion welding processes and micro-joining/lasers. The goal of the symposium was to provide a forum for researchers to present recent developments in joining of advanced materials.

The success of the symposium derived from the hard work of many people. We would like to thank the symposium organizers for their contributions. We would like to express also our gratitude to the rest of the organizing committee as well as the session chairs for their time and effort. Most importantly, we would like to extend our appreciation to the presenters for participating.

T.J. Lienert
V. Acoff
J.E. Indacochea
J.N. DuPont

Program Organizers and Editorial Committee

Thomas J. Lienert
Los Alamos National Laboratory
Los Alamos, New Mexico

Viola L. Acoff
The University of Alabama
Tuscaloosa, Alabama

J. Ernesto Indacochea
University of Illinois at Chicago
Chicago, Illinois

John N. DuPont
Lehigh University
Bethlehem, Pennsylvania

Contents

Keynote Address

Session Chair: T.J. Lienert, Los Alamos National Laboratory, Los Alamos, New Mexico

Computer Modeling –An Emerging Tool for Welding Engineers to Achieve Structurally Sound and Reliable Welds 1
T. DebRoy, The Pennsylvania State University, University Park, Pennsylvania

Friction Stir Welding

Session Chair: T.J. Lienert, Los Alamos National Laboratory, Los Alamos, New Mexico

Experimental and Numerical Study of Thermal Process in Friction Stir Welding 10
J. Xu, S.P. Vaze, R.J. Ritter, K.J. Colligan, J.R. Pickens, Concurrent Technologies Corporation, Johnstown, Pennsylvania

Friction Stir Welding of Copper Plates 20
A. Polar, Universidad Nacional de San Agustin, Arequipa, Peru
T. Shah, J.E. Indacochea, University of Illinois, Chicago, Illinois

Friction Stir Welding of DH-36 Steel 28
T.J. Lienert, Los Alamos National Laboratory, Los Alamos, New Mexico

Friction Stir Welding of Copper Canisters – A Weld that Lasts for 100,000 Years 35
L. Cederqvist, Swedish Nuclear Fuel and Waste Management Company (SKB), Oskarshamn, Sweden
R.E. Andrews, The Welding Institute (TWI), Cambridge, United Kingdom

Solid-State Processing

Session Chair: V. L. Acoff, The University of Alabama, Tuscaloosa, Alabama

The Effect of Rolling Strain on the Processing of Intermetallic Sheet Materials Using Roll Welding and Reaction Annealing 40
G.P. Chaudhari, V.L. Acoff, The University of Alabama, Tuscaloosa, Alabama

Brazing and Soldering

Session Chair: R.W. Smith, Materials Resources International, Lansdale, Pennsylvania

Brazing Using 2.45 GHz Microwaves 45
W.M. Drake, D.D. Kautz, R.U. Vaidya, P.W. Hochanadel, V.R. Dave, Los Alamos National Laboratory, Los Alamos, New Mexico

Thermal Cyclability of Reactive Air Braze Seals in Anode Supported Solid Oxide Fuel Cells 51
J.S. Hardy, C.A. Coyle, J.C. Birnbaum, K.S. Weil, Pacific Northwest National Laboratory, Richland, Washington,
J.T. Darsell, Washington State University, Pullman, Washington

Practical Aspects of Joining

Session Chair: J.E. Gould, Edison Welding Institute, Columbus, Ohio

Developments in Friction Stir Welding 59
W.M. Thomas, The Welding Institute (TWI), Cambridge, United Kingdom
G. Sylva, MTS Corporation, Eden Prairie, Minnesota

TLP/Interfacial Reactions

Session Chair: J.J. Stephens, Sandia National Laboratories, Albuquerque, New Mexico

Hot-Stage Light Microscopy Studies of the Wettability of Nickel-Base Superalloys during Transient Liquid Phase Bonding 67
L.N. Payton, S.V. Chitti, D.R. Taarea, N.I. Sofyan, W.F. Gale, D.A. Butts, R. Aluru, R.D. Love, Auburn University, Auburn, Alabama

Reactions Between Constituents of Two Molten Nickel-Base Superalloys and Ceramic Materials 71
D.R. Taarea, W.F. Gale, J.W. Fergus, Auburn University, Auburn, Alabama

The Influence of Ceramic and Metallic Substrates on the Oxidation Behavior of Gold ABA 76
K.S. Weil, J.P. Rice, Pacific Northwest National Laboratory, Richland, Washington

Microstructural Development and Relationship to Mechanical Properties in Polycrystalline to Single Crystal Nickel-Base Superalloy Transient Liquid Phase (TLP) Bonds 85
L.N. Payton, R. Aluru, W.F. Gale, N.I. Sofyan, V.G. Krishnardula, D.A. Butts, S.V. Chitti, T. Zhou, Auburn University, Auburn, Alabama

NDT of Weldments

Session Chair: M. Moles, R/D Tech, Mississauga, Ontario, Canada

Nondestructive Evaluation for Characterization of Welding Defects and Development of Welding Technology 91
N. Meyendorf, University of Dayton, Dayton, Ohio

Comprehensive Inspection of Friction Stir Welds Using Arrays 96
A. Lamarre, O. Dupuis, M. Moles, R/D Tech, Mississauga, Ontario, Canada

Nondestructive Inspection Techniques for Friction Stir Weld Verification on the Space Shuttle External Tank 104
M.W. Suits, J. Leak, NASA Marshall Space Flight Center, Huntsville, Alabama
D. Cox, Lockheed Martin Michoud Space Systems, Huntsville, Alabama

Ultrasonic Phased Arrays for Thick Section Weld Inspections 112
M. Moles, R/D Tech, Mississauga, Ontario, Canada
N. Dubé, F. Jacques, R/D Tech, Québec, Canada

Fusion Welding Processes

Session Chair: T.J. Lienert, Los Alamos National Laboratory, Los Alamos, New Mexico

Effect of Purging Gas Oxygen Levels on Structure and Mechanical Properties of GTA Welded Type 304 Stainless Sanitary Tube 119
L. Li, Utah State University, Logan, Utah
T. Davis, AWD Sales & Service, Inc., Payson, Utah

Plasma Welding of NiTi to NiTi, Stainless Steel and Hastelloy C276 125
C. van der Eijk, H. Fostervoll, Z.K. Sallom, O.M. Akselsen, SINTEF Materials Technology, Trondheim, Norway

Theoretical Models to Describe Properties of Magnesium Alloy Fusion Welds 130
M. Marya, G.R. Edwards, D.L. Olson, Colorado School of Mines, Golden, Colorado

Investigation of the Fusion Zone Microstructure for Gas Tungsten Arc Welded Gamma Titanium Aluminide 142
M.F. Arenas, V.L. Acoff, The University of Alabama, Tuscaloosa, Alabama

Micro-Joining and Laser Welding

Session Chair: Y. Zhou, University of Waterloo, Waterloo, Ontario, Canada

Bonding Mechanism in Gold Wire Crescent Bonding 150
Y. Zhou, X. Li, N. Noolu, I. Lum, University of Waterloo, Waterloo, Ontario, Canada

Electrode Tip Life Improvement in Resistance Microwelding of Nickel-Plated Steel Thin Sheets 159
N. Scotchmer, Huys Industries Limited, Weston, Ontario, Canada
S.J. Dong, Hubei Automotive Industries Institute, Shiyan, Hubei, People's Republic of China
Y. Zhou, University of Waterloo, Waterloo, Ontario, Canada

Proceedings from Joining of Advanced and Specialty Materials, 13–15 October 2003, Pittsburgh, Pennsylvania, ASM International, 2004

Computer Modeling – An Emerging Tool for Welding Engineers to Achieve Structurally Sound and Reliable Welds

T. DebRoy
Department of Materials Science and Engineering
The Pennsylvania State University, University Park, PA, USA

Abstract

Welded joints contain spatial variations of composition, structure and properties resulting from heat transfer, flow of molten metal in the weld pool and solid state transformations during welding. Computational models, based on scientific principles, help in quantitative understanding of complex welding processes and welded materials and provide solutions for a variety of welding problems. Numerical models of heat transfer, fluid flow and thermal cycles can now be relied upon to provide fusion zone geometry, temperature fields and thermal cycles with a fair degree of confidence. Significant progress has also been made in building a quantitative understanding of various structural features such as phase composition, grain structure and inclusion structure, mostly for several relatively simple alloys.

As progress is made through the application of existing models, advances in computational hardware and software are now providing unprecedented opportunities to build more ambitious, computationally more intensive and more reliable models. An important recent development has been the construction of heat transfer and fluid flow models that can determine uncertain welding variables from a limited volume of experimental data using multivariable optimization. Predictions from these models are more reliable because they learn from experimental data. Since the computed results reveal detailed insight about the welding processes and welded materials that cannot be obtained otherwise, the modeling efforts are already contributing to the growing quantitative knowledge base in fusion welding. The expanded knowledge base is necessary to transform welding to a mainstream engineering branch.

Introduction

In recent decades, new welding processes such as the friction stir welding and advanced technology such as computer aided welding process control have found increased acceptance in the industry. However, our past success cannot be relied upon as insurance for continued good health and prosperity of our field, since we live in an era of unprecedented technological change. So, what can we do now to maintain and enhance the vitality of welding technology? Certainly, innovations cannot be organized or predicted because of their very nature. However, research and development (R&D) can be organized to provide less radical, but more predictable technological solutions. In practice, reliable welds are often fabricated by trial and error without the use of any organized research. So, does it make sense for the industry to invest in organized R&D? In many instances, systematic R&D can be very useful in seeking solutions where traditional empirical methods fail. For example, the trial and error solutions ignore the potential competitive advantage that is attainable through understanding of the underlying science. Furthermore, multiple work-pieces are not always available. Finally, sometimes the value of the work piece is considerably enhanced prior to welding, making trial and error paradigm expensive.

Welded joints typically contain spatial gradients of composition, structure and properties that are affected by processes such as the heat transfer, flow of molten metal in the weld pool and solid state transformations during welding. Organized R&D seeks to better understand these complexities and control characteristics of welded materials and processes.

As a result of organized R&D, there is now a growing quantitative knowledge base in welding. Can this knowledge base provide solutions of today's welding problems? The answer depends on the specific solution desired. It is fair to say that it is already providing solutions for a variety of problems[1]. Computer models of heat transfer, fluid flow and thermal cycles can now be relied upon to provide fusion zone geometry, temperature fields and thermal cycles with a fair degree of confidence. Progress has also been made in quantitative understanding of various structural features such as phase composition, grain structure and inclusion structure, mostly for several relatively simple alloys. The goal of this paper is to review a selection of examples where computational modeling has made a significant difference. These examples show that phenomenological modeling can provide insight that cannot be achieved otherwise and the modeling effort can be used to improve weld quality.

1. Weld Metal Geometry

1(a) Role of Surface Active Elements

Beneficial effects of surface active elements, such as sulfur and oxygen, in improving weld penetration have been known for well over a decade.[2] However, in many cases, the presence of sulfur has not actually resulted in the expected high depth of penetration. For example, cross sections of steel welds containing 20 and 150 ppm of sulfur and laser welded under different powers[3] are shown in Fig. 1(a-d). The weld pool geometry in each steel is similar in appearance with a laser power of 1900 W. However, when the samples are welded at a laser power of 5200 W, the weld containing 150 ppm sulfur has a much greater depth of penetration than that containing 20 ppm sulfur. Thus, the concentration of sulfur may or may not have a significant effect on the weld geometry for the given range of welding variables.

Why didn't the higher sulfur concentration improve penetration at 1900 W? In general, for counterintuitive results, possible lack of reproducibility of the data cannot be ruled out. However, in this investigation, over eighty experiments were carefully conducted and the results were reproducible. These results lead to an important question: how can we predict and control weld penetration?

The answer to this question requires discussion of numerically computed results and is deferred until the end of this section. We start by examining the relative magnitudes of heat transfer by convection and conduction within the weld pool, which is expressed by the Peclet number for heat transfer, Pe:

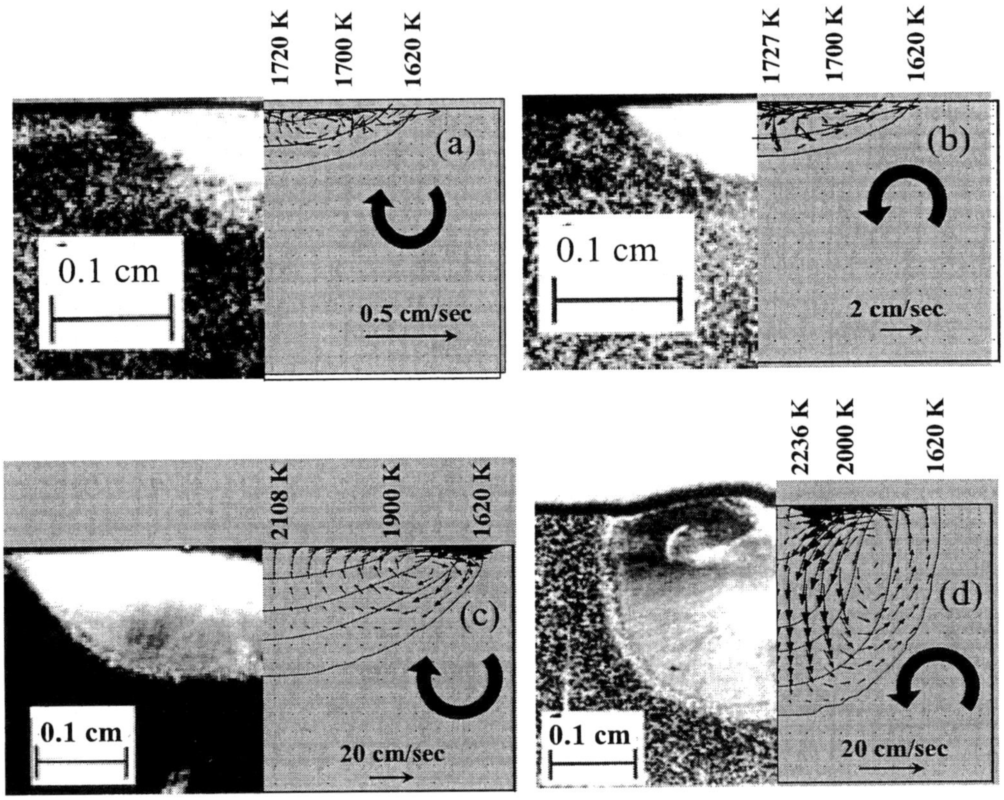

Fig. 1 Comparison between the computed and experimental weld pool geometry[3] at a laser power of 1900 W for steels containing (a) 50 ppm and (b) 150 ppm sulfur and at a laser power of 5200 W for steels containing (c) 50 ppm and (d) 150 ppm sulfur.

$$Pe = \frac{uL}{\alpha} \quad (1)$$

where u is the velocity, α is the thermal diffusivity of the liquid metal, and L is the characteristic length that can be taken as the depth of the weld pool. In the calculation of Pe, the value of the velocity is required. Fig. 1 shows the computed velocity and temperature fields and the weld geometries.[3] These calculations are based on the solution of the equations of conservation of mass, momentum, and energy in a transient, two-dimensional, axi-symmetric form. As shown in each case in Fig. 1, the computed weld geometry agrees well with the corresponding experimental geometry.

At a laser power of 1900 W, the peak temperatures reached on the weld pool surface for both experiments are about 1720 K. The relatively low temperature gradients on the weld pool surface lead to low surface velocities. The maximum values of the Peclet number for the steels with 20 and 150 ppm sulfur are 0.18 and 0.91, respectively. These low values of Pe (<1) indicate that heat transfer by conduction is more important than that by convection. As a result, the direction of fluid flow is not important in determining the melting of the base metal and the weld pool shape at this laser power. Consequently, there is no significant difference between the weld pool geometries for steels containing 20 and 150 ppm sulfur. In contrast, at a laser power of 5200 W, the calculated peak temperatures are higher than 2100 K, which produce much more significant temperature gradients and higher surface velocities. The computed Peclet numbers are large (>200), making convective heat transport the primary mechanism for heat transfer.[3] Thus, the fluid flow field and the direction of convective heat transport have a pronounced effect on the weld pool geometry.[3]

The direction of the fluid flow within the weld pool depends on the spatial gradient of interfacial tension, which is the product of the spatial gradient of temperature and the temperature gradient of surface tension, $d\gamma/dT$. For a sulfur content of 20 ppm, the temperature gradient of surface tension[4,5] is negative at all temperatures. The negative values of $d\gamma/dT$ over the weld pool surface result in radially outward flow over the entire weld pool surface and a shallow weld pool, as shown in Figs. 1(a) and 1(b). When the sulfur content is 150 ppm, $d\gamma/dT$ is positive, which results in a radially inward flow and the convective heat transport in the downward direction near the heat source results in deep pools, as observed in Fig. 1(d). Since $d\gamma/dT$ is negative at temperatures higher than 1980 K for the steel containing 150 ppm sulfur, it generates a radially outward secondary flow in a small region near the middle of the pool. However, this secondary flow is mild and a deep weld pool is obtained at 150 ppm sulfur because of the dominant radially inward flow and high Peclet number.

Only when convection is the dominant mechanism of heat transfer can surface active elements play an important role in enhancing weld penetration. This example shows that numerical calculations of heat transfer and fluid flow can provide detailed insight into welding processes that cannot be achieved otherwise. A well-tested mathematical model, that considers the role of surface active elements such as oxygen and sulfur, can accurately predict weld pool geometry.

1(b) Modeling of "Finger" Penetration in GMA Welding

Weld pools formed during gas metal arc (GMA) welding display a unique shape not commonly observed in other welding processes. These weld pools are characterized by a much deeper penetration,[6] resembling a finger, and is shown in Fig. 2. This feature is the result of droplets detaching from the consumable electrode and impinging on the weld metal surface. These droplets aid in the transport of heat, supplementing the energy transfer from the arc. Knowledge of the weld penetration is useful in developing welding procedures to achieve reliable welds without excessive trial and error.

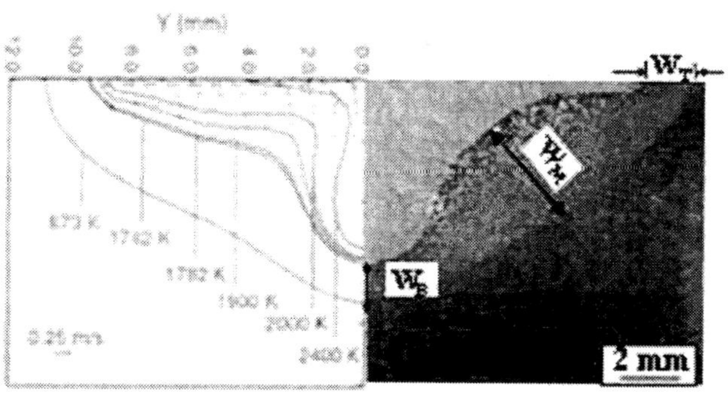

Fig. 2 Comparison of the calculated and experimental geometry of the fusion zone and the HAZ at a heat input per unit length of 2.4 kJ/mm. The welding conditions are defined elsewhere.[6] The symbols W_T, W_M, and W_B are the widths at the top, middle, and bottom of the HAZ.[6]

As an example, the evolution of the weld pool geometry during the GMA welding of HSLA-100 steel is examined using the fundamentals of transport phenomena. A three-dimensional turbulent heat transfer model is used here to solve for the equations of conservation of mass, momentum, and energy. In GMA welding, a high level of agitation in the weld pool is aided by both the large mean velocities present in the relatively small weld pool and the impact of the metal droplets from the consumable electrode. Experimental observations[7] and theoretical calculations[8,9] in gas tungsten arc (GTA) welding have indicated that the fluid flow is turbulent in nature. The widely used K-ε model[10] has been adopted here to simulate turbulent heat transfer and fluid flow in the weld pool.

The difficulty encountered in the simulation of the GMA welding process involves the transfer of metal drops from the consumable electrode to the weld pool and the resulting heat and momentum transfer. These droplets deliver their excess heat to the molten weld pool and are responsible for the finger penetration observed during the GMA welding process. The heat transfer from the metal droplets has been simulated by considering the existence of a cylindrical volumetric heat source in the weld pool,[6] whose dimensions are determined by the drop size, shape, velocity, and frequency. The procedure for the calculation of the volumetric heat source is available in the literature.[11]

An example of the computed temperature and velocity fields in three dimensions considering turbulent transfer of heat and momentum[6] is shown in Fig. 3. The general features of the calculated temperature fields are consistent with the expected results. For example, temperature gradients are greater in front of the heat source than behind it. These temperature gradients in front of the heat source result in slightly greater liquid metal velocities here than behind the heat source. The computed weldment geometry[6] is compared with the experimental weld pool shape and size in Fig. 2. It is observed that the width of the fusion zone, the depth of the finger penetration, and the shapes of both the fusion zone and HAZ predicted by the model agree with the corresponding experimental results. Furthermore, "finger" penetration can be satisfactorily predicted using this model. The modeling effort, apart from providing a detailed insight about the physical processes in GMA welding, can reduce the volume of experiments needed to develop welding protocol for a given application.

The modeling efforts can provide geometric information in both linear[12-15] as well as spot welding,[16-19] in complex geometries[20-22] such as fillet welding and in systems with considerable surface deformation.

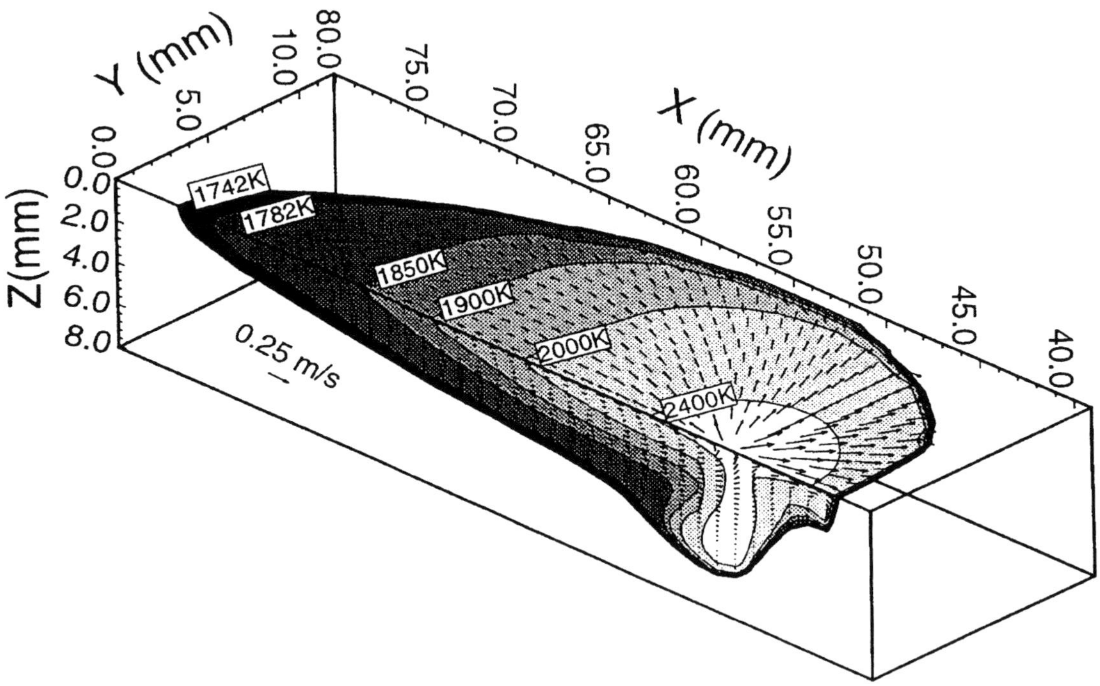

Fig. 3 Calculated temperature and velocity fields in GMA welded HSLA-100 steel with a welding current of 405 A, a welding voltage of 31.3 V, a travel speed of 5.29 mm/s and no preheat.[6]

2. Quantitative Understanding of Weldment Structure

Quantitative prediction of the role of welding variables on the weld metal structure is now possible for many important metals and alloys. For example, modeling can be useful in revealing the phase transformation mechanisms in commercially pure titanium.[24] Modeling is also applied to understand the evolution of grain structure under steep temperature gradients during welding of titanium and its alloy. Computer modeling has also been used to simulate the inclusion composition and size distribution in the weld metal considering the nucleation, growth, and dissolution rates of inclusions.

2(a) Understanding Mechanism of Phase Transformations

Phase transformation rates can be expressed by the Johnson-Mehl-Avrami (JMA) equation when the transformations are controlled by the classical nucleation and growth mechanism. The JMA equation can be modified for non-isothermal kinetics, and the detailed procedure to derive the modified JMA equation is available in the literature.[23] This method uses a discretization of the time dependence of temperature by approximating the continuous temperature-time curves by many isothermal steps. The final expression for the transformed fraction obtained from m subsequent isothermal steps is expressed in the following relationship:

$$f(t(T)) = 1 - \exp\left\{-\left[\sum_{l=0}^{m-1} k(T_o + l\Delta T)\Delta t\right]^n\right\} \qquad (2)$$

where f(t(T)) is the transformed phase fraction under non-isothermal conditions, Δt is the time step, ΔT is the temperature change corresponding to each Δt, T_o is the initial temperature, k is a time dependent constant, and m is the total number of the isothermal steps. The parameters, k and n, depend on the mechanism of the phase transformation.

During welding of commercially pure titanium,[24] the fusion zone undergoes a transformation from the α to the β to the liquid and back through the β to the α phase. The heat affected zone (HAZ) goes from the α to the β and back to the α phase. The extent of the phase transformation in the HAZ and the times for the completion of the $\alpha \rightarrow \beta$ transition for each assumed transition mechanism can be calculated from the knowledge of the thermal cycles. These calculated times for each mechanism are then compared with the times determined from the experimental phase distribution map in Fig. 4. There is good agreement between the times taken from the phase distribution map and those calculated assuming that the $\alpha \rightarrow \beta$ transition is controlled by the transport of Ti atoms across the interface and it may be assumed here that the $\alpha \rightarrow \beta$ transition is primarily controlled by the transport of Ti atoms across the interface (mechanism I).[24] The use of mathematical modeling here produces knowledge of phase transformation mechanism, which can not be achieved with experiments alone.

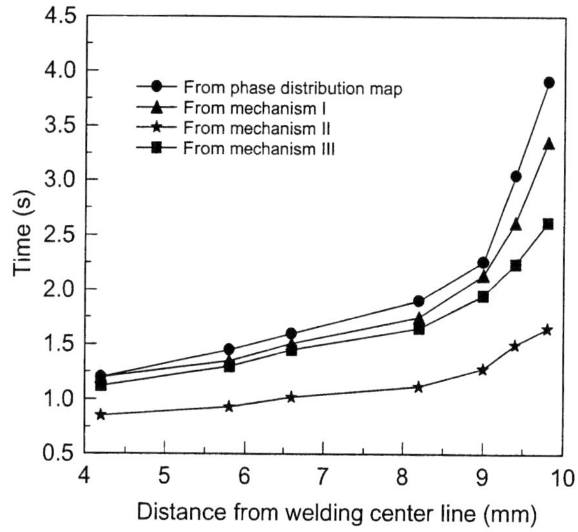

Fig. 4 Comparisons of times need for the $\alpha \rightarrow \beta$ transformation from the JMA equation and from the phase distribution map. Different mechanisms are assumed for the application of the JMA equations are discussed elsewhere.[24] Mechanism I: Ti short range diffusion, II: oxygen long range diffusion and III: oxygen short range diffusion.

2(b) Modeling of Grain Growth in the HAZ

In recent years, modeling of grain growth has greatly aided our understanding of grain structure evolution during welding. In the past, Monte Carlo (MC) techniques have been used to simulate the evolution of grain structure under isothermal conditions[25-27] and recently the technique has been applied to the HAZ of welds[24,28-30] where the steep temperature gradients near the fusion zone have been taken into account. By incorporating the calculated thermal history data obtained from a well-tested heat transfer and fluid flow model, the simulated grain growth kinetics at each site can be calculated. The simulated final grain structure in the HAZ[24] of titanium is shown in Fig. 5. Significant spatial grain size gradients are observed in the simulated HAZ structure. At sites closer to the fusion line, the grains are coarser,

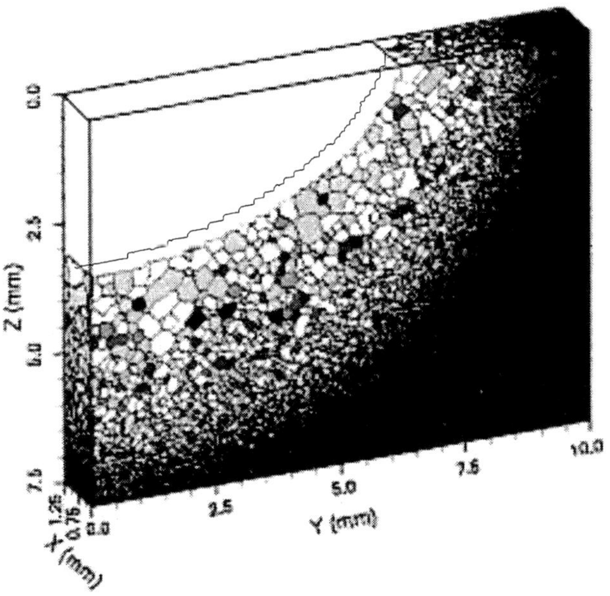

Fig. 5 Three dimensional grain structure of the HAZ in a GTA welded titanium weldment computed using 3D temperature field from the heat transfer and fluid flow model.[24]

since the grain size change depends on both the temperature and time period for grain growth. It is also observed that grain growth on the top surface of the HAZ (line ab) is less than that observed in the vertical direction (line cd). The calculated and experimental grain sizes were found to be comparable.[24] In recent years, advancements in computational modeling have provided significant insight into the evolution of structure in welded materials. The feasibility of calculating the evolution of grain growth in three dimensions, demonstrated by the example presented here, represents a new tool for quantitative understanding of the structure of welded materials.

2(c) Modeling of Inclusion Characteristics

The rate of precipitation of inclusions from a supersaturated solution can be determined from the rate of diffusion of their constituent elements to the inclusion surface where the growth takes place.[31,32] In the weld pool, the inclusions experience considerable recirculatory motion[33,34] and strong temperature gyrations along their path. It has been recently shown that[28,29] our current knowledge of heat transfer and fluid flow in the weld pool can be utilized to understand the growth and dissolution of solid inclusions in the weld pool. The loci and the temperature gyrations experienced by a large number of inclusions have been calculated numerically to seek a statistically meaningful residence time distribution for the inclusions and the number and the intensities of the temperature cycles experienced by these particles.

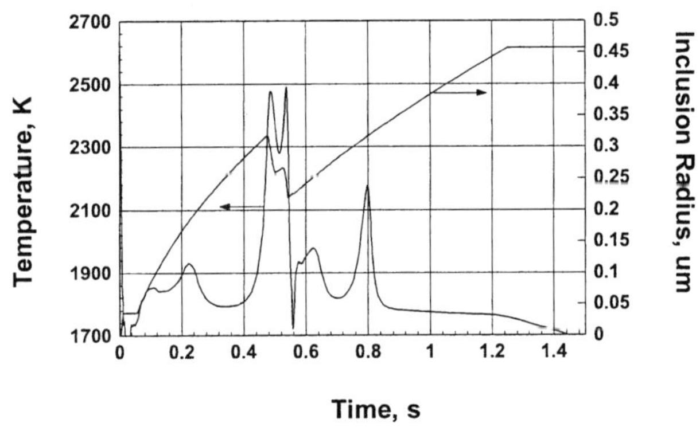

Fig. 6 A typical temperature gyration experienced by an inclusion and the corresponding size change in the weld pool.[34]

A typical computed temperature versus time plot and the corresponding Al_2O_3 inclusion radius versus time plot are shown in Fig. 6 for the SMA welding of a low alloy steel.[34] It can be seen that these inclusions begin to dissolve above about 2210 K and grow again when the temperature falls below this level. Its radius reaches about 1 µm before it is lodged in the solidified weld metal. Calculations show that the majority of the inclusions undergo some dissolution within the weld pool during their life cycle. The coarsening of the inclusions is facilitated by the longer times available at low welding speeds. Furthermore, the size distribution curves become flatter when the welding velocity decreases. The computed and measured size distributions are presented in Fig. 7 in the next page as a plot of radius versus percentage of particles. Good agreement[34] between the computed and the measured values indicates that important aspects of weld metal structure can be understood from the fundamentals of transport phenomena and kinetics.

3. Toward A More Reliable Model of Heat Transfer and Fluid Flow

Although the preceding examples demonstrate that the heat transfer and fluid flow calculations can reveal important features of welding processes and welded materials that cannot be obtained otherwise, the current models suffer from a major problem. The model predictions of weld geometry and cooling rates do not always match the corresponding experimental results. This vulnerability now precludes more widespread use of the transport phenomena based powerful computational models in welding.

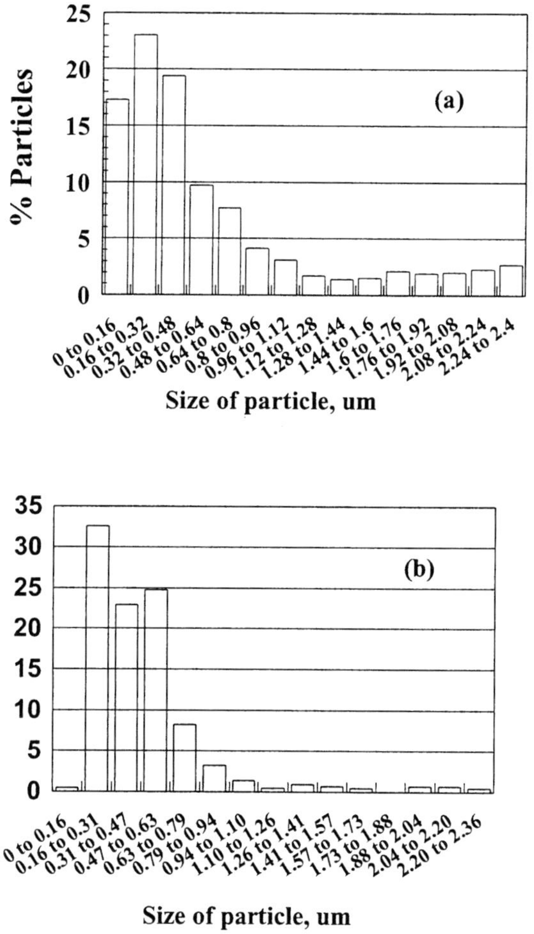

Fig. 7 (a) Computed inclusion size distribution.[34] (b) Experimental inclusion size distribution in a controlled experiment.[34]

Why don't the model predictions always match the experimental results perfectly? The model embodies the fundamental equations of conservation of mass, momentum and energy and the boundary conditions are fairly straightforward in most cases. A minimum requirement for any serious modeling effort is that the computer program is adequately debugged. So the numerical solution of the equations of conservation of mass, momentum and energy is accomplished with adequate fidelity. Because the model uses fundamental equations of physics as governing equations and straightforward boundary conditions, the source of discrepancy between the experimental results and the model predictions cannot be attributed to use of the transport phenomena based model. In contrast, some of the model input variables, such as the arc efficiency, cannot be prescribed from the fundamental principles. For example, although the arc efficiency can be experimentally measured, the measured values are dependent on the nature of the surface, the exact condition of welding such as the torch angle, the electrode tip angle and many other variables.[35] The discrepancy between the experimental results and the model predictions can be attributed, to a large extent, to the significant uncertainties in several model input variables. For the powerful numerical models to be useful to the welding engineers, the problem of uncertain welding parameters must be fully addressed in the model structure.

A potential solution is to formulate models[35,36] that can learn the values of those uncertain variables that cannot be prescribed from fundamental principles from a limited volume of experimental data. How can such a model be developed? Mapping of the entire window of welding variables to achieve a target weld geometry using the current fluid flow and heat transfer models and a trial and error procedure is impractical because of the large number of variables involved. Clearly, the search for the uncertain welding variables must be guided by an appropriate multivariable optimization scheme that ensures compliance with the basic laws of heat transfer and fluid flow and at the same time allows for the determination the values of the uncertain or unknown welding variables.

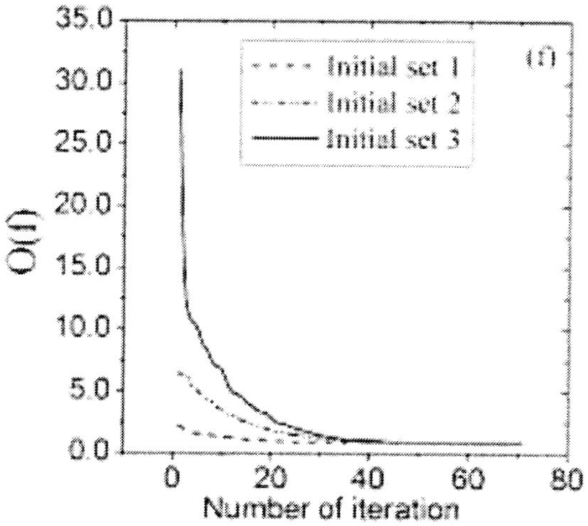

Fig. 8 Decrease of the objective function with the number of iterations indicating progress of the calculations. The results show that the value of the objective function after a certain number of iterations depend on the initial values of the uncertain parameters assumed.[35]

De and DebRoy[35,36] recently developed a methodology to determine uncertain welding parameters from a limited volume of experimental data using two interactive computational modules, one for the direct analysis of heat transfer and fluid flow in weldment and the other for the optimization of the uncertain welding variables. They[35] argued that the values of several parameters that affect calculation of heat transfer and fluid flow in fusion welding such as the arc efficiency, arc radius, and effective thermal conductivity were not always readily available. Following an inverse modeling approach, a method for the optimization of multiple unknown variables was developed within the framework of phenomenological laws that govern heat transfer and fluid flow in the weld pool. An objective function that represented the difference between the computed and the experimental depth and width of the fusion zone was minimized by Levenburg-Marquardt multi-variable optimization technique. The model could correctly learn values of multiple unknown parameters from only a few measurements and, based on the knowledge of these parameters, provide realistic predictions of heat transfer, fluid flow and weld geometry.[35] A set of experimental data on weld penetration and width for several welding conditions were used to estimate five unknown parameters, i.e., arc efficiency, arc radius, arc power distribution parameter, effective thermal conductivity and effective viscosity of the liquid weld metal. The optimization scheme required calculation of how the computed weld geometry was affected by these unknown parameters. The computed weld penetration was found to be sensitive to all these five parameters. The weld width was influenced mainly by arc efficiency and arc radius and, to a much lesser extent by other parameters. The optimum values of the five unknown parameters were independent of their initial guessed values, although the volume of numerical calculations depended on their initial values as shown in Fig. 8. The accuracy of the estimated set of the unknown parameters was verified using the numerical heat transfer and fluid flow model and the experimental data as shown in Fig. 9. Good agreement between the experimental data and the model predictions indicate the high reliability of these new models.

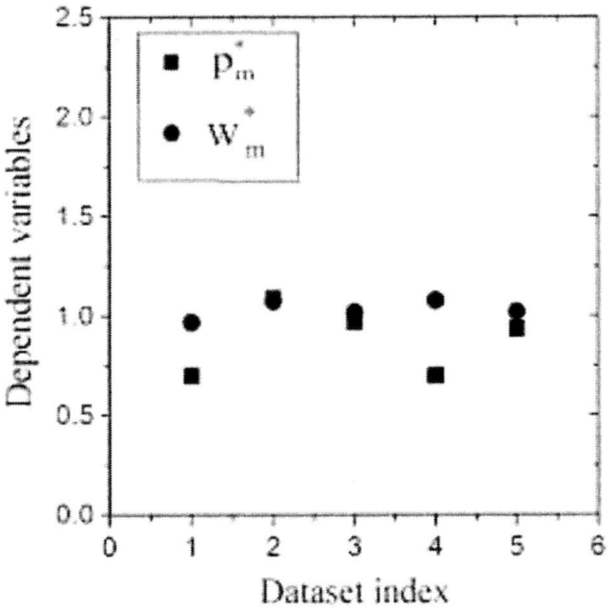

Fig. 9 Computed values of dimensionless penetration, p^*_m, and dimensionless width, w^*_m, using the optimized set of five unknown parameters. The dimensionless variables represent computed values divided by the respective experimental values.[35]

Concluding Remarks

In recent years, phenomenological modeling has provided significant insight into fusion welding processes and weldment characteristics which could not have been obtained solely through experiments. The examples presented in this paper demonstrate how significant insight can be achieved by using large phenomenological models that contain detailed description of important physical processes.

Comprehensive phenomenological models are not widely used by practicing welding engineers today since they are complex and computationally intensive. A practical solution for facilitating the use of numerical models is to develop relatively simple process control models, such as neural networks. These models generally do not consider any of the physical processes in welding, and as a consequence, are computationally simple and can be used in real time. However, development of these simple models requires an extensive data base. In addition to experimental results, some of the necessary data can be generated by the comprehensive phenomenological models. The large models can be used to calibrate, train, and verify, the simple, real-time, process control models. Thus, apart from providing insight into welding processes, phenomenological models are also potentially useful for the development of process control models. Improved user access to the large models at low cost would also be helpful in gaining wider acceptability of the models. As progress is being made through the use of the existing transport phenomena based computer models, new exciting models are also being built and tested taking advantage of the emerging powers of advanced computational hardware and software. An important new capability of the recent heat transfer and fluid flow models is the ability to learn values of uncertain welding parameters from a limited volume of experimental data through multivariable optimization. However, the higher reliability of these models comes from more intensive calculations.

In addition to being potentially more useful to practicing engineers, phenomenological models are a powerful tool for research today. Modeling of weldment geometry and structure, presented here, represents a small contribution to an expanding quantitative knowledge base in welding. Significant expansion of this knowledge base is necessary, if not essential, for transforming welding from almost an empirical art to an engineering science.

References

1. T. DebRoy and S.A. David, Reviews of Modern Physics, 1995, 67(1), 85.
2. C. R. Heiple and J. R. Roper, Welding Journal, 1982, 61, 92s.
3. W. Pitscheneder, T. DebRoy, K. Mundra, and R. Ebner, Welding Journal, 1996, 75(3), 71s.
4. P. Sahoo, T. DebRoy and M.J. McNallan, Metallurgical Transactions B, 1988, 19B, 483.
5. M. J. McNallan and T. DebRoy, Metallurgical Transactions B, 1991, 22B, 557.
6. Z. Yang and T. DebRoy, Metallurgical and Materials Transactions B, 1999, 30, 483.
7. M. Malinowski-Bodnicka, G. den Ouden and W.J.P. Vink, Welding. Journal, 1990, 69, 52s.
8. R.T.C. Choo and J. Szekely, Welding Journal, 1994, 73(2), 25s.
9. K. Hong, D.C. Weckman and A.B. Strong in Trends in Welding Research, H.B. Smartt, J.A. Johnson, and S.A. David, eds., ASM International, Materials Park, OH, 1996, 399.
10. B.E. Launder and B.D. Spalding, Mathematical Models of Turbulence, Academic Press, New York, NY, 1972.
11. S. Kumar and S.C. Bhaduri, Metallurgical and Materials Transactions B, 1994, 25B, 435.
12. H. Zhao and T. DebRoy, Journal of Applied Physics, 2003, 93(12), 10089.
13. J. W. Elmer, T. A. Palmer, W. Zhang, B. Wood, and T. DebRoy, Acta Materialia, 2003, 51(12), 3333.
14. S. A. David, R. Trivedi, M. E. Eshelman, J. M. Vitek, S. S. Babu, T. Hong and T. DebRoy, Journal of Applied Physics, 2003, 93(8), 4885.
15. A. Kumar and T. DebRoy, Journal of Applied Physics, 2003, 94(2), 1267.
16. X. He, P. W. Fuerschbach and T. DebRoy, Journal of Physics, D: Applied Physics, 2003, 36, 1388.
17. W. Zhang, G. Roy, J. Elmer and T. DebRoy, Journal of Applied Physics, 2003, 93(5), 3022.
18. X. He, P. Fuerschbach and T. DebRoy, Journal of Applied Physics, 2003, 94(10), 6949.
19. X. He, P. Fuerschbach and T. DebRoy, Journal of Physics D: Applied Physics, 2003, 36(23), 3079.
20. C.-H. Kim, W. Zhang and T. DebRoy, Journal of Applied Physics 2003, 94(4), 2667.
21. W. Zhang, C. L. Kim and T. DebRoy, Journal of Applied Physics (in press).
22. W. Zhang, C. L. Kim and T. DebRoy, Journal of Applied Physics (in press).
23. P. Kruger, Journal of Physics and Chemistry of Solids, 1993, 54(11), 1549.
24. Z. Yang, J.W. Elmer, J. Wong, and T. DebRoy, Welding Journal, 2000, 79(4), 97s.
25. M.P. Anderson, D.J. Srolovitz, G.S. Grest, and P.S. Sahni, Acta Materialia, 1984, 32, 783.
26. Y. Saito and M. Enomoto, Iron and Steel Institute of Japan International, 1992, 32, 267.
27. J. Gao and R.G. Thompson, Acta Materialia, 1996, 44, 4565.
28. Z. Yang, S. Sista, J. W. Elmer, and T. DebRoy, Acta Materialea, 2000, 48(20), 4813.
29. S. Sista, Z. Yang and T. DebRoy, Metallurgical and Materials Transactions B, 2000, 31B, 529.
30. S. Mishra and T. DebRoy, Acta Materialia, 2004, 52(5), 1183.
31. T. Hong and T. DebRoy, Ironmaking and Steelmaking, 2001, 28(6), 450.
32. T. Hong and T. DebRoy, Metallurgical and Materials Transactions B, 2003, 34B, 267.
33. T. Hong, W. Pitscheneder and T. DebRoy, Science and Technology of Welding and Joining, 1996, 3(1), 33.
34. T. Hong, T. DebRoy, S. Babu and S. A. David, Metallurgical and Materials Transactions B, 2000, 31B, 161.
35. A. De and T. DebRoy, Journal of Physics, D: Applied Physics, 2004, 37, 140.
36. A. De and T. DebRoy, Journal of Applied Physics (in press).

Proceedings from Joining of Advanced and Specialty Materials, 13–15 October 2003, Pittsburgh, Pennsylvania, ASM International, 2004

Experimental and Numerical Study of Thermal Process in Friction Stir Welding

Junde Xu, Suhas P. Vaze, Robert J. Ritter, Kevin J. Colligan, Joseph R. Pickens
Concurrent Technologies Corporation, 100 CTC Drive, Johnstown, PA 15904, USA

Abstract

This paper summarizes an experimental and numerical study of heat generation and dissipation during the friction stir welding (FSW) process. The work was motivated by the optimization of the welding tool design and the selection of the FSW process parameters for the manufacturing of combat vehicle structures. An infrared camera and embedded thermocouples were employed to study the evolution of the temperature distribution during the FSW of one-inch thick aluminum magnesium alloy 5083-H116 plate. The testing matrix covered variations in process conditions, such as tool shoulder diameter, tool advancing speed and thermal insulation conditions in FSW. The collected temperature-time data enabled a detailed quantitative analysis of the thermal process in FSW. The heat generation due to the tool-workpiece interaction, the heat loss through the tool and the anvil, and the temperature history in the weld stir zone were determined using finite element analysis. The model developed accurately predicts temperature profiles in FS weldments, as confirmed by thermocouple and infrared camera temperature measurements.

Introduction

Friction stir welding (FSW) is a solid-state joining technique that can fabricate high quality welds in an environmentally friendly fashion. In FSW, the heat generated by the traversing and rotating tool softens the workpiece material. As the welding tool advances, welds are made by displacing the plasticized workpiece material from ahead of the tool to behind the tool. Compared to the gas-metal-arc welding (GMAW) process currently employed to weld aluminum alloys for combat vehicle applications, FSW eliminates the use of electrode material (filler), shielding gas, and a high intensity heat source (weld arc). Since the workpiece material experiences a much lower temperature in FSW than in GMAW, weld defects (such as porosity, slag and cracking) commonly associated with solid-liquid transformation can be essentially eliminated. Due to all these advantages, the Combat Vehicle Research Program operated by Concurrent Technologies Corporation (*CTC*) proposed friction stir welding as an alternative for the joining of combat vehicle structures. To this end, a full-scale FSW machine system was designed and constructed to adapt FSW processing technology for the fabrication of combat vehicle structures, demonstrate the technology at a research and industrial scale, and establish an engineering knowledge database of aluminum armor alloys and the associated friction stir welded configurations.

In a typical FSW operation, as shown in Figure 1, the workpiece is firmly fixed on an anvil to preclude any significant movement during the welding process. The FSW tool consists of a shoulder and a pin that are typically fixed relative to each other. The FSW process starts with the plunging of the rotating pin tool into the workpiece. The length of the protruding pin is slightly less than the desired depth of the weld nugget (thickness of the workpiece in a butt joint) so that the annular end surface of the shoulder can have full contact with the workpiece and not leave a considerable groove behind the advancing tool. The function of the end surface of the shoulder is to generate frictional heat and to confine the softened and deformed base material within the weld nugget. The helical thread on the pin is usually left-handed. The rotating direction of the tool is chosen to drive the cut base material down into the workpiece and enable "vertical mixing." When the pin advances in the workpiece, base material surrounding the pin experiences severe plastic deformation: another source of heat. Because the rotation of the pin controls the deformation of the base material as the tool advances, as shown in Figure 1, the FSW process is not symmetric. The relative velocity between the rotating pin and the workpiece reaches its maximum on one side of the workpiece (designated the retreating side) and the minimum relative velocity occurs on the opposite side (advancing side) (Figure 1).

To guide the process design of FSW, much effort has been devoted to better understand the fundamental physics of the process. As described in several publications [1, 2], typically four distinct microstructures can be observed in friction stir welds. The material centermost in the "weld nugget" consists of fine, relatively equiaxed grains, indicating that the plastically deformed material has dynamically or continuously recrystallized. This region is also called the stir zone (SZ). Outside of the recrystallized fine grains is the thermo-mechanically affected zone (TMAZ) where dislocations are pumped from the stir zone and alter the microstructure. In butt welds, the TMAZ is characterized by the change of the grain orientation with respect to the original plate while the general size of the grains remains unchanged. Between the TMAZ and the unaffected parent material is the heat-affected zone (HAZ), which does not experience deformation in the process, but where the microstructure is affected by the heat generated in the SZ. The microstructural change in the HAZ depends on the alloy type, its original heat-treatment condition and the temperature history it experiences. Finally, there is the base metal, where no significant heating or deformation takes place.

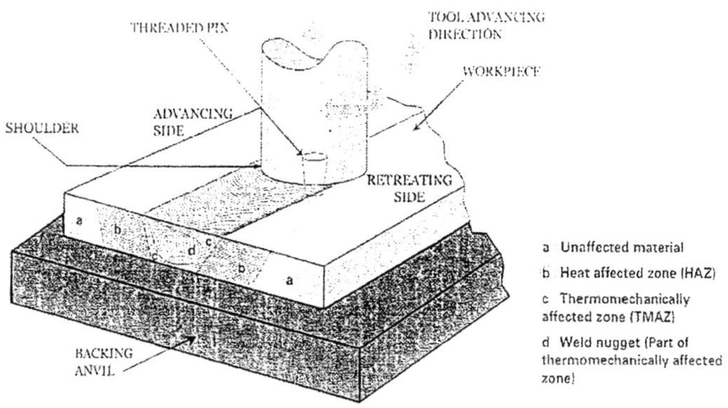

Figure 1. Schematic diagram of friction stir welding.

Although the friction stir welding process is mechanically driven, it is assisted by the softening of the workpiece in response to the generated heat. An important aspect of friction stir welding is the relatively complex thermal field in the plates or structures being welded. For a given base material, the heat generated in FSW is determined by the tool geometry and process parameters such as the tool rotating speed and the advancing speed. Part of the generated heat is transferred to the tool, the anvil and the surrounding air whereas the rest is retained and distributed in the workpiece. With the advancing tool, incoming base material is heated through its proximity to the large deformation stir zone, while the material behind the tool, either friction stir welded or not, cools down gradually to the ambient temperature. This thermal cycle can have a substantial influence on the microstructure of the welded material and the resulting mechanical properties. Because of this, the thermal process in FSW has been subjected to a variety of investigations from the infancy of the joining technique [3–6].

Compared with most of the previous work on friction stir welding, the manufacture of combat vehicle structures is characterized by the joining of thick (one inch and above) gauge workpieces. The temperature gradient through the workpiece thickness can affect the quality of the weld. To better control the process, it is desired to quantitatively understand how the welding tool shoulder and the pin contribute to heat generation in the FSW process and their individual dependence on the process parameters. Another feature of the FSW of combat vehicle structures is that the thermal insulation conditions in the FSW process vary with the type of the weld to be made. For example, in weld structures such as rabbet corner welds, the weld zone is usually distant from the backing anvil. This is different from the typical FSW of butt-weld in which the anvil intimately contacts the workpiece (hence, the weld zone) and dissipates considerable heat during the process. A better understanding of the heat loss through the anvil will also help optimize the selection of process parameters in FSW.

A welding pin has been developed by *CTC*, based on the concepts originated by The Welding Institute (TWI), for the joining of combat vehicle structures [7]. The pin's most distinctive features are the frustum shape and three or four flats on its threaded surface. Macrosections of friction stir welds (Figure 2) show that the profile of the SZ almost coincides with the profile of the pin and no significant TMAZ is observed, revealing that minimal material has been displaced during the FSW process. Outside the weld stir zone, vertical mixing is not observed. In a trial focused on a smaller tool shoulder for the aforementioned welding pin in the FSW of a relatively thick workpiece, it was found that, opposite to reducing the advancing speed of the tool to accommodate the lower heat generation due to a smaller shoulder, sound welds could only be made at relatively high advancing speed. This raises the issues of how heat is generated in FSW and how to determine the optimal combination of tool shoulder diameter, tool rotating speed and tool advancing speed for a given FSW workpiece. To address these issues and establish a framework for the optimization of the FSW process, a test matrix was designed to investigate the influence of the tool shoulder diameter and tool advancing speed on heat generation during the process. The heat loss through the tool, the anvil and the air was also quantitatively determined.

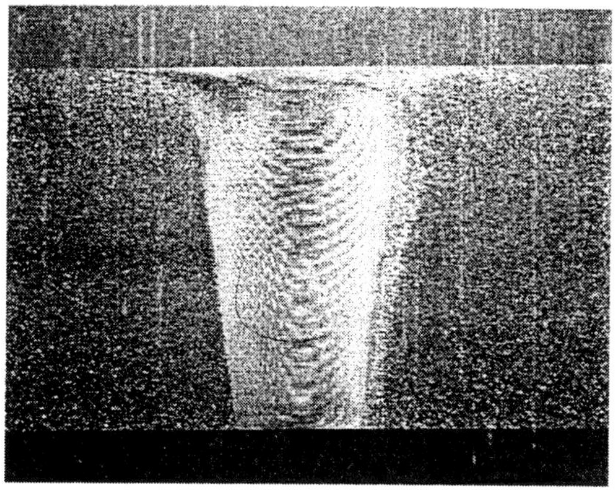
Figure 2. Macrograph of the transverse section of the joint welded using the welding tool developed by *CTC*.

Objectives

The present work focused on the friction stir welding of aluminum alloy 5083, an important armor material for combat vehicle structures, as well as marine applications. It is a non-heat-treatable Al-Mg alloy that derives its strength from cold work, Mg solid solution strengthening and grain/subgrain strengthening. The alloy is widely used in the H131 condition for armor application. One-inch thick 5083-H116, a similar work-hardened temper, plate was available and was used to conduct "bead-on-plate" friction stir welding. The objectives were:

- Measure the in-situ development of the temperature field on the surface and in the interior of the workpiece during friction stir welding
- Investigate the influence of the tool shoulder diameter and tool advancing speed on the heat generation in the workpiece

- Measure and confirm the salient thermophysical properties of 5083
- Calculate the heat generation due to the tool shoulder and pin, and the heat loss through tool and anvil.

The results of the present work can be used to improve the tool design and select the process variables. The measured temperature history can be correlated to the resulting microstructure and mechanical properties of the workpiece. Based on this framework, further process optimization work can be conceived.

Experimental Setup

In this work, 1-in thick 5083-H116 plates were used for the bead-on-plate welding tests. The length and the width of the specimens were 30-in and 8-in, respectively. An NTE Model H30K 5-Axis Stirwelder, manufactured by MCE Technologies, Inc. (Seattle, Washington) was used to perform all the welds for this work (Figure 3). The Stirwelder is equipped with 12 thermocouple data acquisition ports. In addition, the machine is capable of recording the torque and force on the welding tool during the FSW process. The machine was designed to fabricate large structures such as a full-size combat vehicle. The pin used to produce the welds was made from MP159, a Co-Ni-Cr alloy. It has a frustum shape with a 0.6-in root diameter, a 0.35-in tip diameter, and a 0.996-in length. The thread pitch is 1/18-in. Four flats were made on the threaded surface. The spiral tool shoulder was made of Viscount 44 steel (a free-machining H-13 tool steel). Two different shoulders were used in this study: one with a 1.65-in diameter and one with a 1.2-in diameter.

Figure 3. Large scale FSW system at *CTC*, Johnstown, Pennsylvania.

Temperature Measurement

An infrared (IR) camera and thermocouples were employed to measure and characterize the temperature field in the plate during FSW.

A Prism DS IR camera made by FLIR Systems, Inc., was used to detect emitted thermal energy during the friction stir welding process. Thermal video images were recorded via an 8-mm camcorder attached to the IR camera. The temperature range for this camera using the F05 filter was up to 450 °C (842 °F) with an accuracy of ±2 °C or ±2%, whichever is greater. The chosen spectral range was from 3.6 μm to 5 μm. In order to maximize the emissivity of the surface of the plate and thus enable better recording from the IR camera, the surface of the plate was painted with a Dupli-Color flat black spray paint that is heat resistant up to 1200 °F. A fixture was fabricated to mount the IR camera aligned with the weld centerline but underneath the tool head. The fixture was bolted onto the base plate of the welding tool (Figure 4). The lens was focused on the top surface of the workpiece. A shield was designed and fabricated to block the reflected radiation from the welding tool to the surface of the plate (not shown in Figure 4).

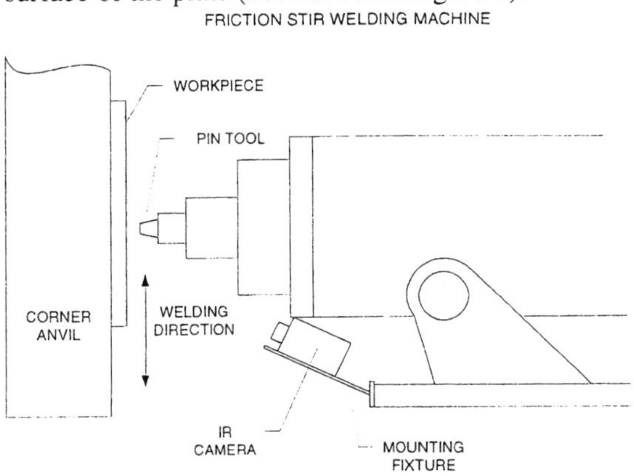

Figure 4. Sketch of the IR camera setup.

In order to determine the correct emissivity setting for the IR camera for the painted plates, a preliminary experiment was conducted. The air cooling of a hot plate was monitored using both the IR camera (by aligning the crosshair in the camera viewport with the thermocouple location) and the thermocouples by taking readings every 60 seconds. The emissivity setting on the camera was adjusted until the readings obtained from both instruments coincided. It was found that an emissivity of 0.55 is appropriate for the IR camera setting for the flat black painted surface and 0.26 for an unpainted oxidized surface. Figures 5 and 6 show the data from the calibration of the emissivity for the above two cases, respectively. In Figure 6, results of two tests with different starting tenperature (at time=0) are reported to demonstrate the insensitivity of the IR temperature data to the incline angle of the camera.

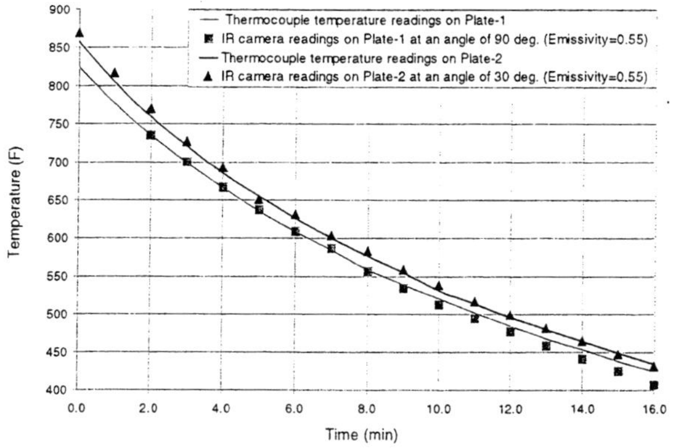

Figure 5. Calibration of emissivity for surface painted flat black.

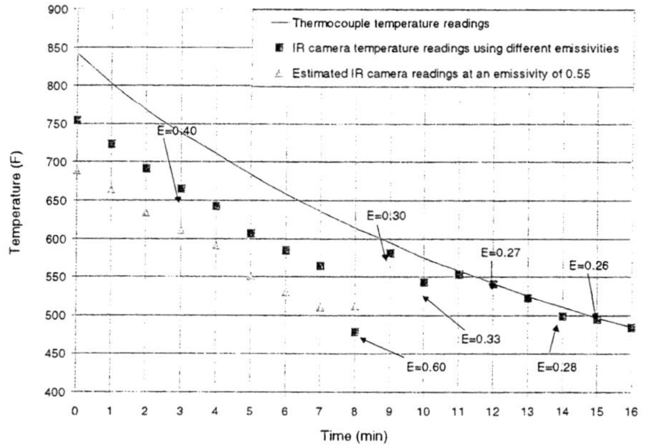

Figure 6. Calibration of emissivity for unpainted surface.

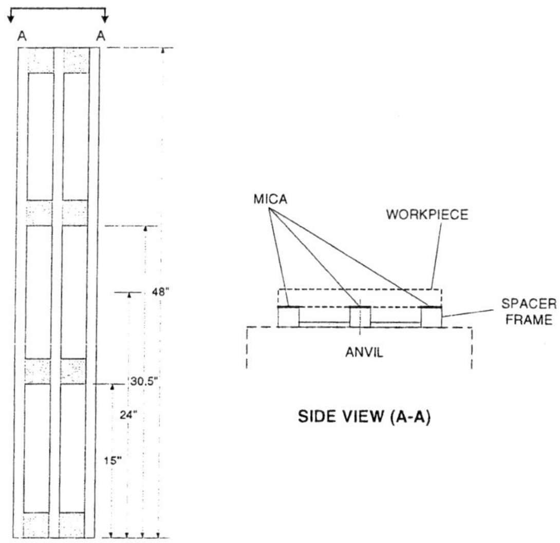

Figure 7. Spacer frame used to insulate the workpiece from the anvil.

Due to their relatively high strength, 1/16-in diameter Chromega Alomega 304 SS Sheath Type K Quick Disconnect thermocouples were chosen to measure the temperature in the interior of the workpiece during welding. In designing the thermocouple placement, the focus was on obtaining thc temperature history in the vicinity of the stir zone and at different depths in the workpiece. Holes were drilled into the plate for the insertion of thermocouples. For those tests in which mica was used to insulate the workpiece from the anvil, thermocouple holes were drilled from the bottom surface of the plate, in most cases at an angle of 50° relative to the surface of the plate. A spacer was fabricated using A36 steel strips (Figure 7). One side of the spacer was glued with strips of mica to prevent the heat in the plate from being transferred to the backing anvil. The spacer also allowed enough space for extension wires to connect all the thermocouples to the data acquisition ports on the friction stir welding machine. For those plates directly fixed on the anvil, thermocouple holes were drilled from the top surface of the plate. The diameter of the thermocouple hole was 0.0635 in.

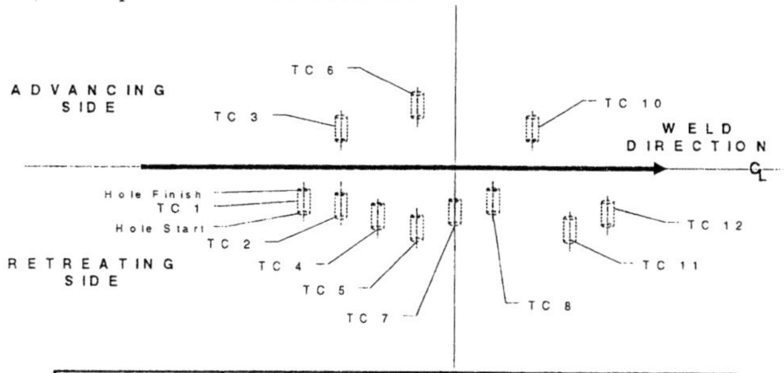

Thermocouple	Distance to Centerline (in)	Depth from the Bottom (in)
1.00	0.24	0.25
2.00	0.27	0.50
3.00	0.27	0.50
4.00	0.45	0.50
5.00	0.70	0.50
6.00	0.70	0.50
7.00	0.30	0.75
8.00	0.24	0.25
10.00	0.27	0.50
11.00	0.70	0.50
12.00	0.30	0.75

Figure 8. Typical thermocouple layout.

Figure 8 shows a typical thermocouple placement. Thermocouples were inserted at two sides of the weld centerline, at different distance to the centerline, at different depths in the plate, and at different axial locations (location in the advancing direction of the welding tool). Some thermocouples were symmetrically arranged. Some were inserted on the same side of the plate but differed in the axial location. For example, TC2 (Thermocouple 2) and TC3 were symmetric with respect to the weld centerline, both at a distance of 0.27 in to the centerline and at a depth of 0.5 in. TC5 and TC6 formed another symmetric pair. These thermocouples were used to verify whether the actual weld line was precisely on the centerline and also to confirm the symmetry of the temperature field about the centerline observed by other researchers. In addition, both TC7 and TC12 were located on the retreating side of the plate (0.3 in from the weld centerline and 0.75 in from the bottom surface of the plate) with the latter 1.0 in behind the former in the tool advancing direction. TC1 and TC8 were similarly arranged. These thermocouples were used to verify the repeatability of the measured temperature history. During the FSW process, the temperature data were collected at a frequency of 1 Hz.

Thermal Model Development

To investigate the heat generation and dissipation in the FSW process, the following strategy was employed in the model development. At first, the temperature measurements were conducted on the workpiece attached to the spacer frame with mica inserted between the workpiece and the backing anvil to minimize the heat loss via the anvil. This allows a better calibration of the heat loss to the air and through the tooling. Then, test specimens were fixed directly to the anvil as is normally done during butt-welding. Temperature changes during FSW were measured to determine the heat loss through the anvil. Finite element analysis was combined with the measured temperature-time data to quantitatively determine the following parameters for the FSW process.

- Power needed to generate the heat in the workpiece
- Heat generation due to tool shoulder-workpiece interaction
- Heat generation due to pin-workpiece interaction
- Heat loss through air
- Heat loss through tooling
- Heat loss through anvil

To determine the above parameters, the general purpose non-linear finite element package ABAQUS [8] was chosen for the thermal analysis. Two finite element models were developed for the analysis. As reported by many previous studies and confirmed in this effort, the asymmetry of the temperature field between the advancing and retreating sides is present but is negligible for all practical purposes. Therefore, most of the finite element simulation was conducted on a half-scale model, meaning that only one side of the plate was modeled. The welding tool (the shoulder and the pin) and anvil were not included in the finite element model. However, their contributions to the heat generation and heat loss were taken into account as described in the following sections. In addition to the half-scale model, a full-scale finite element model was also constructed to numerically simulate the possible asymmetric temperature distribution due to the porosities, lack-of-fusion defects or voids generated under improper process conditions and residing on the advancing side of the stir zone. For 5083, which possesses very good thermal conductivity, no significant asymmetric temperature distributions were observed in the finite element simulation. For this reason, the details of the full-scale finite element model are omitted, although the model is ready to be used in the future to analyze the possible asymmetric temperature distribution for less conductive materials, such as titanium alloys.

In order to accommodate the steep temperature gradients produced by the FSW process, 20-node triquadradic cuboidal elements with eight integration stations were used. The half-scale model consists of 3,840 elements and 20,041 nodes. Before the finite element model was used to calibrate the process parameters of interest, it was first verified using the two-dimensional Rosenthal analytical solution [9] for a point heat source having a constant strength and moving at a constant velocity. The results confirm that the discretization introduced in the finite element analysis is acceptable.

The thermophysical properties of 5083 are temperature dependent. Since very limited data could be found in the open literature, the thermal diffusivity of 5083 was measured using a Netzsch Laser Flash Analyzer. A differential scanning calorimeter measured the specific heat while the conductivity was derived from these measurements. The results are shown in Figures 10 and 11.

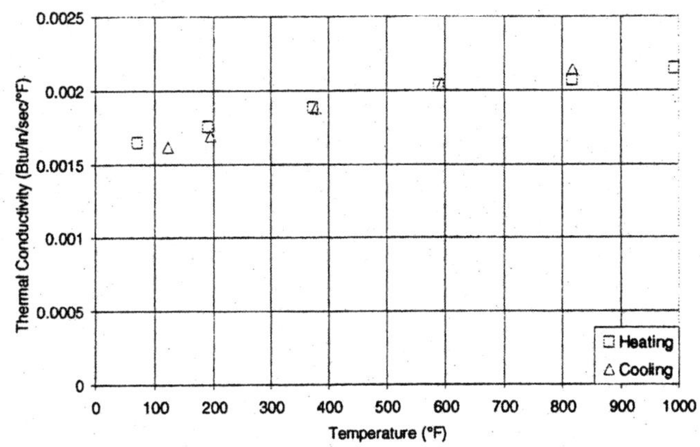

Figure 10. Thermal conductivity of 5083 at different temperatures.

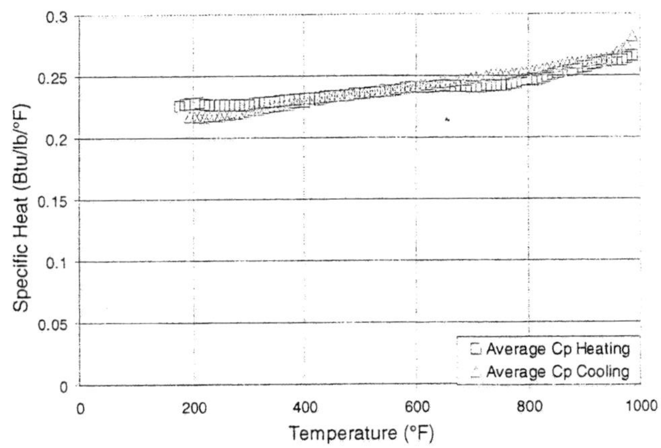

Figure 11. Specific heat of 5083 at different temperatures.

Heat Generation Due to Tool Shoulder-Workpiece Interaction

Previous work [5] has derived the following relation for the total heat input, q, from the contact area between the tool shoulder and the workpiece.

$$q = \int_A f \, dA \tag{1}$$

In the above, A is the area of the contact and f is the local strength of heat input. The parameter f depends on the relative motion between the tool shoulder and the workpiece. Letting r denote the distance to the tool axis, it is assumed that

$$f = \alpha \cdot r \tag{2}$$

where, for a tool shoulder having a radius r_o, the parameter α can be expressed as [5]

$$\alpha = \frac{3\,q}{2\,\pi\,r_o^3} \tag{3}$$

In the finite element analysis, the heat generated by the tool shoulder is simulated by imposing a body heat flux. In constructing the finite element models, the thickness of the top layer elements was chosen as 0.125 in, which makes the four (4) top integration points only 0.0264 in beneath the top surface. A user-coded subroutine, DFLUX, was developed to implement the heat input scheme in Eq. 2 on the top layer integration points when their distance to the pin axis is less than the radius of the tool shoulder.

Heat Generation Due to Pin-Workpiece Interaction

During the FSW process, severe deformation (hence, a temperature rise) occurs around the pin; the heated base material then is quickly transported to the backside of the pin. As mentioned before, the welding tool developed by *CTC* does not generate a considerable TMAZ outside the weld stir zone, which is contrary to the result with conventional pins that consist of a cylindrical surface with no flats or reentrant features. Since the profile of the stir zone nearly coincides with the profile of the pin, the heat generation between the pin and the workpiece is assumed to occur in the volume occupied by a fictitious pin in the finite element models. The generated heat is modeled by a transient uniform heat source in the aforementioned volume. Again, this is implemented using a body heat flux in the finite element analysis. The finite element mesh was designed to match the profile of the pin and the stir zone as well as the locations where the centermost thermocouples were inserted. When the distance between an integration point and the pin axis is less than the radius of the pin, heat flux is imposed. This scheme was also implemented in the subroutine DFLUX. A parameter was introduced in the subroutine DFLUX to control the fractional heat generation due to shoulder-workpiece interaction and pin-workpiece interaction. In the finite element simulation of the temperature distribution in the plate, the value of this parameter was adjusted through comparison with experimental data to achieve the proper balance between shoulder heating and pin heating.

Heat Loss through Air and Anvil

In the finite element simulation, the heat loss through the air and the anvil was modeled by specifying different film coefficients on different portions of the outer surface of the finite element models. The values of the film coefficient were determined by finite element simulation.

Results and Discussion

Figure 12 shows the cross section of a workpiece on which two holes were drilled to insert thermocouples to measure the plate mid-thickness temperature at a distance of 0.45 in from the weld centerline. It can be seen that the two holes are symmetric to the weld nugget. In most cases, temperature measurement made using thermocouples produced data scatter less than 30~50 °F. In Figure 13, the solid, dash, and dotted lines show the temperature history of three locations at different distance to the weld centerline and at different depth in Plate 02. The process conditions are as follows: 1.65-in tool shoulder diameter, 250-rpm tool rotating speed, and 5-in/min tool advancing speed. The test conditions of all the plates are summarized in Table 1. In all the tests, the tool rotating speed was kept at 250 rpm. In Table 1, the machine power consumption was calculated by multiplying the measured torque on the welding tool by the tool rotating speed. The power input to the workpiece and percentage heat generation due to shoulder and pin were determined through the finite element simulation.

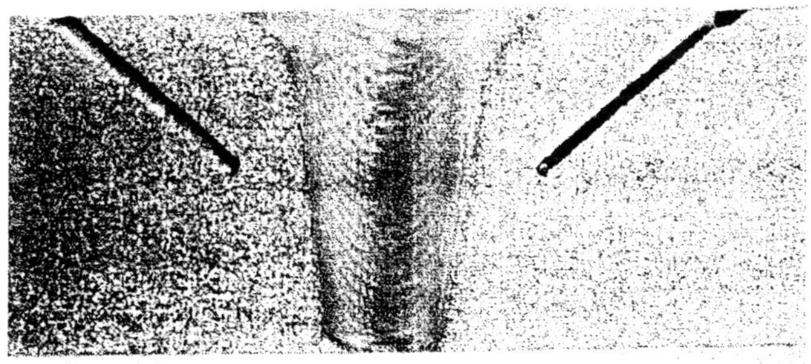

Figure 12. Macrophotograph showing the accuracy of thermocouple insertion.

The finite element analysis started with the simulation of FSW in which mica was used to block the heat loss to the anvil. For different shoulder diameters and tool advancing speeds, the measured temperature profile was simulated by adjusting the total power input, the heat generation at the tool shoulder and the pin, and the film coefficient between the workpiece and the air in the finite element analysis input data files. It was found that a film coefficient of 10^{-7} Btu/in^{2}s°F yields the best simulation results. Figure 13 also compares the half-scale finite element simulation results against the measured temperature profile shown. The finite element results are denoted using symbols. For the weld in Plate 02, the simulation results indicated that the total power input is 5.9 Btu/s. In this case, 50% of the heat is generated by the shoulder-workpiece contact and 50% by the pin-workpiece interaction. Having calibrated the film coefficient between the workpiece and the air, the temperature history of the workpiece directly fixed on the anvil was simulated. Analysis indicated the film coefficient between the workpiece and the anvil is 10^{4} Btu/in^{2}s°F. In Figure 14, a comparison is made between the temperature profile measured on plate 09 and the finite element simulation results. In this case, the tool shoulder diameter is 1.2 in, the tool advancing speed is 7 in/min and the workpiece is directly fixed on the anvil. To match the measured temperature profile, the finite element simulation determined that the power input to the plate is 5.6 Btu/s and the fractional heat generation due to the shoulder and pin needed to be adjusted to 30% and 70%, respectively.

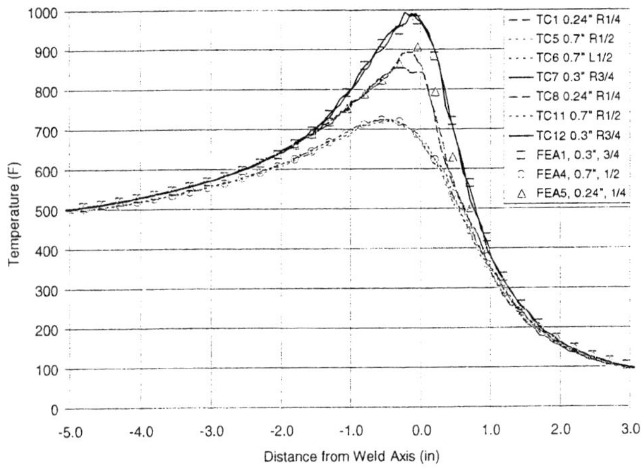

Figure 13. Temperature profile of Plate 02 (shoulder diameter 1.65 in, advancing speed 5 in/min, and mica used to insulate from the anvil)., showing the comparison between the temperature measurement and the finite element simulation.

Table 1. Summary of Test Conditions and Results

Plate Number	Shoulder Size (in)	Advancing Speed (in/min)	Insulation Conditions	Machine Power Consumption (Btu/sec)	Power Input to Workpiece (Btu/sec)	Percentage Heat Generation due to Shoulder : Pin
Plate04	1.65	4	with mica	7.3	5.4	60% : 40%
Plate02	1.65	5	with mica	8.1	5.9	50% : 50%
Plate05	1.65	6	with mica	8.9	6.4	40% : 60%
Plate08	1.65	5	no mica	9.1	5.8	30% : 70%
Plate03	1.2	5	with mica	6.5	4.9	50% : 50%
Plate06	1.2	7	with mica	7.9	5.9	50% : 50%
Plate07	1.2	8	with mica	7.7	6.1	40% : 60%
Plate09	1.2	7	no mica	8.8	5.6	30% : 70%

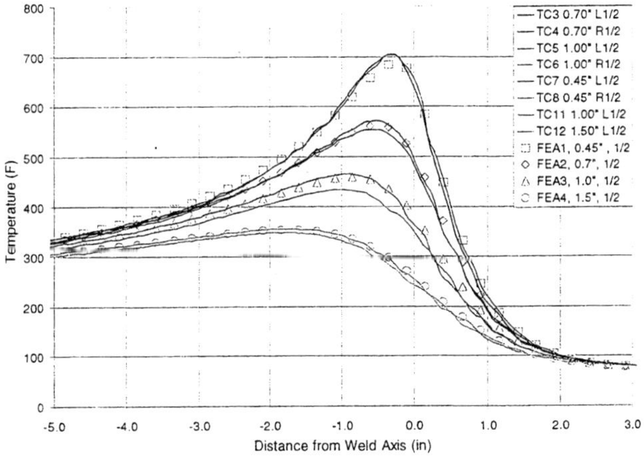

Figure 14 Comparison between the temperature measurement and the finite element simulation for Plate 09 (shoulder diameter 1.2 in, advancing speed 7 in/min, and workpiece directly fixed on the anvil).

Figure 15 shows the IR camera image on the top surface of Plate 02. The camera lens was set to focus on the top surface of the workpiece at a spot 2.0 in behind the advancing pin. The top surface of the plate and the shield were all painted flat black. An emissivity of 0.55 was used for the IR camera setting to record the thermal image. Since the weld mark is equivalent to an unpainted surface that requires an emissivity of 0.26, the estimated temperature on the weld mark must be corrected. As shown in Figure 7, a temperature increase of approximately 150 °F is associated with an emissivity of 0.26 compared with 0.55. Figure 15 shows the IR camera temperatures at two locations on the weld mark. At location 1 (approximately 2 in behind the weld pin axis), the thermal image records a temperature of 502 °F. With the 150 °F correction, the estimated actual temperature is 652 °F. Location 2 is 1 in behind location 1 or 3 in behind the welding pin. Based on the color at location 2, the temperature there was estimated as the average of the temperature at location 1 (502 °F) and the lowest detected temperature (332 °F), which is 417 °F. With the correction, the temperature at location 2 was estimated as 567 °F. Figure 16 compares these two corrected temperatures with the finite element calculation results; notice the close agreement.

Effect of Tool Shoulder Diameter

For both Plate 02 and Plate 03, the same tool advancing speed (5 in/min) was used in the testing. The former used a tool shoulder diameter of 1.65 in whereas the latter used 1.2 in. Comparing the predicted power input of these two plates (5.9 Btu/s versus 4.9 Btu/s) shows that a larger shoulder generates more heat and consumes more energy in the FSW process (Table 1). This higher heat generation results in higher temperatures distributed in the workpiece.

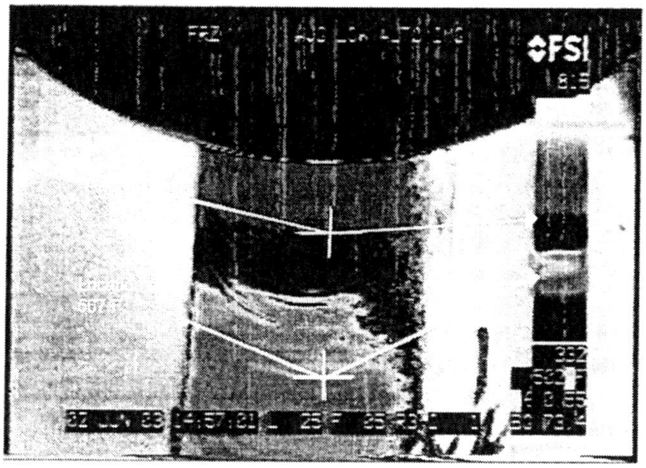

Figure 15. IR camera image on the weld mark of Plate 02.

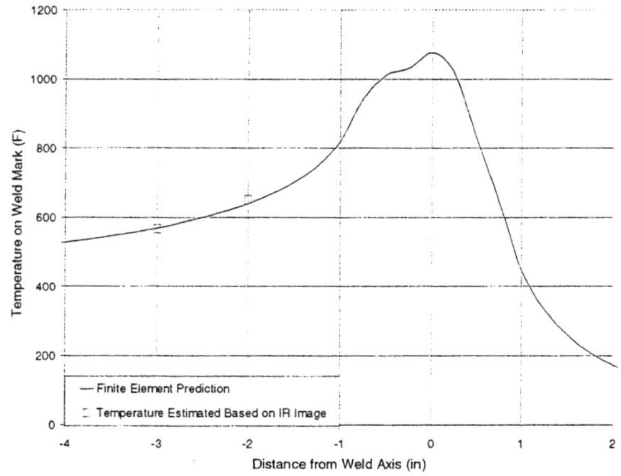

Figure 16. Comparison between the temperature estimated from the IR camera image with the finite element simulation results.

Effect of Tool Travel Speed

As shown in Table 1, for Plates 04, 02, and 05, each with a 1.65-in shoulder diameter, and Plates 03, 06, and 07, each with a 1.2-in shoulder diameter, increasing the tool advancing speed increases the machine power consumption and the power input to the workpiece. Figure 17 plots the power input to the workpiece versus the welding tool advancing speed. For example, for the 1.65-in shoulder, the power input increases from 5.4 Btu/s to 6.4 Btu/s when the tool travel speed increases from 4 in/min to 6 in/min. It is instructive to examine the energy consumption per unit length of weld. Using the data in the columns "Advancing Speed" and "Power Input to Workpiece" in Table 1, the "specific energy" (energy consumption per unit weld [7]) versus the advancing speed for two different shoulder diameters is plotted in Figure 18. The observations are as follows.

- For the same tool shoulder diameter, specific energy decreases with increasing tool advancing speed.
- For the same advancing speed, specific energy increases with increasing tool shoulder diameter.

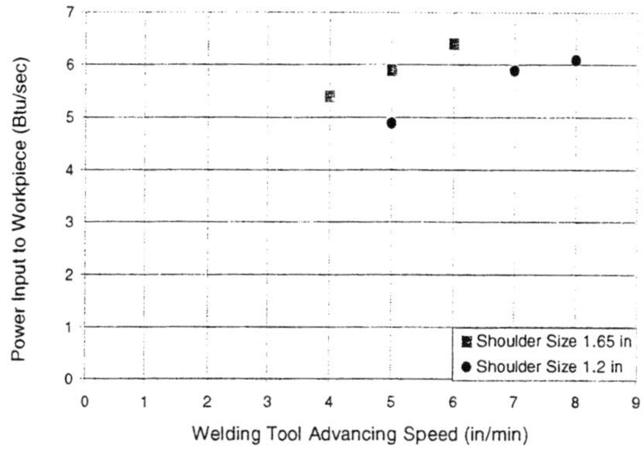

Figure 17. Effect of tool advancing speed on power input to workpiece for different tool shoulder diameter.

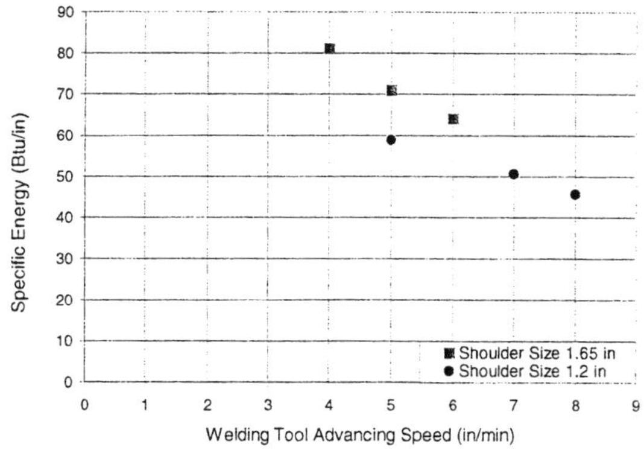

Figure 18. Effect of tool advancing speed on energy input per unit length of weld for different tool shoulder diameter.

Macrophotographs from welds produced using the 1.65-in tool shoulder show that sound welds were made under all three different travel speeds. However, the weld produced using a 1.2-in shoulder and 8-in/min travel speed shows unacceptable defects in the weld nugget (Figure 19). Figure 18 indicates that the "specific energy" for this process condition is the lowest, suggesting that insufficient heat was generated during the FSW process. It should be noted that these welds were made with the mica insulation on the root side of the weld. Different defect formation results would be expected with a conventional anvil boundary condition, but the trends in the data are expected to be the same.

According to the finite element simulation, increasing the tool travel speed also increases the percentage of heat generation due to the pin (Table 1). This is more prominent for the larger tool shoulder (1.65-in diameter) than the smaller tool shoulder (1.2-in diameter).

Effect of Heat Loss Through the Welding Tool

The heat loss through the welding tool can be easily obtained by subtracting the "Power Input to Workpiece" (found from finite element simulation) from the "Machine Power

Consumption" (calculated from the measured torque) in Table 1. When mica was used to insulate the workpiece from the anvil, the heat loss through the welding tool was about 20~30% of the total machine power consumption. In contrast, when the workpiece is directly fixed on the anvil, the heat loss through the welding tool and anvil is about 36%.

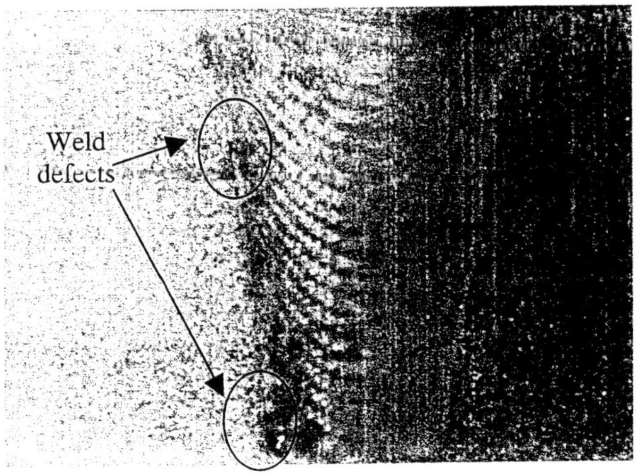

Figure 19. Transverse section of Plate 07 (shoulder diameter 1.2-in, advancing speed 8-in/min, and mica used to insulate from the anvil).

Effect of Heat Loss through the Anvil

The film coefficient for the heat loss through the anvil has been determined. From Table 1, it is also found, for the same tool shoulder diameter and tool advancing speed, the heat loss through the anvil slightly reduces the required power input to the workpiece but increases the machine power consumption (or the heat loss through the welding tool). For example, comparing Plate 06 (1.2-in tool shoulder, 7-in/min travel speed, and workpiece insulated from the anvil) with Plate 09 (same shoulder diameter and travel speed, but plate directly fixed on the anvil), the heat generated in the plate drops from 5.9 Btu/s to 5.6 Btu/s, but the machine power consumption increases from 7.9 Btu/s to 8.8 Btu/s. Another noticeable influence of the heat loss through the anvil is the increase of the percentage heat generation by the pin. As shown in Table 1, for both Plate 08 and Plate 09, 70% of the heat was generated by the pin.

Prediction of the Temperature in the Weld Stir Zone

Using the results of finite element simulation, the temperature history of the weld stir zone can be estimated. Since the temperature in the stir zone depends on location, an "average" temperature of the stir zone is defined as the mean of the temperatures at the weld centerline and at the edge of the stir zone in the middle layer of the workpiece. Figure 20 plots the history of the above "average" temperature of all four welds produced using the 1.65-in tool shoulder. Comparing Plates 04, 02, and 05 (tool advancing speed 4, 5, and 6 in/min, respectively) where mica was used to minimize the heat loss to the anvil shows that increasing the tool travel speed reduces the time duration for the material to experience elevated temperature but increases the peak temperature in the stir zone. This indicates a faster advancing speed in FSW forces the welding tool to deform colder, higher strength material and generate more heat in unit time, thus yielding a higher peak temperature in the stir zone as the material is transported behind the pin. Figure 20 also shows that the welds done with the workpiece directly on the anvil had higher peak temperatures than those done with the mica insulating the back of the plate. This can be seen by comparing the temperature history of Plates 02 and 08, which have the same advancing speed of 5 in/min but different thermal insulation conditions. To understand this, consider the following two competing scenarios involved in the thermal-mechanical process in FSW. On one hand, the heat loss through the anvil tends to lower the temperature of the material. On the other hand, a lower temperature will cause a higher flow stress for the material in the stir zone, which requires more torque to plasticize it and generates more heat. Yet both conditions were observed to exist at the same time. It is speculated that this is possible because the reduced workpiece temperature occurs in front of the pin, while the higher peak temperatures occur behind the pin in the stir zone. Further study is planned to explain this complex issue. For the welds produced using the 1.2-in tool shoulder, the same trend is also observed in the finite element simulation. Figure 21 plots the peak temperature for all the welds.

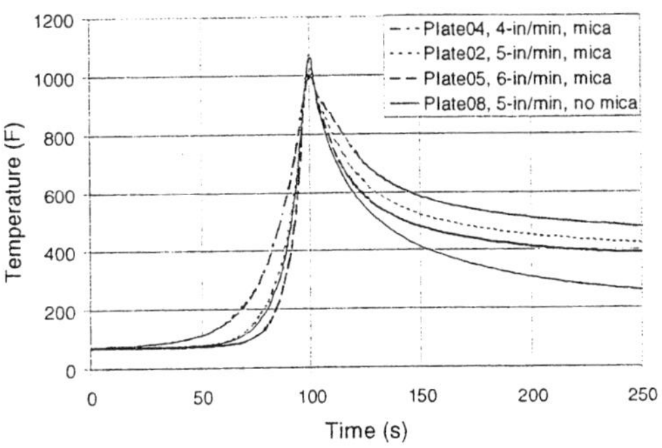

Figure 20. Comparison of the predicted average temperature in the weld stir zone for the welds produced using 1.65-in tool shoulder.

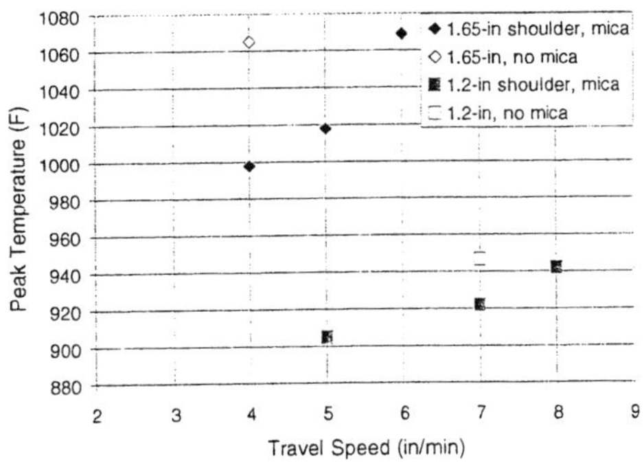

Figure 21. Predicted peak temperature in the weld stir zone for different tool shoulder diameter, advancing speed, and thermal insulation conditions.

Conclusions

The present work provides the following insights for the design of friction stir welding tools and the selection of the process parameters for FSW on aluminum alloys such as 5083-H116.

- Both the tool shoulder and the pin contribute significantly to the heat generation in the FSW process. The shoulder : pin contribution ranges from 60% : 40% to 30% : 70%, depending on the process variables such as tool shoulder diameter, tool advancing speed and insulation conditions.
- Heat generation due to the tool shoulder increases with tool shoulder diameter, but decreases with tool advancing speed. In addition, the recorded torque, thus, the machine power consumption (for a given rotational speed) to produce a viable weld increases with heat loss through the anvil.
- Heat generation due to the pin increases with tool advancing speed. When significant heat loss through the anvil is present, the pin plays a more important role in heat generation.
- Increasing the tool travel speed or the heat loss through the backing anvil reduces the time duration for the material to experience elevated temperature, but increases the peak temperature in the stir zone.
- The simulation system established in the present work also facilitates the estimation of the temperature history of the material in the weld nugget. This, combined with the metallurgical behavior of the workpiece material, will permit a better evaluation of the resulting microstructure and mechanical properties of the workpiece.

Acknowledgements

This work was conducted by Concurrent Technologies Corporation (*CTC*) for the Tank-Automotive Research Development & Engineering Center (TARDEC) through the National Center for Excellence in Metalworking Technology (NCEMT). *CTC* operates the NCEMT under Contract No. N00014-00-C-0544 to the U.S. Navy as part of the U.S. Navy Manufacturing Technology Program. The authors appreciate the support of the Combat Vehicle Research Program Monitor Jamie Florence of TARDEC, who has been a consistent supporter of Friction Stir Welding Technology.

The contributions of John L. Lancione (Intern at *CTC* until May 2003) are gratefully acknowledged. John was responsible for the thermocouple layout, data acquisition during welding, and fixture design for the infrared camera.

References

1. M.W. Mahoney, C.G. Rhode, and W.H. Bingel: *Metal. Mater. Trans. A*, Vol. 29A, pp. 1955-1964 (1998)
2. T.J. Lienert and R.J. Grylls: *Proc. 1st Int. Symp. on 'Friction Stir Welding,* Thousand Oaks, CA (1999)
3. M.J. Russel and Shercliff: *Proc. 1st Int. Symp. on 'Friction Stir Welding'* Thousand Oaks, CA (1999)
4. A. Askari, S. Silling, B. London, and M. Mahoney: *Friction Stir Welding and Processing*, ed. K.V. Jata, S.L. Semiatin and D.P. Field, TMS (2001)
5. Y.J. Chao, X. Qi, and W. Tang: *Journal of Manufacturing Science and Engineering*, Vol. 125, pp. 138-145 (2003)
6. P. Colegrove and H.R. Shercliff: to appear in *Science and Technology of Welding and Joining*
7. K.J. Colligan, J. Xu, and J.R. Pickens: *Friction Stir Welding and Processing*, ed. K.V. Jata, M.W. Mahoney, T.J. Lienert, and R.S. Mishra, TMS (2003)
8. Hibbitt, Karlsson & Sorensen, Inc., *ABAQUS User's Manual*, Version 6.1 (2000)
9. D. Rosenthal and R. Schmerber: *Welding Journal*, Vol. 17, No. 4, pp. 208-214 (1938)

Proceedings from Joining of Advanced and Specialty Materials, 13–15 October 2003, Pittsburgh, Pennsylvania, ASM International, 2004

Friction Stir Welding of Copper Plates

A. Polar*, T. Shah, and J. E. Indacochea
University of Illinois at Chicago, Chicago, Illinois USA.
* Universidad Nacional de San Agustín, Arequipa, Perú/Visiting Professor at UIC

Abstract

Friction stir welding of annealed copper plates was carried out in a standard milling machine with a non conventional tool made of alloy tool steel. Welds were made using different process parameters. Metallographic evaluation was done using optical and scanning electron microscopes. A close relation between metal flow pattern and shape of the pin thread was observed. Only when a critical temperature is reached using most favorable weld parameters, a sound joint is possible. The typical weld nugget found in the friction stir welding of other metals was not observed in this case, the weld consisted primarily of a thermo-mechanical affected zone (TMAZ). Microhardness as well as tensile strengths of the joints are closely related to the weld microstructure. The crossweld specimens all failed in the weld metal with the failure originating at the cold joint at the advancing side of the weld.

Introduction

Friction Stir Welding (FSW) is a solid state welding process where the weld is produced by the rotation of a viscous mass of the metal workpiece pushed against its cold walls due to the action of the pin rotation of the weld tool and forged by the shoulder of the tool. The rotating tool is plunged and pressured against two metal sheets butted together and firmly secure. The heat of friction developed by the contact of the tool with the workpiece softens the material, and the severely deformed metal is also highly stirred by the probe located at the bottom of the turning tool. As the tool moves along the faying interface under a fixed travel speed and fixed applied pressure, the metal from the front of the tool is swept around this plasticized ring to the rear. The benefit of the localized heating is that limits the softening of the components joined.

Since there is no melting with this process oxidation is reduced, the solidification defects are eliminated, and the residual stresses are lessened. The microstructure of the weld consists of a central nugget showing often an onion ring structure, a thermomechanically affected zone (TMAZ) close to the nugget, and a very small heat affected zone (HAZ) [1, 2]. The central zone receives the most attention because true bonding is produced in this region. In FSW of aluminum alloys it is accepted that the process entails a continuous dynamic recrystallization of the nugget [2-16], where the former grains of the parent metal are heavily strained and then recrystallized as a fine grain structure. This fine grain structure is the result of the large straining and relatively low temperature of the process. It has been established that the grain size of the nugget increases as the temperature of the process also increases [10]. The TMAZ is also influenced by the straining caused by the stirring action of the probe; but large recrystallization is not observed in this region. This lack of recrystallization is likely affected by the relatively low peak temperature in this region.

Evaluations of precipitation hardened Al alloys welded by FSW include microhardness surveys across weld joints, which define the different microstructures and matching hardness variations [4, 8-10, 13, 19, 21, 22]. As expected, there is decrease of microhardness in regions of the weld nugget and the TMAZ, because of the metallurgical changes associated with the dissolution or coarsening of the precipitates following exposures to temperatures above the solvus curve during welding. The tendency is for the microhardness to be higher at the weld nugget in view of the larger straining experienced by this region, and due to its finer microstructure. The microstructural changes will have an impact on microhardness, tensile and fatigue properties of the weld joint [19-26]. Studies are also being undertaken on the corrosion resistance of FSW joints [27-29].

Peak temperatures during FSW of aluminum and aluminum alloys reach about 500°C, equivalently to 80% of its melting temperature. The peak temperatures expected on FSW of copper or steel are likely to be 55% and 43% of their melting temperatures respectively. These low temperatures would be expected to have an impact on the metal flow, nugget formation, and recrystallization.

The objective of this study is to determine the weldability of copper by FSW as an alternative to arc welding of copper and copper alloys. It is recognized that fusion welding of cooper and its alloys may encounter several problems related to the process itself; for example oxygen may enter into the weld pool during arc welding, dissolve in the liquid metal and segregate to grain boundaries on solidification; this can lead to embrittlement of the weld joint. Precipitation hardened copper alloys may loose their alloying elements by oxidation during fusion welding and thus compromise their strength. In other cases fusion is incomplete due to the high thermal conductivity of copper; this reduces the amount of heat necessary for melting of a critical mass of metal that ensures a complete filling of the weld cavity. This study analyzes the effect of weld parameters on the metal flow characteristics during welding, and the resulting microstructures and mechanical properties of FSW copper C10100.

Experimental Procedure

Electrolytic tough pitch copper sheets, C10100, were employed in this study. The dimensions of the coupons were 203.0 x 41.3 x 1.6 mm (8.0 x 1.625 x 0.062 in), the composition of the sheet metal was 99.99 wt. % copper and 0.01 wt. % oxygen. The sheets were received in the as-rolled condition and were annealed at 450°C for two hours. The sheet blanks were cleaned with acetone and the edges machined to give a close fit when butted together for the weld fabrication.

A standard two-axis milling machine was utilized in the fabrication of FS welds. The weld tool was machined from heat treated O2 tool steel with a shoulder diameter of 16.0 mm (0.622 in); the pin diameter of 5.0 mm (0.210 in) was off-centered by 0.43 mm (0.017 in) with respect to the main axis of the welding tool. The pin was right hand threaded using a 22-thread/in specification (9.5 threads/cm). The welds were fabricated by butt joining the square matting edges of both sheet-coupons; all resting on top of an alloy steel backing plate and the whole system firmly secured with holding wedges, as shown in Figure 1.

Figure 1. Experimental set-up for making the friction stir welds.

In making the FS welds the milling machined was first started at the set rotation and travel speeds; the tool was then gradually plunged into the workpiece to produce a hole on the plate and held in place momentarily to allow the heat of friction to heat up and soften the metal adjacent to the tool. The tool was then was allowed to travel along the faying interface of the butted joint until the end of the weld was reached. The travel motion was stopped, the tool was then retracted from the workpiece while still in rotation and the milling machine was then shut off. In the present work, the plunge depth was controlled by maintaining the shoulder of the welding tool forcibly in contact with the top of the joining sheets. The pin length is slightly shorter, by 0.1mm, than the thickness of the sheet for each metal. The tool axis had a lead angle of 2°. Different weld parameters were used to obtain a complete weld, the most successful parameters are shown in Table 1.

After fabrication the welds were first examined by naked eye for surface weld appearance. Portions of the weld were later cross sectioned at locations near the start and towards the end of the weld for metallographic evaluation. The samples were prepared by means of standard ASTM procedures. In addition, weld cross sectioned tensile specimens were machined from towards the end region of the weld for tensile testing. Microhardness measurements were obtained of the weld cross sections.

Table 1

Welding Parameters for FSW of 10100 Copper

Trial	Travel Speed in/ min (mm/sec)	Rotation Speed (rpm)
1	1.0 (0.42)	2000
2	0.5 (0.21)	
3	1.0 (0.42)	1500
4	0.5 (0.21)	

Results and Discussion

Weldability of Copper

Copper has physical and mechanical properties different than those of aluminum that need to be carefully considering in designing the FSW procedure for making the weld. The developments of FSW have come about using aluminum as the primary metal. Although both metals have a face centered cubic crystalline structure and similar tensile strengths, the differences become notable when melting point, thermal conductivity, and stress range between the yield and tensile strengths at 200°C are compared. Table 2 shows some of the physical and mechanical properties corresponding to two aluminum alloys and copper.

Table 2

Physical and Mechanical Properties of Selected Aluminum alloys and Copper

Properties	2024-T6	6061-T6	Cu
Thermal Conductivity (W/mK)	120	167	391
Liquidus Temperature (°C)	638	652	1083
Tensile Strength at 200 °C	179	138	160
Yield Strength at 200 °C	131	103	152

The larger thermal conductivity of copper, about three times larger than that of the Al-alloys, is a factor of concern since heat needs to be build up in the metal next to the weld tool to become soft. Furthermore, Cu has a higher melting point than the Al-alloys, approximately 1.7 times greater; consequently, higher temperatures are also needed to soften the material, which implies a larger heat flux and heat accumulation per mass of metal. The higher thermal conductivity of copper makes this task more difficult, and hence a higher heat input is required and this is controlled by the welding parameters. In addition, materials that are to be FS welded should have a large uniform plastic region, usually demonstrated by a large yield to tensile stress difference. The reductions in yield and tensile strengths with temperature should be significant even at low temperatures to make the material more plastic and thus improve its forging and manipulation during welding. Table 2 shows that the drop in yield strength for copper at 200ºC is not as large when compared to the Al-alloys. Figure 2 displays the drop in tensile strengths with temperature for copper and aluminum. Essentially, higher temperatures are required for the FSW of Cu-alloys.

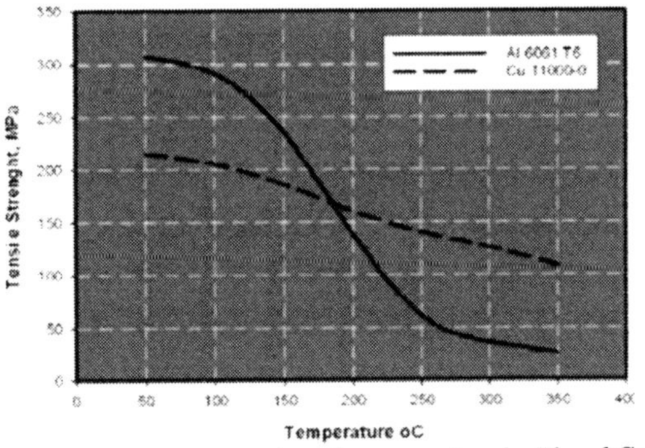

Figure 2. Tensile strength with temperature in Al and Cu alloys [30].

The preliminary trials on FSW of the Cu sheets showed that the welding parameters selected were adequate to produce a weld between the two Cu specimens. Figure 3 displays a sample of the initial welds, which also shows a thin film of oxide coating parts of the weld run.

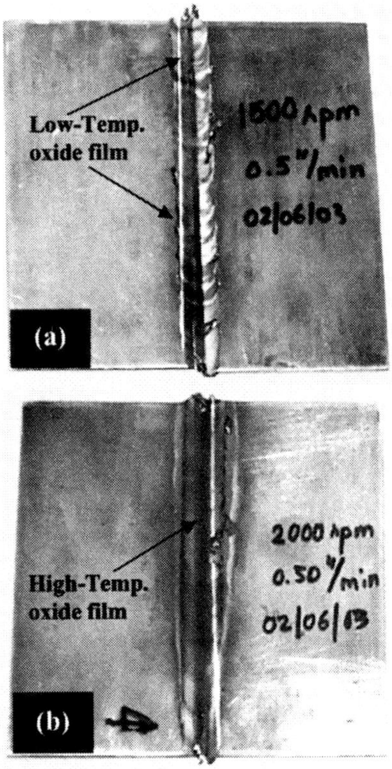

Figure 3. Cu-FSW for two welding parameters. (a) 1500 rpm and (b) 2000 rpm.

Metal Flow Characteristics

It is recognized in FSW that the plasticized metal is displaced from the front of the tool to the trailing region of the weld tool. The metal flow is strongly influenced by the thread design, as observed in Figure 4. It is observed in this case that the metal flows from the top of the retreating side clockwise to the bottom of the weld cavity behind the moving tool. The three bands displayed in the figure were seen to correlate with the thread design of the pin used in this investigation. The left half of the Figure 4 shows the metal flow from the bottom of the weld cavity towards the top at the advancing side. It is evident that the metal is squeezed in place by the tool shoulder. Figure 5 shows a larger magnification of this region and demonstrates the turbulence and vigorous smashing of the plasticized metal against the shoulder by the advancing side. The numbers observed in the micrograph correspond to the microhardness measurements performed in this region; these numbers reflect the level of straining.

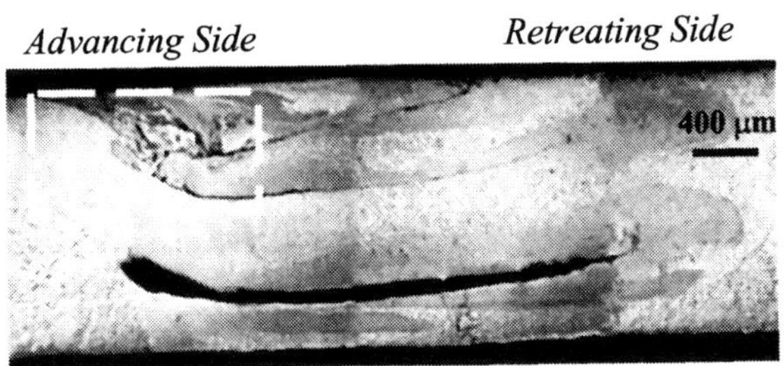

Figure 4. Photomicrograph displaying the metal flow pattern at the start of the weld for a rotation speed of 2000 rpm and a travel speed of 0.5 in/min.

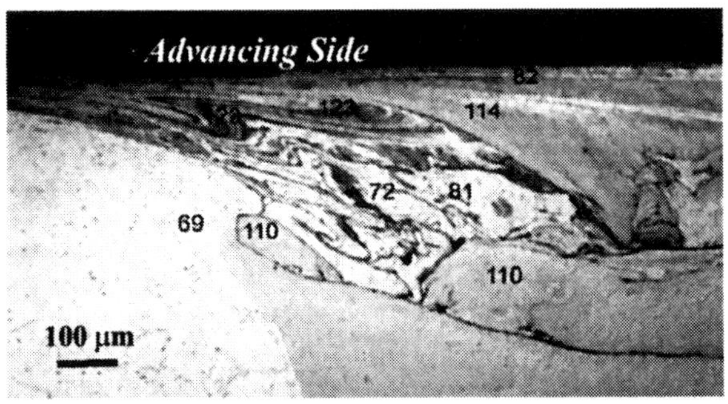

Figure 5. Illustration of turbulence and strain of metal on advancing side. Region of rectangle in Figure 4.

The micrographs shown in Figures 4 and 5 correspond to a cross section specimen machined at the start of the weld. The weld is observed to be incomplete were the metal did not fill complete certain regions in the weld cavity, this is observed at the bottom of the weld in Figure 4. Some other regions of the weld show cracks and cavities; for example there is a definite boundary or incomplete bonding between the metal flow bands, and the turbulent region on the advancing side (Figure 5), show chunks of metal squeezed together without bonding. This defective FS weld implies that the temperature reached at this point was not sufficiently high to soften enough metal and make it more plastic to fill all the weld cavities. A high temperature would also make the forging of the metal easier.

By contrast the cross section specimen, for the same weld (2000 rpm and 0.5 in/min travel speed), machined from the end of the weld displayed a more complete appearance. That is, the long cavity at the bottom of the weld nugget disappeared and the turbulent region found at the advancing side is not present either. This gives further evidence to the fact that at later stages of the FSW process the temperature of the copper was high enough to able to stir and forge the soften metal into remote places. Figure 6 shows the micrograph corresponding to the later sample.

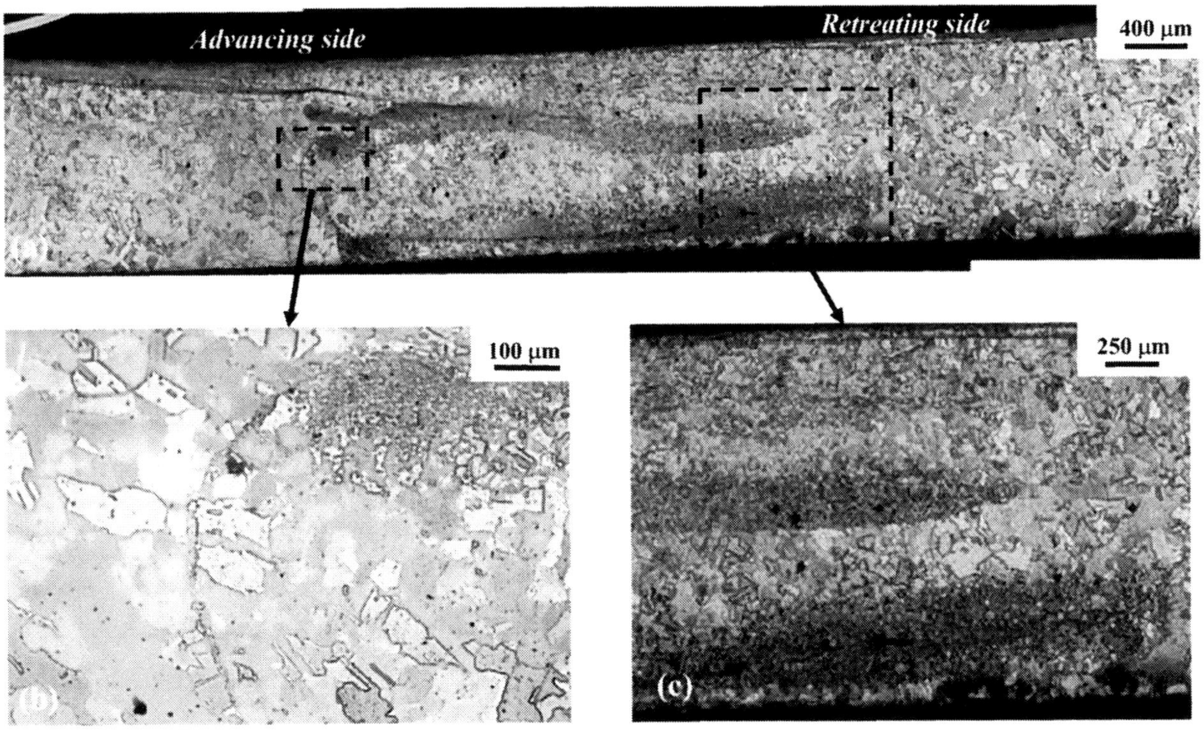

Figure 6. Micrograph of the cross section towards the end of the weld for the specimen fabricated at 2000 rpm at a travel speed of 0.5 in/min. Amore complete weld was observed at this location.

Microstructural Characterization of Copper FS Welds

Aluminum friction stir welds are characterized for having a weld nugget with the "onion structure", because of the concentric and layered distribution of the metal flow lines that remind the appearance of the peels in an anion. Since deformation and heat is the largest at the center of the weld nugget, this region is expected to undergo dynamic recrystallization and dynamic recovery, depending on the location of the weld nugget region with respect to the tool threads. The copper welds fabricated in this study displayed microstructure features different than those encountered in aluminum; first of all, there was no onion structure in these welds. In the case of the weld produced 1500 rpm and 0.5 in/min travel speed, the cross section examined from the end of the run shows an incomplete weld as shown in Figure 7. This denotes that the weld parameters used in producing the weld did not generate the heat needed to plasticize a critical mass of metal to produce a complete weld. Large cavities and extensive cracking was observed through out the weld nugget, but more extensively by the advancing side.

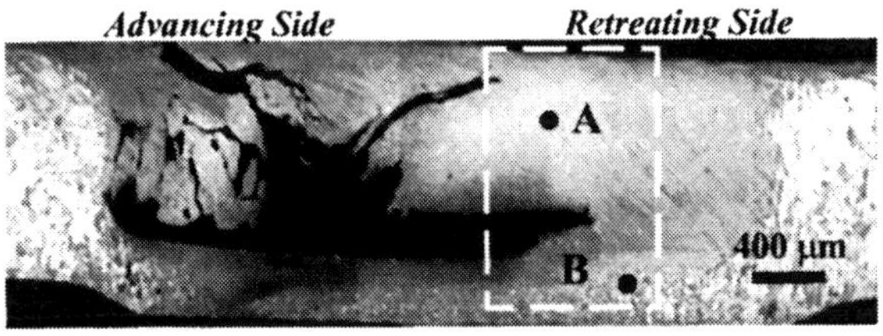

Figure 7. Weld nugget of a sample sectioned from the end of the run. The weld was fabricated at 1500 rpm and at a travel speed of 0.5 in/min.

A closer examination of the weld nugget shows regions undergoing different levels of recrystallization. For instance, regions "A" and "B" inside the rectangle drawn on Figure 7, show significant differences in microstructures. Region "A" has a fine grain structure; this is most likely the result of dynamic recrystallization. In describing the FSW, it is recognized that hot working phenomena is an integral aspect of the process. During hot working of metals, when considerable amount of strain is produced, two types of behavior could be observed. For materials with high staking fault energy, like BCC metals and some FCC like Al, the strain produced leads to an annihilation and rearrangement of dislocations but mainly in the subgrains. This is what is called *dynamic recovery* and is shown as a plateau for a constant flow stress after an amount of strain. In low stacking fault energy metals, such as copper and copper alloys, it is reported that the rearrangement of dislocations produces regions with almost no dislocations and separated from the surrounding area by high angle boundaries. This is called *dynamic recrystallization* [31]. In addition this area undergoes significant straining that aids nucleation, and since the temperature is relatively low, grain growth is limited. It is recognized that *dynamic recrystallization* is often followed by *static recrystallization* and some times grain growth. However, in this study region "A" appears to just have undergone dynamic recrystallization. Region "B" on the other hand, underwent static recrystallization. The deformation at this location is not as severe as in region "A", the straining is caused by the compression of the metal by the probe bottom as the tool moves along the faying interface. The final grain size at this spot was still finer than that found in the annealed base metal, but it was coarser than in region "A". Obviously the nucleation rate at "B" is less than "A". These differences are better appreciated in Figure 8, since this a larger magnification of the rectangular portion highlighted in Figure 7.

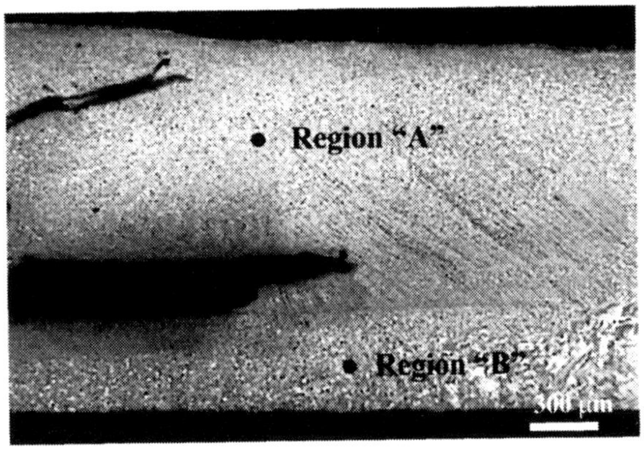

Figure 8. Micrograph of the selected rectangular region in Figure 7 to show regions "A" and "B", by the retreating side of the weld. The weld was fabricated at 1500 rpm with a travel speed of 0.5 in/min.

Increase in weld heat input is expected to produce higher temperatures and larger heat fluxes that will soften a larger volume of metal, as well as making the stirring and forging of the metal more manageable. This effect was observed as the rotation speed was increased to 2000 rpm and the travel speed to 1.0 in/min. The quality of the weld improved significantly, as seen in Figure 9. The higher temperatures made also the differences in microstructures in the weld more visible. The weld cross section presented in Figure 9 corresponds to a sample machined towards the end of the weld length. The amounts of cavities and cracks have been reduced, and once again most of the defects were located towards the advancing side of the weld. The weld presents metal flow bands sloping down from the retreating side towards the advancing side in a shape that matches the probe threads.

The microstructure at the center and edges of the metal flow bands were examined. The higher temperature in this weld has definitely produced a coarser microstructure at selected locations. Figure 10 is a larger magnification of Figure 9 and focuses on the microstructures of the bands at

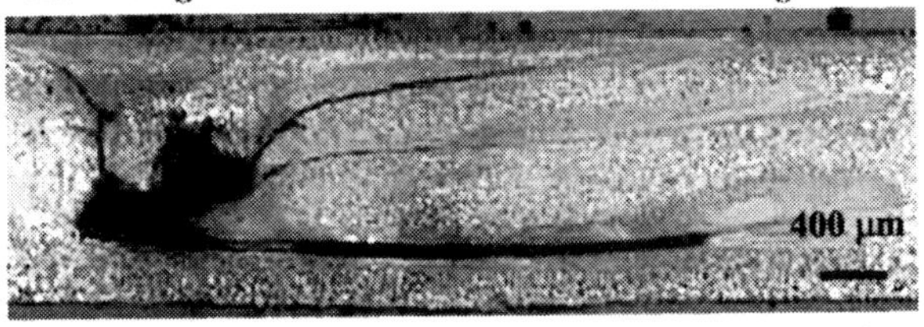

Figure 9. Micrograph of a weld cross section removed from before the end of the run. Weld produced at 2000 rpm and at a travel speed of 1.0 in/min.

the center and at interface between bands as they bond. The microstructure at the band center showed complete recrystallization compared to the edges where a finer recrystallization and some simultaneous recovery is apparently present.

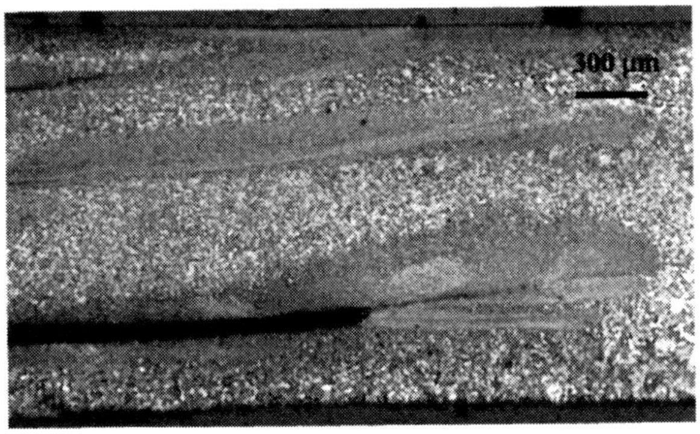

Figure 10. Larger magnification of microstructure of Figure 9 by the retreating side showing flow bands.

The best FS weld produced in this investigation was the one corresponding to a rotation speed of 2000 rpm and a travel speed of 05. in/min. Figure 4 corresponds to a specimen from this weld obtained near the start of the run. Obviously from this figure, it is seen that defects are still present; a long cavity is observed near the bottom of the weld, as well as unbonded and strained chunks of metal at the advancing side of the weld. But the cross section extracted from the same weld coupon but towards the end of the run, shows a more complete weld with no cavities, as shown in Figure 6. As the travel speed was reduced from 1.0 to 0.5 in/min, keeping the rotation speed constant at 2000 rpm, a higher heat input was introduced and hence, a larger temperature was reached that soften the metal further and made it easier to stir and filled the weld cavity. The metal flow bands have practically disappeared; and only dark brown fringes are found between the layered-like microstructure seen in Figure 6a. The higher processing temperature also impacted the recrystallization of the weld by producing coarser grain structures. Although no detailed grain size measurements were taken at this stage of the study, Figures 6a and 10 show qualitative differences in grain size between these two welds. The weld fabricated at the lower travel speed (Figure 6a) had a coarser grain structure than the weld produced at a faster travel speed (Figure 10). Moreover, there are differences in grain size between the center of the weld and the brown fringes for the weld section displayed in Figure 6a. The dark brown regions consisted of a finer grain size structure; inclusions were found to have segregated to these sites too, as seen in Figure 11. It is possible these inclusions play a role in controlling the growth of the grains here.

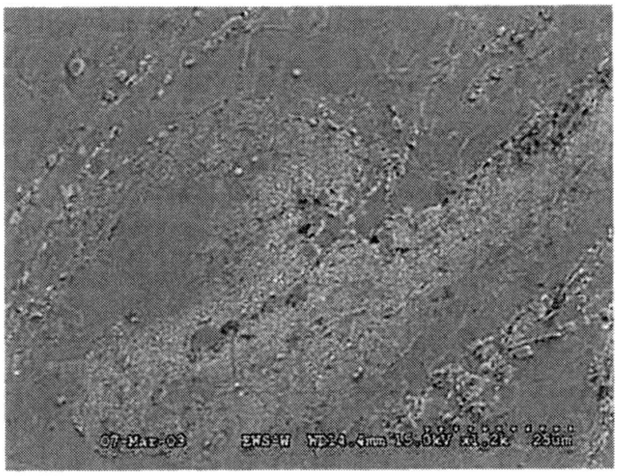

Figure 11. Micrograph of the dark brown regions between metal flow layers (Figure 6c) inside the weld.

Further evaluation of the microstructure for the last weld sample was done of areas between the recrystallized region of the nugget and TMAZ, at both the advancing side (Figure 6b) and retreating side (Figure 6c). At the advancing side a discontinuity or boundary between these two regions is observed; however in the case of the retreating side there was continuity of the microstructure were the grains are coarse and uniform across the two regions, except for some of the narrow dark brown discussed above. This difference must be a consequence of the fabrication procedure itself, i.e. the mechanics of the welding process. In analyzing the rotational movement of the tool probe in conjunction with the travel velocity during weld production (Figure 12), the following assessment could be made. At the advancing side the tool rotation is in the same direction as that of the travel speed, so the contact of the probe with the workpiece is accelerated. On the other hand, by the retreating side the rotation speed opposes the travel speed; this means that the net velocity slows down at this location, and thus the time of contact between the tool and the workpiece is prolonged. In addition the material removed ahead of the tool is slowed down by the travel speed. Such circumstances point towards

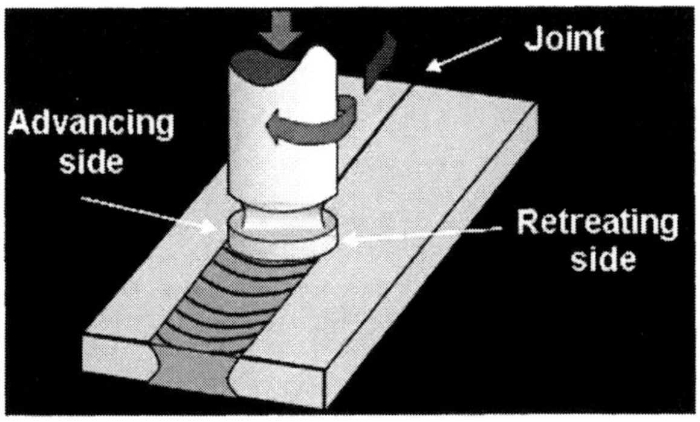

Figure 12. Schematic illustrating the friction stir welding process and interaction of the rotation and travel speeds.

different thermal experiences on either side of the weld. Although, it was difficult to measure temperatures at these spots, the grain microstructure was examined. This analysis shows a finer grain microstructure at the advancing side (Figure 13a) compare to a coarser grain microstructure at the retreating side (Figure 13b).

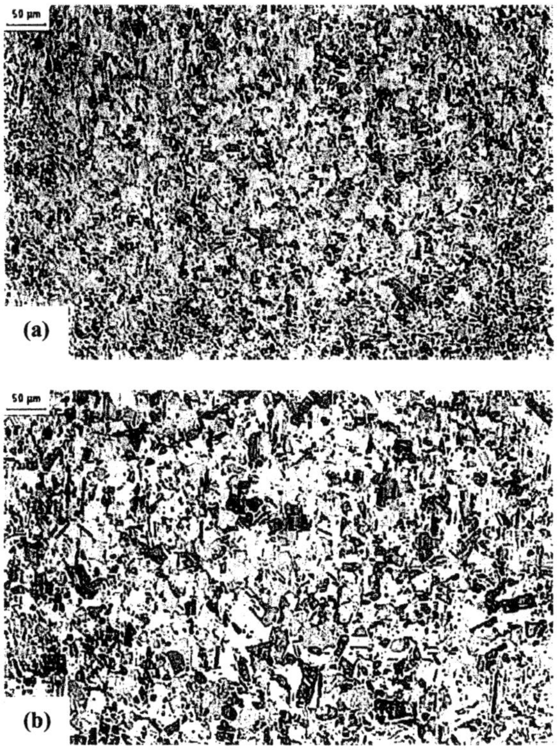

Figure 13. Microstructures of the (a) advance side and the (b) retreating side. Weld produced at 2000 rpm and 0.5 in/min.

The minor differences in microstructures encountered between both sides of the last copper friction stir weld justify the thinking that higher temperatures are encountered at the retreating side of the weld.

Mechanical Evaluation

Microhardness across the weld metal was performed, as well as tensile testing of cross weld samples extracted from locations near the start of the weld and from about the end of the weld run. This evaluation was done on the sample made with a rotation speed of 2000 rpm and a travel velocity of 0.5 in/min. Figure 14 displays the microhardness readings at the two locations. As expected the microhardness values at the start of the weld were higher than the values of the base metal because the temperature did not reach a critical temperature to recrystallize the structure and thus relieve some of the straining cause by the welding operation. This correlates with the fact that the weld was incomplete because of the lack of plasticity of the metal in view of the low temperature. The weld cross section from the end of the run shows a uniform microhardness throughout because towards the end of the weld run the higher temperature allowed for recrystallization to be more complete.

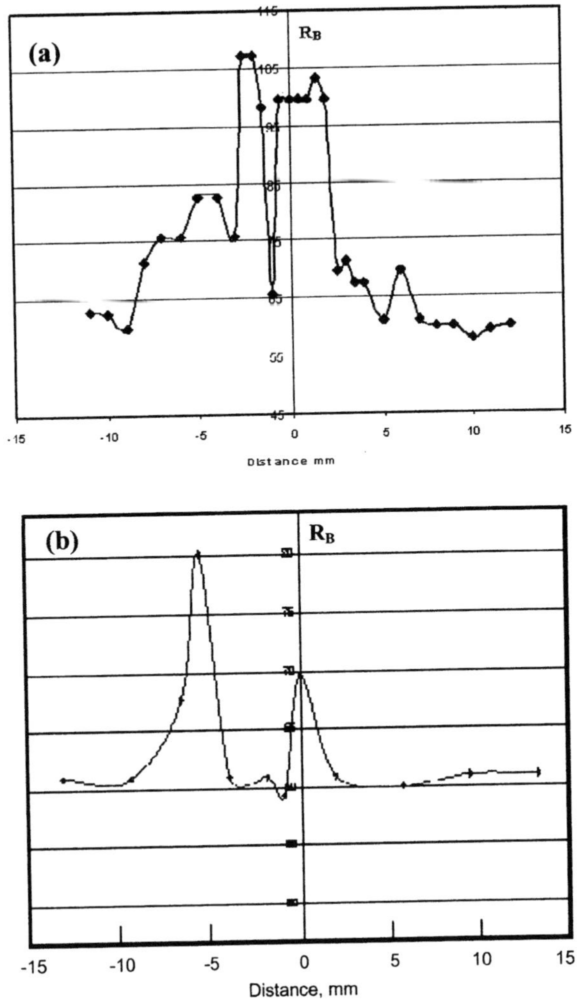

Figure 14. Microhardness survey across the weld for the sample fabricated at 2000 rpm and 0.5 in/min. (a) start of weld, and (b) end of weld.

The tensile tests showed higher strength and elongation values for the samples extracted from the end of the weld. These results were expected based on the results of the metallographic evaluation and microhardness survey. Table 3 summarizes the tensile test results. The difference in strengths between these locations are in part related to the differences in weld defects present, that is, there were few or no defects towards the end of the weld run, compare to the start of the weld. The tensile samples all failed in the weld metal. To determine the initiation of failure a crossweld tensile sample was tested and stopped before failure and it was found that the fracture started at a cold joint located at the advancing side of the weld.

Table 3
Tensile Test Results

Sample	UTS (MPa)	% Elongation
Start of weld 1	111.6	3.77
Start of weld 2	173.5	21.66
End of weld 1	185.5	21.44
End of weld 2	187.5	24.27

Conclusions

1. No *onion ring structure* was detected in the friction stir welds of 10100 Cu; instead a homogeneous TMAZ is formed with regions of high strain and fine grains.
2. Plasticity of the metal is limited at the start of the weld due to lack of heat built up. This is characterized by cavities, cracking, and lack of bonding between layers of the extruded material at the weld.
3. Higher heat inputs by adjusting the welding parameters resulted in higher temperatures that improved the metal plasticity and flow.
4. Higher processing temperatures lead to coarsening of recrystallized grain structure that may lower the mechanical properties.
5. All crossweld tensile samples fractured in the weld metal, with the fracture originating in the cold region at the advancing side.

Acknowledgements

The authors appreciate the financial support provided by the Campus Research Board of the University of Illinois at Chicago to undertake this research.

References

1. P. Threadgill, TWI bulletin, March – April, 30-33 (1977).
2. M.W. Mahoney, C.G. Rhodes, et al. *Properties of FSW 7075 T651 aluminum.* Met. and Mat. Transactions A, Vol. 29A, July 1955-1964, (1998).
3. C.G. Rhodes *Effects of FSW on microstructures of 7075 aluminum.* Scripta Materialia vol 36,1 69-75, 1997
4. G. Liu, et al. *Microstructural aspects of the FSW of 6061 – T6 Aluminum.* Scripta Materialia, Vol. 37, No. 3, pp. 355-361 (1995).
5. L.E. Murr, et al. *A TEM study of precipitation and related microstructures in FS welded Aluminum.* Journal of Materials Science. Vol. 33, pp. 1243-1251 (1998).
6. Flores, V. et al. *Microstructural issues in FS welded aluminum alloy.* Scripta Materialia. 38,5 703-708, 1998. Elsevier
7. Li Ying et al. *Solid state flow visualization in the FSW of 2024 Al to 6061.* Scripta Materialia 40, 9 1041-1046, (1998).

8. Ying Li et al. *Flow visualization and residual microstructures associated with the FSW of 2024 Al to 6061 Al.* Materials Sci. & Eng. A271 (1999) 213-223
9. Murr, L. et al. *Microstructures in FS welded metals.* Journal of Materials Processing & Manufacturing Science vol 7 Oct 1998.
10. Benavides, S. et al. *Ultrafine grain structure in the FSW of aluminum alloy 2024 at low temperature.* The Minerals, Metals & Materials Society. 2000
11. Benavides, S. et al. *Low temperature FSW of 2024 aluminum.* Scripta Materialia, 41,8 809-815. 1999.
12. Sutton, M. A. et al., Microstructural *studies of FS welds in 2024-T3 aluminum.* Materials Science and Engineering A323 (2002) 160-166.
13. Heinz Beate et al. *Microstructural characterization of a friction stir welded Al alloy.* Materials Science Forum, vols. 331-337 (2000), pp 1757-1762.
14. Norman, A. F. et al. *High resolution EBSD analysis of the grain structure in an AA2024 Friction Stir Weld.* Materials Science Forum, vols. 331-337 (2000) pp 1731-1718.
15. Yutaka, S. et al. *Microstructural evolution of 6063 aluminum during FSW.* Met. and Mat. Transactions A, Vol. 30A, pp. 2429-2437 (1999).
16. Jata, K. V. *Friction stir welding of high strength aluminum alloys.* Materials Science Forum, vols. 331-337 (2000), pp 1701-1712.
17. J. Chao, et al *Effect of FSW on dynamic properties of AA 2024 and AA 7075 T7351.* Welding Journal, August 2001, 196s-299s.
18. Svensson, L. E. et al. *Microstructure and Mechanical properties of friction stir welded aluminum alloys with special reference to AA 5083 and AA 6082.* Science and technology of welding and Joining, vol 5, 2000. 285-296.
19. K.Karlsson, L.-E.Svensson, H.Larsson.: Characteristics of friction stir welded Aluminum alloys. 5th Intl. Conf. on "Trends in welding research", ASM International, 574-579, (1998).
20. Sutton, M. et al. *Mode I fracture and microstructure for 2024-T3 friction stir welds.* Materials Science and Engineering, A354 (2003) 6-16
21. Bussu, G. et al. *The role of residual stresses and the heat affected zone properties on fatigue crack propagation in FS welded 2024.* International Journal of Fatigue, 25 (2003) 77–88.
22. Lockwood, et al. *Mechanical response offs welded AA2024: experiment and modeling. Materials Science and Engineering.* A323 (2002) 348-353.
23. Liming Ke, Li Xing, J. E. Indacochea, *The material Flow Pattern and the Cavity Model in Friction Stir Welding,* accepted by TMT A
24. Ouyang, J.H. et al, *Material flow microstructure in the friction welds of the same and dissimilar Al alloys.* Journal of Materials Engineering and Performance. 11(1) February 2002. 51.
25. Pao, P.S. *Corrosion fatigue crack growth in FS welded Al 7050.* Scripta Materialia 45 (2001) 605-612.
26. Krishnan, K. N. *On the formation of the onion ring in FS welds.* Matls Sci. and Eng. Vol. A327, pp. 246-251 (2002).
27. C. Dalle Donne. *Investigations on Residual Stresses in Friction Stir Welds, Proceedings III Symposium on Friction Stir Welding,* Kobe Japan Sept. (2001).
28. Field, David et al *Texture effects on corrosion behavior of friction stir welded 7075 Aluminum.* Web Site. School of Mechanics and Materials Engineering Washington State University
29. Zucchi, F. et al. *Pitting and Stress Corrosion cracking resistance of friction stir welded AA 5083.* Materials and Corrosion, Vol. 52, pp. 853-850 (2001).
30. ASM Handbook, Vol. 2, 1990.
31. G. Dieter. *Mechanical Metallurgy, Third Edition,* McGraw Hill (1986).

Proceedings from Joining of Advanced and Specialty Materials, 13–15 October 2003, Pittsburgh, Pennsylvania, ASM International, 2004

Friction Stir Welding of DH-36 Steel

T.J. Lienert
Los Alamos National Laboratory
Los Alamos, NM 87545

1.0 Abstract

The objective of this work was to determine the feasibility of FSW for joining DH-36 steel for shipbuilding applications. Welds were produced on 0.18 inch-thick plates of DH-36 steel using tools of W-25%Re with a shoulder diameter of ¾ inches. The tool was rotated at 400 or 500 rpm and travel speeds of either 8 or 10 ipm were employed. Subsequent to welding the welds were evaluated using optical microscopy, microhardness and tensile testing. No measurable changes in tool dimensions were found after welding suggesting that no appreciable tool wear or deformation occurred. Examination of metallographic samples revealed microstructurally distinct regions including the stir zone, heat-affected zone (HAZ) and base metal. Many of the welds contained small defects near the bottom of the stir zone resulting from incomplete bonding. The base metal microstructure was composed of equiaxed grains of ferrite approximately 15 to 20 μm in size with pearlite colonies. The microstructure of the base metal also showed evidence of banding. The stir zone microstructure contained grains of acicular ferrite about 5 um in size. No regions of martensite were observed in the stir zone. The HAZ contained regions of partially-spherodized pearlite (fuzzy pearlite) but no grain-coarsened or grain-refined regions. The banding extended from the base metal into the stir zone. The microhardness of the stir zone averaged approximately 215 VHN, while that of the base metal was about 190 to 195 VHN. Consistent with the lack of martensite, no evidence of hardening was found in the microhardness results. The weld samples showed acceptable strain to failure with yield and tensile strengths exceeding those of the base metal. With one exception, all samples failed through the stir zone owing to the presence of lack-of-bonding defects. These defects can be eliminated with proper tool design. FSW of DH-36 appears feasible, and further work is warranted.

2.0 Introduction

2.1 DH-36 steel

DH-36 steel is a higher strength mild steel used widely in construction of ship hulls. It is produced as a fully-killed, grain refined material to ensure toughness and is supplied in the normalized condition. The nominal composition of DH-36 steel is given in Table 1 (Ref 1). DH-36 is specified to have minimum yield strength of 50 ksi and minimum tensile strength of 70 ksi with minimum elongation of 21 percent (Ref. 1).

2.2 Friction Stir Welding of Steels

2.2.1 Previous Work on FSW of Steels

The pioneering efforts on friction stir welding (FSW) of steels were reported in back-to-back presentations by Thomas et. al of TWI (Ref. 2) and Lienert and Gould of EWI (Ref. 3) at the First International Conference on FSW in 1999. These initial studies focused on determining the feasibility of FSW for joining of steel alloys. A subsequent publication by Thomas and co-workers (Ref. 4) outlined current capabilities of FSW of steel with regard to alloy type, possible plate thicknesses and achievable properties. Successful welds were produced on a low carbon steel as well as a 12% chromium steel. Tensile and bend test on the weld samples showed that acceptable mechanical properties can be achieved.

More recently, Lienert et. al (Ref. 5) have reported on an extensive study of FSW of a 1018 steel alloy. Welds were produced on 6.35 mm-thick plates using a CP W tool held firmly in the tool holder by a collet system. The tool holder was cooled by a mixture of ethylene glycol and water from a chiller system. Weld power and heat input were reported, and a process efficiency of ~0.75 was determined using simple heat flow calculations. A peak temperature just above 1000°C was calculated for the tool shoulder/stir zone interface by extrapolating tool temperatures at steady-state. Microstructures of the stir zone and HAZs were characterized and microstructural evolution of the various weld regions was discussed. Weld samples showed acceptable tensile properties with failure occurring in regions corresponding to the base metal.

Successful friction stir welds have also been produced on two austenitic stainless steels: 304L and AL-6XN (Refs. 6,7). Welds were reported to exhibit high quality with regard to strength and ductility. Residual stresses in the 304L welds were found to be substantial.

2.2.2 Advantages of FSW of Steels.

Recent developments in tool materials and design now make successful FSW of steel alloys possible. FSW offers several possible advantages relative to arc welding for joining of steels. The lower apparent heat inputs of FSW (with respect to arc welding) are expected to minimize grain growth in the HAZ, and to limit distortion and residual stresses. Use of FSW also eliminates concerns over welding fumes, especially those containing Cr^{VI}, to allow compliance with anticipated changes in OSHA standards. Finally, problems with hydrogen cracking are eliminated since FSW is a solid-state process. The objective of this work was to determine the feasibility of FSW for joining DH-36 steel for shipbuilding applications.

3.0 Experimental Procedures

3.1 Friction Stir Welding

All welds were produced on the MTS Process Development System machine at the University of South Carolina using a single tool of W-25%Re (Rhenium Alloys, Elyria, OH) with a shoulder diameter of ~¾ inches. The tool shoulder had a flat geometry rather than a domed geometry. The pin of the tool was not threaded, but was slightly tapered. A 2.5° pushing angle was used for welding. The tool was held in the tool holder with a single set-screw. No direct or indirect cooling of the tool or tool holder was employed. The tool and weld area were protected from oxidation by inert gas shielding in a plexiglass chamber with a sliding top to allow tool traverse. Wear and deformation of the tool were assessed by comparing tool dimensions before and after each weld using an optical comparator.

The welds were made parallel to the rolling direction on plates of DH-36 steel approximately 30 inches long, eight inches wide and ~ 0.18 inches thick in a square butt configuration (30" dimension parallel to the rolling direction). Individual welds of approximately five to six inches in length were produced. The tool was rotated at either 400 or 500 rpm for all welds, and travel speeds of either 8 or 10 ipm were employed. Welds were produced using control of the axial load with loads in the range of 3000 to 5000 lbf. Weld torque was measured in real time during welding.

3.2 Characterization

Subsequent to welding, samples for microstructural evaluation and for tensile testing were removed by water jet cutting. Transverse metallographic specimens were prepared using standard procedures, etched with 2% nital and examined using light optical microscopy (LOM). Vickers microhardness traverses were produced across the weld region at the plate mid-thickness using a 500-gram load and 15-second dwell time. The surfaces of the transverse tensile samples were ground to remove flashing and tested to failure at room temperature.

4.0 Results

4.1 Process Results

The tool glowed red-hot during welding indicating that the tool reached a temperature above 1000° C. A single tool was used to produce approximately six feet of welds. No measurable changes in tool dimensions were found after welding indicating that no appreciable tool wear or deformation had occurred.

Process results are shown in Table 2. Weld energy was calculated using the RPM and measured torque for each weld. Weld torques varied from about 1000 in-lb to 1400 in-lb, and weld energies spanned a range from approximately 1183 to 1755 J/mm. After initial screening experiments, a total of six welds were produced, as shown in Table 2. The first number of each weld trial corresponds to the plate number (either 618 or 619) whereas the second number specifies the order of welds on each plate (1, 2 or 3). The a/b designation indicates the first part of the weld (a) or the second part (b).

An interesting pattern of events was observed for many of the welds. For weld 619-1, for example, the average torque value remained relatively constant after an initial transient period, as shown in Figure 1a. However, for other welds, for example welds 619-2 and 619-3, the torque for the second part of the weld dropped relative to that of the first part, as shown in Figure 1b, despite the fact that input weld parameters were identical. This trend was common to many of the welds.

4.2 Microstructural Characterization

The microstructure of the DH-36 base metal is shown in Figure 2. The base metal microstructure was composed of equiaxed grains of ferrite approximately 15 to 20 μm in size with pearlite colonies. The microstructure of the base metal also showed evidence of banding running normal to the plate thickness.

Metallographic examination of weld samples revealed several microstructurally distinct regions including the stir zone, heat-affected zone (HAZ) and base metal. Microstructures of the various weld regions are presented in Figure 3. Figures 3a and b show the stir zone/HAZ boundary. The HAZ contained regions of partially-spherodized pearlite (fuzzy pearlite) but no grain-coarsened or grain-refined regions. The banding extended from the base metal through the HAZ and into the stir zone without interruption. Figures 3c and d show the microstructures of the stir zone. The stir zone microstructure contained grains of polygonal ferrite about 5 um in size. No regions of martensite were observed in the stir zone. Several of the welds contained small defects near the bottom of the stir zone resulting from incomplete bonding.

4.3 Mechanical Testing

A plot of microhardness as a function of position across the weld region is shown in Figure 4. The microhardness of the stir zone averaged approximately 215 VHN, while that of the base metal was about 190 to 195 VHN. Consistent with the lack of martensite, no evidence of significant hardening was found in the microhardness results. The slight increase in hardness seen in the stir zone is thought to stem from the local refinement in microstructure.

Results of the tensile testing are summarized in Table 3. The weld samples showed acceptable strain to failure with yield and tensile strengths exceeding those of the base metal. With the exception of weld 619-2b, all samples failed through the stir zone as a result of the presence of incomplete bonding defects. Note that mechanical properties of samples taken from the first part of welds 619-2 and 619-3 differed from those of the second part despite the use of constant weld parameters for each weld. The differences in properties may stem from differences in heat input, as explained earlier. However, this interpretation is complicated by the presence of the incomplete bonding defects, and more work is needed to rationalize the possible effects of the variation of heat input on mechanical properties.

5.0 Discussion

Comparisons between the results of this work and those from previous work by the author with a similar steel (Ref. 5) can provide interesting insights into FSW of steels. In the previous work, FSWs were produced on ¼" plates of 1018 steel using commercial purity (CP) tungsten (W) tools at weld speeds ranging from 1 to 4 ipm (Ref. 5). Additionally, weld torques, loads and temperatures were measured, and microstructures and mechanical properties were evaluated. Consequently, comparisons of process results, microstructures and mechanical properties can be made in the following sections.

5.1 Process Results

The change in torque, and hence power and heat input, between the first and second half of many of the welds appeared to coincide with the point in time when the tool reached a constant temperature as indicated by a constant shade of red. Initially, the tool was cold and acted as a heat sink drawing some of the weld energy into the tool. At some point during the weld, the tool reached a temperature equal to that of the top of the stir zone. Since the temperature gradient was then flat, the tool no longer acted as a heat sink and no longer removed weld energy as before. Instead, all of the weld energy was transferred to the workpiece causing an increase of workpiece temperatures. The increased workpiece temperatures in turn promoted greater softening of the workpiece and a lowered torque. The softening also allowed the tool to push deeper into the workpiece (under the constant axial load) producing greater volumes of weld flashing. Indeed, the plot of actual tool position (not shown) monitored during welding indicated that the tool did move downward into the workpiece during the second half of the welds showing a torque drop. Hence, the thermal mass of the tool/chuck assembly plays an important role in welding. The tool time constant for reaching thermal equilibrium should be engineered to be much smaller than the weld time if steady-state conditions are to be achieved.

In the previous work using CP W tools, deformation and wear of the tool caused a loss in pin length of approximately 0.012" for a ten-inch long weld. Most of the loss was found to occur during the plunging period. In the work reported here using W-25%Re tools, no measurable changes in tool dimensions were found after producing about six feet of weld using a single tool. Note that these trials also included at least six plunges. These results suggest that much longer lengths of steel may be friction stir welded successfully with W-25%Re, and that further work on FSW of steels is warranted.

5.2 Microstructural Characterization

The continuity of the banding that extends from the base metal through the HAZ and into the stir zone (Fig. 3) is very interesting. The bands persist largely undeviated through much of the stir zone. This observation suggests that limited vertical flow of material occurred during welding. The limited flow may have resulted from the use of a flat-shoulder tool in the current work. The flat-shoulder tool apparently does not produce the downward motion of material at the trailing edge of the shoulder.

In the current work, small defects resulting from lack of bonding were commonly found in the lower part of the stir zone on the advancing side of the weld, whereas similar defects were not encountered in the previous work. The defect problem also likely stemmed from the different tool designs employed and insufficient downward flow at the trailing edge of the shoulder to close the defects. Hence, use of proper tool designs incorporating domed shoulders are recommended for FSW of steels.

While weld and tool temperatures were not measured directly in the work reported here, the temperature fields are probably similar to those determined previously by the author. The temperatures measured and microstructures observed in the previous work indicated peak temperatures in the stir zone in the austenite phase field. More specifically, the red glow of the tool observed during welding and the microstructure are consistent with temperatures of at least 1200° C.

Unlike the previous study (Ref. 5), little evidence of grain growth was found in the HAZ and stir zone despite the high temperatures. The lack of grain growth likely derived from the lower heat inputs (mainly due to faster travel speed) of the current work and from the differences in deoxidation practices used to produce the two steels. The heat inputs from the current study were approximately 25% to 30% of that from the previous work when the different thicknesses are taken into account. In addition, the DH-36 steel is killed with Al, whereas the 1018 steel of the previous study was killed with Si. The Al additions are known to promote AlN precipitates that pin grain boundaries and limit grain growth in both the stir zone and HAZ. Note that the fine polygonal ferrite microstructure found in the stir zone is indicative of higher toughness. Finally, it is important to point out that cooling rates were insufficient for martensite formation.

5.3 Mechanical Properties

Despite the presence of the lack of bonding defects, the yield and tensile strengths of the weld samples exceeded those required of the base metal by standards (Ref. 1). Note that the percent elongation of the weld samples was lower than that required by standards. However, elongation of defect-free welds normally also appears low relative to the base metal owing to non-uniform elongation throughout the gage length stemming from microstructural gradients created by the welding process. For welded samples, the strain is usually carried by narrow regions of lower strength such as the HAZ, and assumptions made concerning uniform elongation throughout the gage length are not valid. Hence, interpretation of the lower elongation experienced by samples produced in this study is complicated by the coincidence of two factors, the defects and non-uniform elongation. Further work using techniques aimed at measuring strain over the entire weld region (such as applying a grid to the sample and measuring grid spacing before and after straining) is required to separate the two factors.

6.0 Conclusions

The friction stir welds on DH-36 showed acceptable tensile properties with no measurable tool wear. The absence of tool wear stemmed from the use of an improved tool material relative to previous work by the author (W-25%Re vs. CP W). The lack-of-bonding defects found in the stir zone can be eliminated with proper tool design incorporating a domed shoulder. The weld region displayed a microstructurally distinct stir zone consisting of grains of acicular ferrite about 5 um in size and a HAZ containing partially-spherodized carbides. No grain-coarsened region nor grain refined region was observed in the HAZ, and no evidence of martensite was found anywhere in the weld region. Retention of microstructures apparently resulted from the lower heat inputs employed in this study relative to the previous work by the author. The average microhardness of the stir zone was increased relative to the base metal owing to the refined microstructure. Failure of the tensile samples through the stir zone resulted from the presence of the defects. FSW of DH-36 appears feasible, and further work is warranted.

7.0 Acknowledgments

The authors wish to thank L.G. Kvidahl of Northrup Grumman Ship Systems for supplying the material used for this study and for valuable insight. Appreciation is also extended to W. Tang, D. Wilhelm, W. Bailey and R. Beeready of the University of South Carolina (USC) for help with the experiments and to Prof. A.P. Reynolds of USC for helpful and stimulating discussions. Finally, T.J. Lienert wishes to thank LANL and USC for financial support during the course of this work.

8.0 References

1. Datasheet for Lloyd's Ship Hull Steel, 1998.

2. W.M. Thomas, 1999. Friction Stir Welding of Ferrous Materials: A Feasibility Study. *Proceedings of the First International Conference on Friction Stir Welding.* Session 10, Paper 2. TWI.

3. T.J. Lienert and J.E. Gould, 1999. Friction Stir Welding of Mild Steel. *Proceedings of the First International Conference on Friction Stir Welding.* Session 10, Paper 3. TWI.

4. W.M. Thomas, P.L. Threadgill and E.D. Nicholas, 1999. Feasibility of Friction Stir Welding of Steel. *Science and Technology of Welding and Joining.* 4(6): 365-372.

5. T.J. Lienert, W.L. Stellwag B.B. Grimmett and R.W. Warke, 2003. Friction Stir Welding Studies on Mild Steel. *Welding Journal.* 82(1): 1s-9s.

6. A.P. Reynolds, M. Posada, J. DeLoach, M.J. Skinner, J. Halpin and T.J. Lienert, 2001. FSW of Austenitic Stainless Steels. *Proceedings of the Third International Conference on Friction Stir Welding.* Session 2, Paper 1. TWI.

7. A.P. Reynolds, W. Tang, T. Gnaupel-Herold and H. Prask, 2003. Structure, Properties and Residual Stress of 304L Stainless Steel Friction Stir Welds. *Scripta Materialia* 48: 1289-1294.

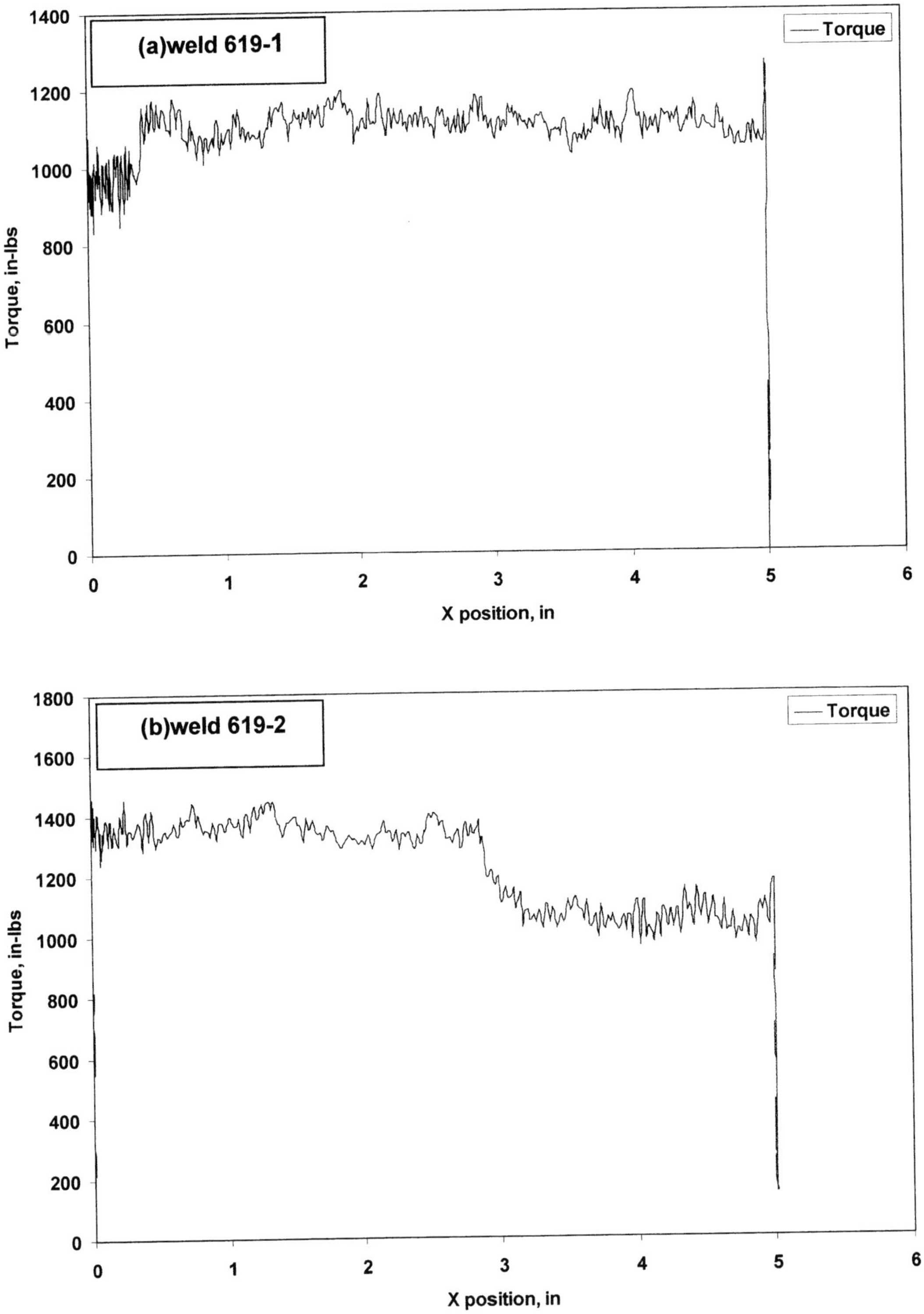

Figure 1 – Torque curves vs. distance for welds (a) 619-1 and (b) 619-2.

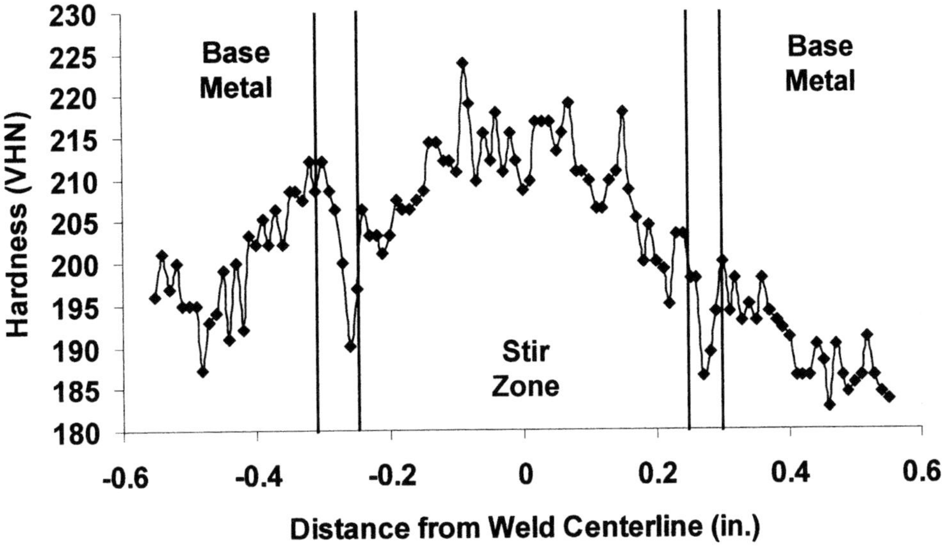

Figure 4 – Microhardness vs position.

Table 1: Nominal Composition of DH-36 Steel (weight percent)

C max	Mn	P max	S max	Si max	Al min	Nb
0.16	0.90-1.60	0.035	0.035	0.50	0.020	0.020-0.050

Table 2: Weld Parameters and Results

Weld #	TS	RPM	Z-load (lbf.)	Torque (in-lb)	Energy (J/mm)	Comments
618-1b	8 ipm	400	3000 lbf.	1003.2	1402.11	Insuff Z
618-2a	8 ipm	400	4000 lbf.	1146.0	1601.69	Good Weld
618-3a	10 ipm	500	4200 lbf.	1255.5	1754.73	Too hot
619-1a	8 ipm	400	4000 lbf.	1112.0	1554.17	Good Weld
619-2a	10 ipm	400	4500 lbf.	1352.0	1511.68	Too hot
619-2b	10 ipm	400	4500 lbf.	1058.3	1183.29	Too cold
619-3a	10 ipm	400	4800 lbf.	1373.6	1535.83	Too hot
619-3b	10 ipm	400	4800 lbf.	1118.3	1250.38	Too hot

Table 3: Mechanical Properties

Weld #	0.2% YS	TS	ε_f
618-1b	68 ksi	81 ksi	8.6 %
618-2a	57 ksi	65 ksi	4.5 %
619-1a	63 ksi	71 ksi	8.0 %
619-2a	64 ksi	73 ksi	3.6 %
619-2b	65 ksi	78 ksi	7.0 %
619-3a	62 ksi	68 ksi	2.6 %
619-3b	66 ksi	78 ksi	6.6 %

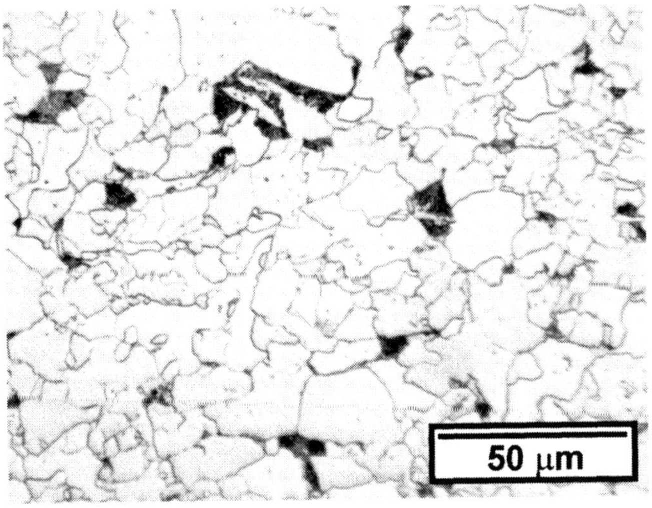

Figure 2 – Base metal microstructure.

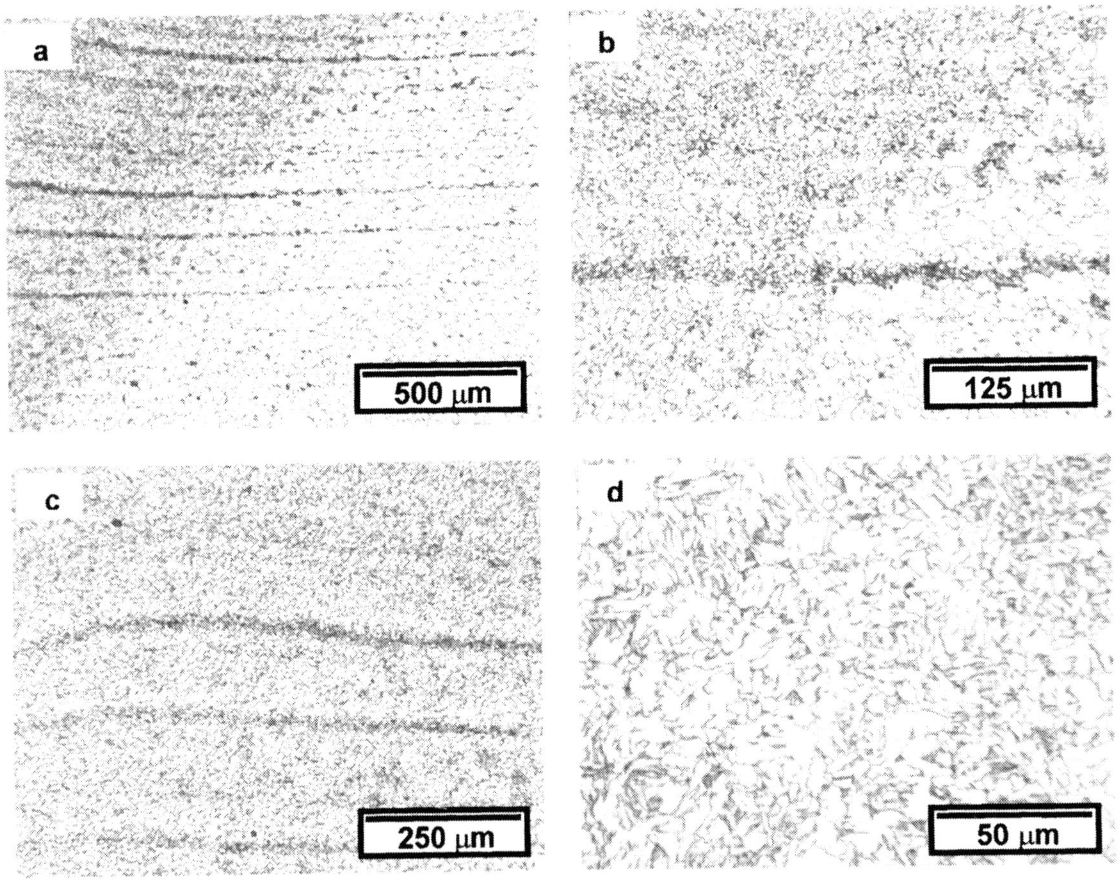

Figure 3 – (a) lower magnification micrograph of base metal (right), HAZ (center) and stir zone (right); (b) higher magnification micrograph of HAZ (right) and stir zone (left); (c) and (d) micrographs of stir zone

Proceedings from Joining of Advanced and Specialty Materials, 13–15 October 2003, Pittsburgh, Pennsylvania, ASM International, 2004

Friction Stir Welding of Copper Canisters - a weld that lasts for 100,000 years

L. Cederqvist
Swedish Nuclear Fuel and Waste Management Co (SKB), Oskarshamn, Sweden

R. E. Andrews
The Welding Institute (TWI), Cambridge, England

Abstract

Friction Stir Welding (FSW) is a novel solid-state joining process where the work pieces are joined together using a rotating non-consumable tool. The process has mainly been used for joining thin section aluminium alloys, and has not yet been used in production for any other metal or thickness above 15mm. However, the results from the development program confirm that FSW can consistently seal 50mm thick copper canisters without creating defects. The Swedish Nuclear Fuel and Waste Management Co (SKB) recently acquired a custom FSW system, which was installed at the Canister Laboratory in April 2003, to further automate the process to production level.

Introduction

The Canister Detailed information about the Swedish system for managing and disposing of radioactive waste can be found at www.skb.se

The canister (~5m long) containing the nuclear waste needs to remain intact for 100,000 years and a corrosion barrier of 50mm thick copper and a cast iron insert for mechanical strength are used to meet this requirement (Fig. 1). The 50mm thick cylindrical body with an outer diameter of 1050mm can be produced (in seamless form) by extrusion or the pierce and draw technique. Alternatively, roll-formed tube halves can be welded together, potentially by FSW (Ref. 1). The canister manufacturing route requires that a bottom and a lid are sealed using a method that produces extremely high joint integrity. The only viable method available in 1982 for welding thick section copper was high power electron beam welding (EBW). In 1997, SKB decided to also investigate FSW as an alternative joining method.

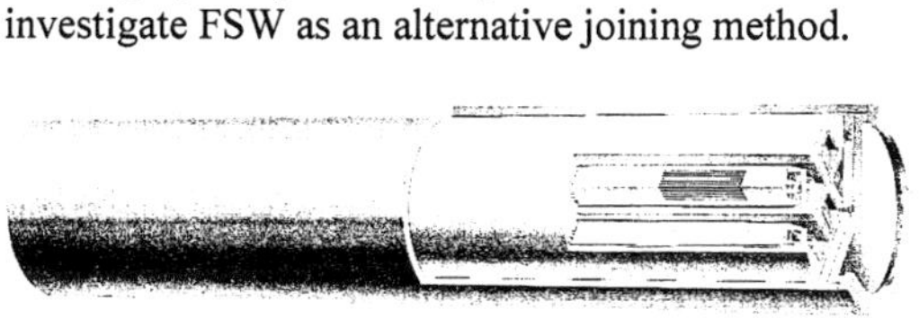

Figure 1 - Copper canister with cast iron insert.

Friction Stir Welding FSW was invented at The Welding Institute (TWI) in 1991 and is a solid-state thermo-mechanical joining process which is a combination of extrusion and forging. A cylindrical, shouldered tool with a profiled probe is rotated and slowly plunged into the material. Frictional heat is generated between the wear-resistant tool shoulder and the material, causing the material to soften without reaching the melting point, and allowing the tool to traverse the joint line. This results in the transfer of plasticized material from the leading edge of the tool to the trailing edge, creating a solid phase bond between the two work pieces. Schematic drawings of the process are shown in Fig. 2, which also provides information regarding the terminology used to describe friction stir welds. Friction stir welds are not symmetric about the centerline due to the tool rotation. The side of the weld on which the rotational velocity of the tool has the same direction as the welding velocity is designated the advancing side of the weld. The side of the weld on which the two velocities have opposite direction is designated the retreating side of the weld.

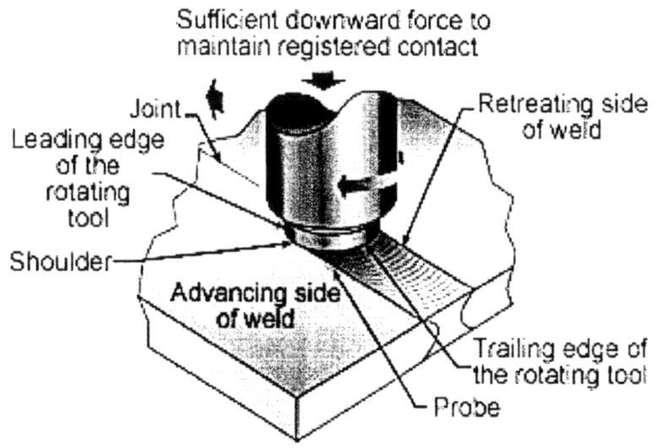

Figure 2 - Schematic of the FSW process (Ref. 2).

In production, the process has only been used for making butt and lap joints in aluminum alloys, mainly for the aerospace industry but also including the marine and automotive industry. In research and development, the process has also been used to join other materials like steel, titanium, and copper. The major difference between aluminum and the other materials is the welding temperature (that is just underneath the melting point). The melting point for aluminium, steel, titanium and copper is 660, 1520, 1680 and 1080°C, respectively. The welding temperature determines the tool requirements and so far it has been difficult to find reliable tool material for steel and titanium. In the tool design and material section, the process to find a suitable probe design and material for FSW of 50mm thick copper is described.

One reason for the rapid success of FSW is the ease of process control due to the relatively few input parameters. The tool is rotated at a specified rotation rate and the tool is traversed along the joint line at a specified traverse rate. The tool is usually also tilted relative to the work piece so that the leading edge is above the surface. The tool shoulder to work piece penetration is then controlled by applying a specified downward force (see Fig. 2) or by specifying the position of the shoulder relative to the work piece. In addition to the input values, the spindle motor torque is measured and provides an indication of the amount of frictional heat generated.

When welding thick section material a pilot hole is drilled to avoid damage to, or fracture of the tool probe. The diameter of the pilot hole is smaller than the top diameter of the tapered probe to enable sufficient frictional heat to be generated. However, starting a weld in a tool probe exit hole could be an important feature in production welding and is under investigation. In this case, a slightly oversized probe might have to be used in order to generate enough heat, and to create a defect-free weld.

Experimental Procedure

Equipment As the FSW process showed great promise when welding 10-50mm thick copper plates using a powerful, modified milling machine at TWI, a full size experimental canister welding machine, where the welding head is stationary and the canister rotates, was designed and built at TWI in late 1999.

Most parameters on the experimental machine were changed manually during the process development which meant reproducibility between welds was not sufficient for production quality. In order to be able to automate the welding process, SKB ordered a FSW machine from ESAB in January 2002 that was ready for welding at the Canister Laboratory in April 2003, and the first lid weld (Fig. 3) were produced in late April 2003. On this machine the welding head rotates up to 400°, while the canister is clamped stationary. The production machine has a capacity of 110kW, which makes it one of the most powerful welding machines in the world. However, in order to be able to control the heat input, only 35-40 kW is used.

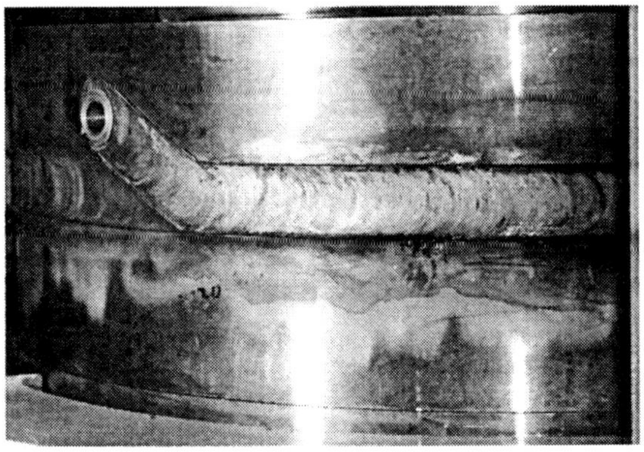

Figure 3 - First lid weld at the Canister Laboratory.

Tool design and material Due to the high welding temperature (up to 950°C), the high welding forces and the duration of the weld (up to 1hr), the demands on the tool probe are very severe.

The tool probe profile that provided the breakthrough when welding up to 50mm thick copper plate was the so-called MX Triflute™ probe (Fig. 4) which increased the maximum welding speed from 40mm/min to 100mm/min (without probe fracture) due to the reduced volume created by the three flutes. Although the tool probe that was developed for welding thick section copper at TWI was essentially an MX Triflute™ variant, its efficiency and reliability at high temperature in combination with the specifically designed probe features warranted a SKB/TWI patent application.

FSW of copper provided a challenge in terms of the selection of a probe material with suitable physical properties. In the assessment of tool probe material for this high temperature application, seventeen different materials were evaluated, and included carbides, ceramics, refractory metals and superalloys.

The current material used is a nickel-based superalloy, Nimonic 105, which is commonly used in aero engines and therefore was readily available. The good creep rupture properties up to 950°C together with the consistent ductility from ambient temperature to 900°C was crucial to the successful application as a tool probe. Increased wear resistance has also been achieved using a proprietary carbo-nitriding process. There is no doubt that superior tool probe materials to Nimonic 105 alloy exist, but at this point further investigation is not a high priority due to the success with the Nimonic 105 alloy.

The majority of FSW tool shoulders have been machined from the sintered tungsten alloy Densimet, which was one of seven refractory metals evaluated as probe material. Its high temperature capability and high thermal conductivity (130W/m°C) make it an ideal shoulder material. Other materials with lower thermal conductivities such as the nickel-based (10-20W/m°C) and cobalt-based (70W/m°C) superalloys have also been assessed but Densimet proved to be superior for maintaining a balanced heat transfer into the weld zone.

Figure 4 - Drawing of the MX Triflute™ design.

Welding material and parameters The use of real canister lids during the development work on the experimental machine could not be justified on the basis of availability, cost and most importantly, the required frequency of welding trials. Instead two 100mm deep ring samples (cut from canisters) were mounted on top of each other and welded. However, the development program at TWI culminated in the production of a real lid weld in January 2003. On the new machine at the Canister Laboratory, only real lids and bottoms have been and will be welded.

The experimental machine could be used in either the position or force controlled mode. The copper rings welded were often not machined to the tolerance required. As a result, the machine was only used in force control to achieve acceptable weld quality. The range of input values that were used successfully during the development stage is as follows: Axial Force 70-100kN, Welding Speed 70-100mm/min, and Tool Rotation 300-500rpm. In addition to the spindle torque, a more meaningful diagnostic parameter was investigated. The welding temperature (measured with a type K thermocouple inside the tool probe) proved to be a more precise weld condition indicator and it was found to be possible to control the weld quality by holding the welding temperature constant by using a controlled cooling system. This thermal management facility has proved to be an extremely valuable addition for production welding, and is the subject of a SKB/TWI patent application. A complementary method of measuring the temperature, infra-red thermography, has been pursued and tested successfully (on the tool shoulder) due to the importance of the welding temperature.

Weld cycle The weld cycle can be divided into three regions, when welding full circumferential copper canisters: start, steady-state and park. For thick section material, the plunge sequence and dwell period are made under force control in order to generate sufficient heat to soften the interface region. During the plunge period the welding temperature is allowed to reach ~500°C, and then the welding speed is accelerated until it reaches the steady-state value, where a constant welding temperature (±10°C) of 800-900°C is attained. Due to the probe exit hole, it is necessary to move the tool upward (~70mm) from the joint line to a position where the exit hole can be machined away.

An automatic start sequence has been developed at the Canister Laboratory, where the welding speed is accelerated as a function of the welding temperature up to the steady-state value, to not create any defects.

After the steady-state welding speed (and temperature) is reached, the welding conditions are kept constant for process stability and reproducibility, and the thermal management facility controls the welding temperature. If the thermal management facility is not able to keep the welding temperature constant at the steady-state value, subtle axial force changes are made to keep the heat input (spindle torque) constant.

Non-destructive testing Three methods are currently being developed and evaluated at the Canister Lab.

Radiographic inspection is performed by a dedicated digital X-ray system consisting of a 9MeV Linear Accelerator that pulses X-rays (60x the power of medical diagnostic x-ray) through the weld zone while the canister rotates. The transmitted x-rays are collected by a linear detector and produce a digital X-ray image. The radiographic inspection will find volumetric flaws but has limited application for non-volumetric flaws.

For ultrasonic testing of EBW, SKB are using the phased array technique. For FSW, this inspection technique is currently being modified and optimised. Ultrasonic inspection is a more suitable method than the X-ray inspection for the detection of non-volumetric flaws.

The eddy current inspection technique is a conductivity-based testing method which is evaluated for inspection of surface flaws. The as welded surface is unsuitable for eddy current inspection as it is too rough. However, in production, 3mm of the surface will be removed by milling to produce a smooth surface, which will make it possible to use eddy current inspection for detection of near surface flaws.

Results and Discussion

Rings at TWI All welds made were analyzed by visual inspection, metallographic examination and/or non-destructive testing. At an early stage, small sub-surface voids (~1mm) were sometimes detected towards the advancing side (Fig. 5a). These voids were found to be associated with changes in welding speed when attempting to control the welding temperature, and were also found using both radiographic and ultrasonic inspection (Fig. 5b-c). It should be noted that these voids could be repaired by a weld overlap, and that no flaws were found when the process was controlled with an automatic thermal management facility, instead of welding speed changes.

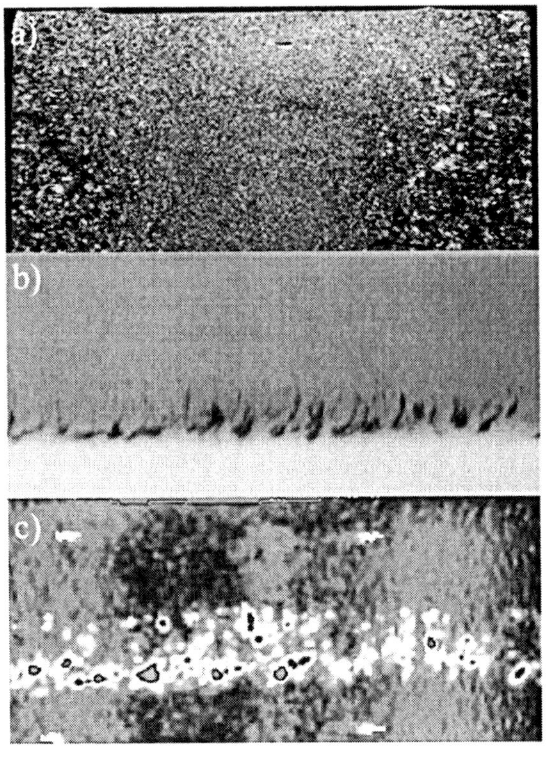

a) Metallographic cross section of sub-surface void.
b) X-ray image of advancing side defects.
c) Ultrasonic image of advancing side defects.

Lid weld at TWI The development program at TWI culminated with a lid weld in January 2003. It was found that the same welding parameters could be used successfully on both rings and lids, however a new problem occurred. Excessive flash was formed on the canister side (Fig. 6a), which results in more heat being generated and makes the process more difficult to control. Since the excessive flash was caused by a hotter canister than lid (thermal expansion), a more symmetric joint location (Fig. 7) were designed.

Figure 6 - Flash generation in steady-state region.
a) Old joint line design, b) New joint line design.

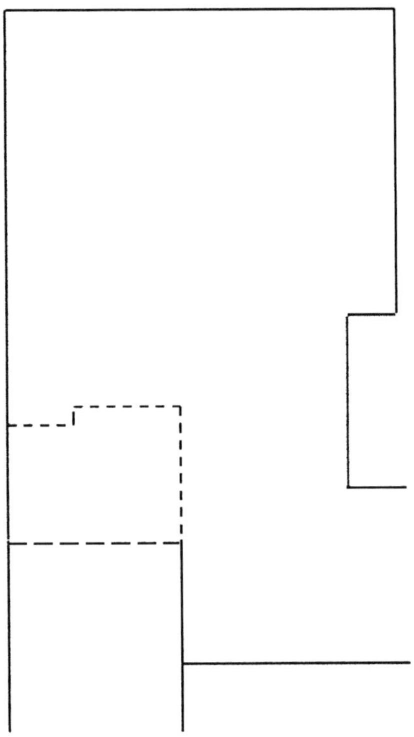

Figure 7 - New joint line design (short dashed line), and old joint line (long dashed line). Picture to scale.

Lid welds at the Canister Lab. Already during the first lid weld (Fig. 3) at the Canister Laboratory, it proved possible to transfer the technology from the experimental machine to the production machine with successful weld production. The result of the new joint line design can be seen in a close-up picture (Fig. 6b) of the same weld. So far, five lid welds have been made, and in four of them no defects have been found at the joint line by radiographic examination. The fifth lid weld was made with a machined defect (31.5x2mm) to see if it could be repaired. During the start region, it was not repaired completely, but during the steady state region, it was repaired.

In addition to the use of an automatic thermal management facility during the steady-state region, an automatic start-up sequence has been developed on the new machine to eliminate the chance of defect formation during welding speed acceleration. Although this automatic start-up sequence has proved successful, the weld cycle is now started above the joint line (Fig. 8) until steady-state welding conditions are reached, due to pilot hole and the potential of defect formation during the start sequence. As a result, only steady-state microstructure (Fig. 9) is produced at the joint line.

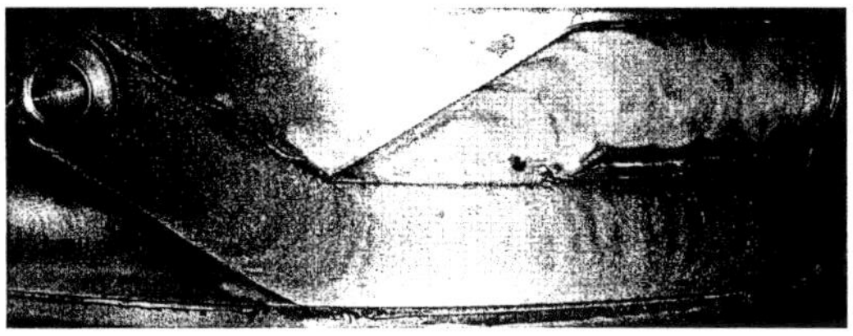

Figure 8 - Start and park above joint line.

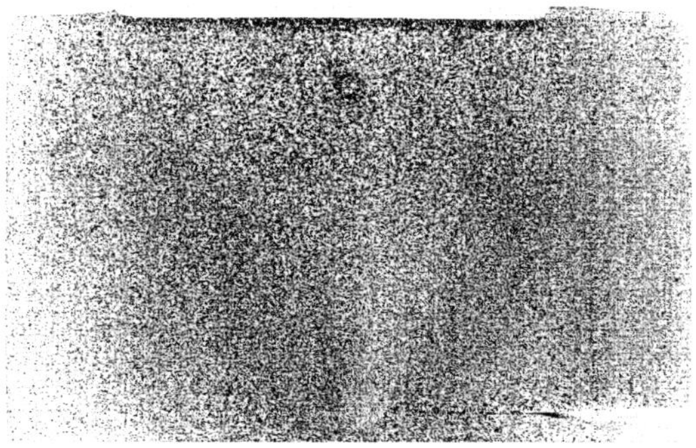

Figure 9 - Metallographic cross section from the steady-state region of the lid weld at TWI.

Tensile results Tensile tests were made from the lid welds, both from TWI and the Canister Laboratory, to examine the quality of the weld joint. All tensile specimens failed outside the weld zone in the so-called heat-affected zone at a stress level of 207-208 MPa, which is at least 99% of the base material value (208-210 MPa). The specimens were 45mm wide.

Conclusion

The results of the program so far have shown that FSW can be used to seal 50mm thick copper canisters with the production of high integrity joints. Even at this stage of the development, the potential for use in the production site (SKB's Encapsulation Plant) is clear.

The development has involved three major components, which are the tool probe material and geometry, the experimental canister welding machine and the creation of an automatic thermal management facility.

The combination of a viable tool probe and machine quickly revealed that the major cause of flaws was the introduction of process instability. Instability was created by changes in the welding parameters. The need to introduce process stability led to the design and build of a thermal management system to control the welding temperature. This system has greatly simplified weld production and improved quality. The advantages of this system include, but are not limited to:

- Welding parameters and conditions are set to a constant value throughout a weld cycle.
- The temperature at which the weld is made can be chosen and, as a result, grain size can be controlled.
- The tool probe and shoulder is never exposed to excessively high temperature.

In conclusion, this type of control system could help improve the quality and reproducibility for production welding of, not only SKB's thick section copper application, but all thicknesses and materials such as aluminium, titanium and steel. In addition, the tool probe feature developed in this program could be essential for quality and speed of thick section welding of other materials.

References

1. Cederqvist, L., and Andrews, R. E. *Proceedings of 4th International Friction Stir Welding Symposium*, Park City, UT (2003).
2. Kallee S. W., and Nicholas E. D. *www.twi.co.uk/j32k/unprotected/band_1/fswintro.html*

Proceedings from Joining of Advanced and Specialty Materials, 13–15 October 2003, Pittsburgh, Pennsylvania, ASM International, 2004

The Effect of Rolling Strain on the Processing of Intermetallic Sheet Materials Using Roll Welding and Reaction Annealing

Gajanan P. Chaudhari and Viola L. Acoff
The University of Alabama, Department of Metallurgical & Materials Engineering, Tuscaloosa, AL 35487

Abstract

Intermetallic sheet materials that are difficult to process by conventional techniques can be processed using a method involving cold roll welding and reaction annealing of elemental foils. This method was used to study the reactions that form at the roll welded interface as a function of rolling strain using the Ti-Al system as the model. Titanium aluminides were successfully produced at the Ti/Al interface by cold roll welding of elemental titanium and aluminum foils and subsequent two-stage reaction annealing. Differential scanning calorimetry (DSC) studies revealed that increased rolling strain increased the exothermicity of the reactions. Increasing the strain was found to accelerate the diffusion rate of the metallic elements, which could lead to a reduction in processing time. The thickness, morphology, and composition of the intermetallic compounds that formed at the Ti/Al interface were examined using light microscopy and scanning electron microscopy equipped with energy dispersive spectrometry. The phases formed were verified using x-ray diffraction.

Introduction

The processing and fabrication of certain alloys and layered composites is inordinately expensive because of the sophisticated methodologies associated with the existing processing techniques for these materials. Among these materials that are difficult to process by conventional techniques are intermetallic compounds. Intermetallic compounds are important technologically because they are light in weight and maintain high strengths at elevated temperatures. As a result, intermetallic compounds have considerable potential in numerous aerospace, structural, and high temperature applications. However, the lack of breakthroughs in cost-effective techniques is partly responsible for the lack of widespread use in industry [1-5].

The processing of gamma-based alloys in the form of sheets and foils has been of interest for certain engineering applications. In particular, structural and high temperature applications such as hypersonic spacecrafts and future gas turbine engines are ideal candidates for gamma-TiAl sheet materials [6]. Presently, sheet processing of gamma-based TiAl alloys is accomplished using two different routes: ingot metallurgy (casting followed by hot isostatic pressing) and powder metallurgy [7,8]. Although these processes have proven to be reliable, neither method can directly process gamma-TiAl with good compositional uniformity and a fully lamellar microstructure. Additional processing steps such as extended heat treatments and mechanical alloying are required to improve the properties to desirable levels [9-13]. The additional operations result in a coarsening of the lamellar microstructure and sometimes disrupt unstable lamellar microstructures. In addition, regardless of their improved properties by alloying, gamma-TiAl alloys are relatively brittle and difficult to shape into sheet materials via conventional forging and rolling methods [14-17]. These existing methods for processing gamma-TiAl sheet materials are also rather expensive [18-19]. Although several studies [20-24] have investigated reactive processing for producing intermetallic compounds and metal-intermetallic composites, all require the application of isostatic pressure during annealing.

The use of roll bonding and reaction annealing can prove to be a viable technique for processing of sheets of difficult-to-process materials. The authors have previously developed a sheet processing method for gamma-TiAl [25-29]. In the following study the effect of rolling strain on the processing rate of titanium aluminide sheet materials is discussed.

Experimental Procedure

Elemental titanium (99.9% pure, 0.074mm thick) and aluminum (99% pure, 0.070mm thick) foils were employed. The foils were given a surface treatment consisting of degreasing with acetone followed by scratch brushing and cleaning with a stainless steel brush having wire diameter of 0.32mm. They were stacked in 21 alternating layers with titanium foils at the top and bottom. These laminated "sandwiches" were cold roll welded using 30-35%

reduction in the first pass. A 2-high rolling mill with 100mm diameter rolls was used, with both rolls revolving at the same surface speed of 0.192 m/sec. All rolling was done under dry friction conditions. Welded specimens were rolled further to different (50, 70, 80, 90, 97.6, and 98.77%) reductions in thickness.

Cyclic rolling experiments were performed to introduce severe strain in the sheet. The roll-welded composite reduced to 50% of the original sandwich thickness, was scratch brushed, cut into half, restacked and rolled again. Rolling reduced its thickness by 50%. This constituted one cycle. These cycles were performed 2, 4, 8, and 11 times to prepare samples for this study. Sound sheets were obtained until 11 cycles after which edge cracking became significant. Overall reductions of up to 99.97% could be achieved.

Specimens were observed in a direction transverse to the rolling direction using scanning electron microscopy (SEM) with backscattered electron imaging. Microstructural observation was also done using transmission electron microscopy (TEM). Cross-sectional specimens were prepared and ion milled. TEM studies were conducted using a Hitachi H-8000 microscope operated at 200kV. Differential scanning calorimetry (DSC) was performed on welded specimens. A constant scan rate of 10°C per minute was used during heating from 50°C to 740°C in argon atmosphere. Energy dispersive spectroscopic analysis was used for composition while X-ray diffraction (XRD) was used for phase determination.

Roll welded multi-layered Ti/Al specimens were given two stage annealing treatment in a tube furnace. The tube furnace was evacuated and then backfilled with argon. The specimens were encapsulated in titanium foil bags prior to heat treatment. The first stage annealing treatment was carried out at 600˚C to convert the elemental aluminum into $TiAl_3$. After converting the aluminum, a second stage heat treatment at 1225˚C was employed to promote the formation of gamma titanium aluminide.

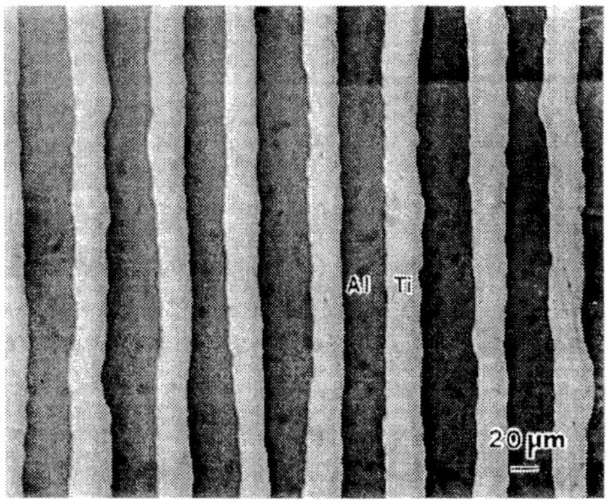

(a)

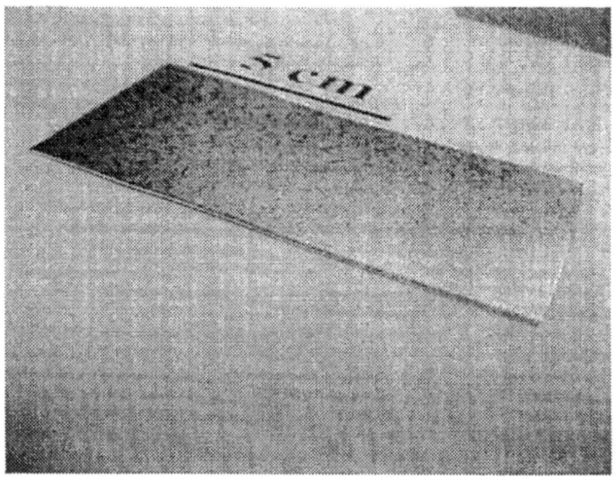

(b)

Figure 1. (a) Light micrograph of the as-cold roll welded specimen subjected to 50% total reduction. (b) Representative Ti/Al composite sheet produced using the cyclic rolling method.

Results and Discussion

Excellent bonding was obtained when the rolling reduction in the first pass was more than 30-35%. Figure 1a shows a light micrograph of the transverse section of a cold rolled specimen subjected to 50% overall reduction, which was obtained in three rolling passes. It shows a well-bonded composite with nearly straight interfaces. Figure 1b shows a representative specimen of the cyclic cold roll welded Ti/Al multi-layered composite obtained after 8 cycles.

Figure 2 shows the DSC traces of specimens cold rolled to 70, 80, and 97% reduction. It shows a broad endothermic heating up to about 500°C for all three specimens indicating the absence of any solid-state reaction. For the 97% reduction specimen, the exothermic reaction begins at 500°C, reaches a maximum at 640°C, followed by an endothermic reaction corresponding to fusion of unreacted aluminum. Comparison of the DSC curves clearly shows that the amount of heat generated in the reaction increases with increased rolling strain.

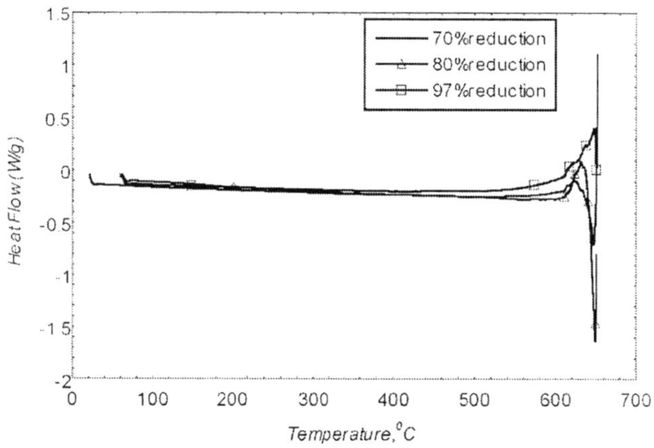

Figure 2. DSC traces of cold rolled specimens subjected to 70, 80 and 97% reduction. (Scan rate 10˚C/min).

The specimens rolled to different reductions in thickness were subjected to one-hour annealing treatment at 600°C. Figure 3 shows the relevant scanning electron micrographs obtained using backscattered electron (BSE) imaging. This figure shows that as the amount of reduction increases (Figure 3a to Figure 3c), the thickness of the reaction layer formed at the interface also increases. SEM/EDS analysis of the layer formed indicated that the composition corresponded to the intermetallic compound $TiAl_3$. The increased formation of $TiAl_3$ at the interface as a function of rolling strain was confirmed by XRD as shown in Figure 4.

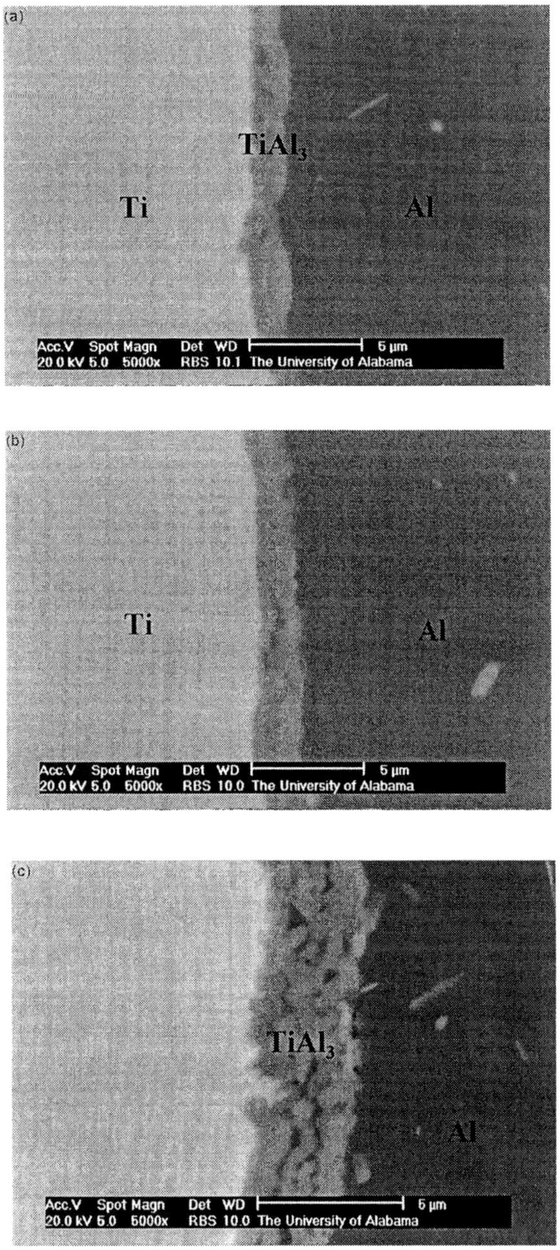

Figure 3. SEM backscattered electron images of (a) 50%, (b) 70%, and (c) 90% reduction specimens subjected to 1-hour heat treatment at 600°C.

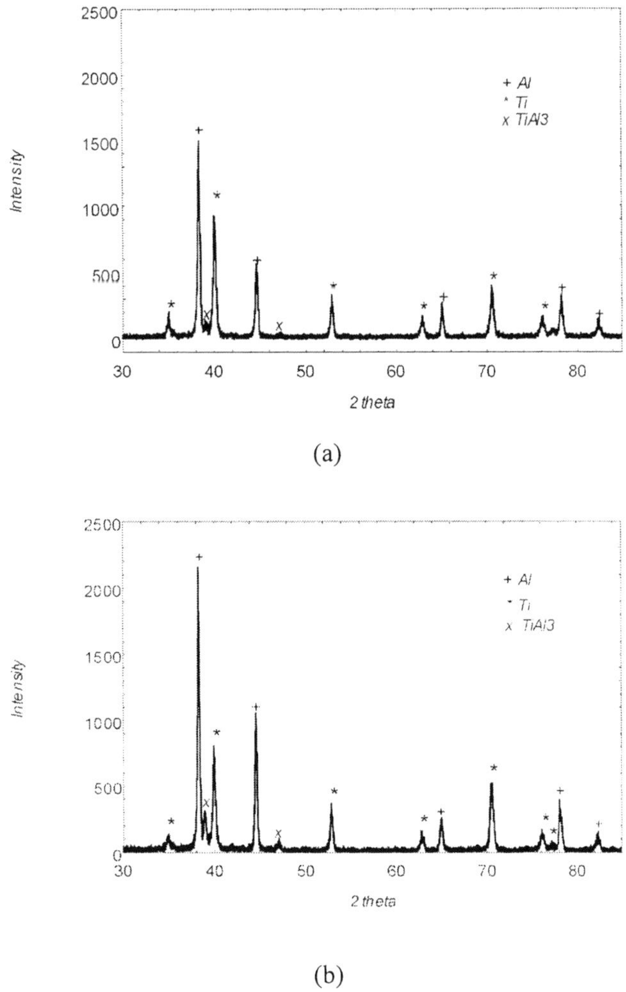

Figure 4. XRD plot for the multi-layered composite subjected to the first stage of annealing at 600°C for 1 hour. (a) 70 % reduction in thickness (b) 90% reduction in thickness.

The cyclic roll welded specimens were also subjected to annealing treatment at 600°C for 1 hour. Figure 5 shows the effect of rolling strain on the microstructural development. As noted previously, the rate of reaction increased as the number of cold rolling cycles increased. It can further be observed that increasing the number of cycles increases the number of interfaces, thereby providing increased contact area for the reaction to occur. It is, however, noted that the occurrence of the reaction product thickness is not uniform.

A cross sectional TEM bright field image of the cold roll welded specimen rolled to 80% reduction is shown in Figure 6. The figure shows that the aluminum layer has been deformed to form sub-micron sized grains. This is also evident in the corresponding selected electron diffraction pattern shown in the inset. This refining of grain size causes an increase in grain boundary area. Other researchers have also observed grain refining as a function of cold rolling [30]. Since grain boundary diffusion is the dominant mechanism of mass transport in the Ti-Al system [31], grain refinement caused by increased rolling strain can enhance the diffusivity substantially. Cold working is also known to introduce lattice defects. The reaction enhancement may occur by short-circuit paths along the defects and grain boundaries created in the deformation. These activate the solid-state reaction to form $TiAl_3$. From microscopic and DSC studies, it is apparent that extensive cold rolling is beneficial in facilitating the processing of intermetallic sheets by improving the kinetics of the reaction.

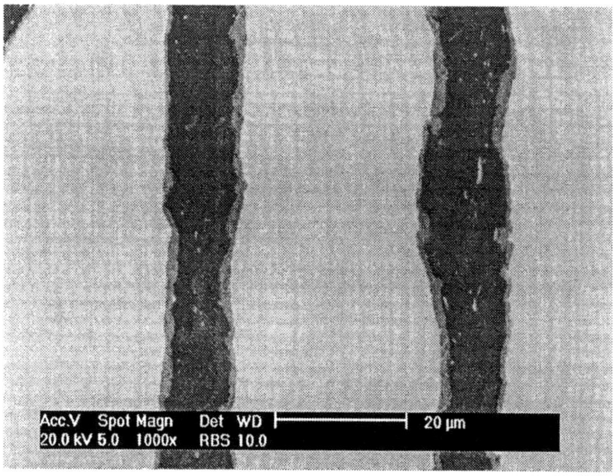

(a)

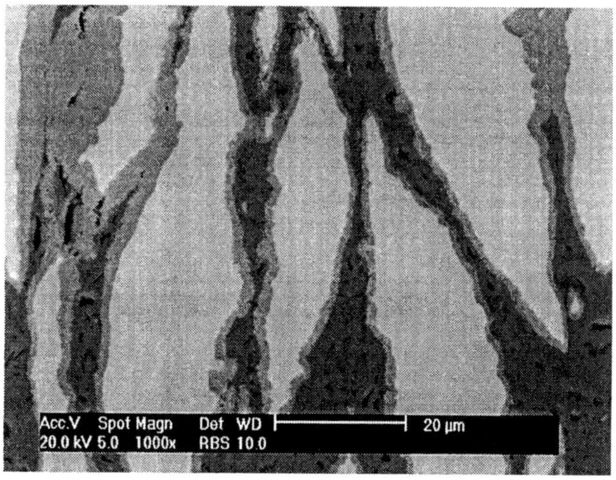

(b)

Figure 5. SEM BSE images of (a) 2 cycles, (b) 8 cycles, and (c) 11 cycles specimens that were subjected to 1-hour heat treatment at 600°C. (The light phase is Ti, the dark phase is Al, and the interface phase is $TiAl_3$)

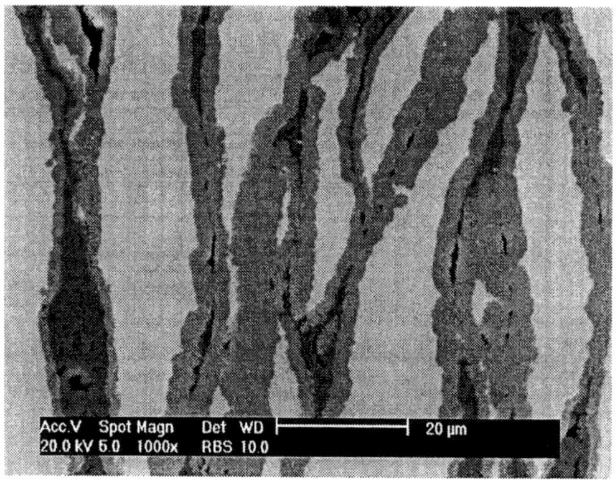

(c)

Figure 5 (continued). SEM BSE images of (a) 2 cycles, (b) 8 cycles, and (c) 11 cycles specimens that were subjected to 1-hour heat treatment at 600°C. (The light phase is Ti, the dark phase is Al, and the interface phase is $TiAl_3$)

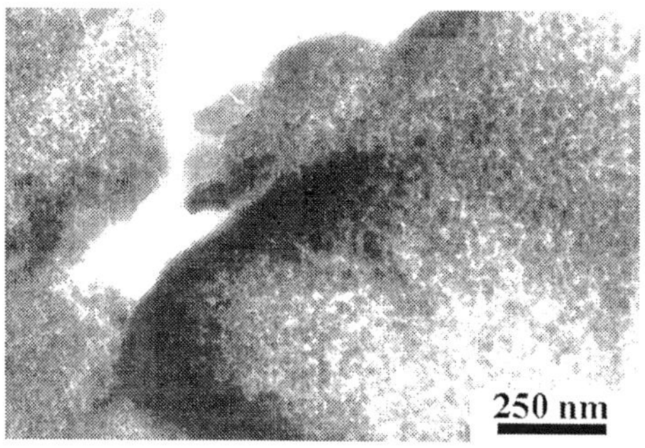

(a)

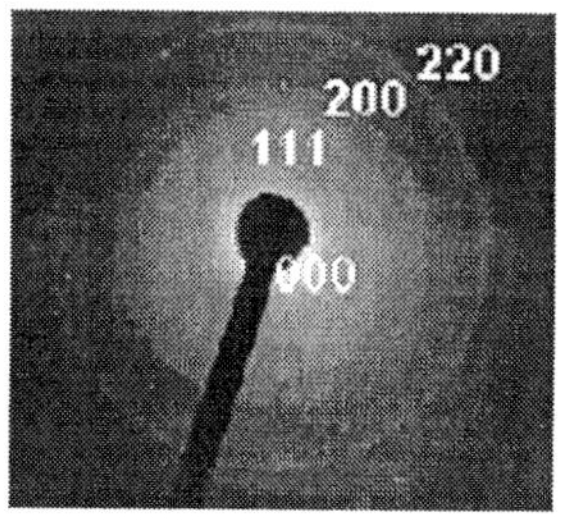

(b)

Figure 6. (a) Cross sectional TEM brightfield micrograph of the aluminum layer in the cold roll welded composite. (b) Corresponding diffraction pattern for TEM image shown in (a).

Extended annealing treatment at 600°C resulted in a composite consisting of layers of $TiAl_3$ and titanium. Second stage annealing treatment was given at 1225°C for 8 hours. The resulting microstructure is shown in Figure 7. Energy dispersive spectroscopy analysis indicated that the microstructure consisted of grains of alpha-2 (Ti_3Al) and gamma (TiAl) phases. The microstructure also reveals the presence of Kirkendall porosity.

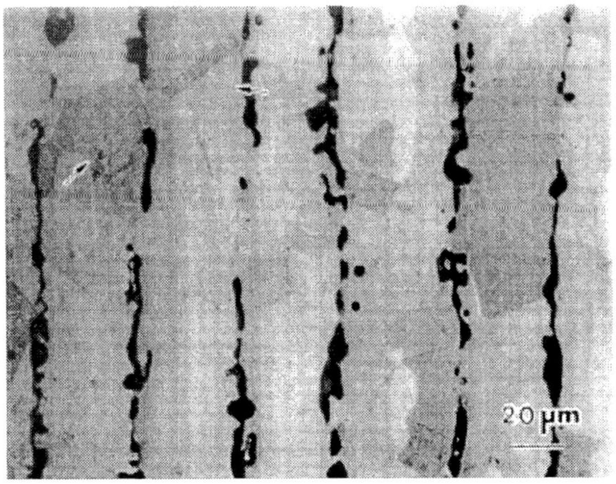

Figure 7. Microstructure after the second stage heat treatment at 1225°C for 8 hours. (Arrows denote TiAl grains)

Conclusions

Titanium aluminide intermetallic sheets were processed using a method involving roll welding of elemental foils followed by reaction annealing. The effect of rolling strain on the formation of $TiAl_3$ was studied. It was observed that employing increased rolling strain using the cold roll welding process accelerated the solid-state reactions that occurred during annealing. This will help in reducing the time required for the processing of the intermetallic sheets by this method.

Acknowledgments

The authors gratefully acknowledge the National Science Foundation for financial support (grant number DMR-0206826) and Dr. C.C. Sorrell for donating the titanium foils.

References

1. M. Yamaguchi, H. Inui and K. Ito, *Acta Mater.*, 48, 307-322(2000)
2. S. Polkin and F.H. Froes, *Int. J. Materials and Product Technology*, 10(1-2),106-120 (1995)
3. D.M. Dimiduk, *Mater. Sci. and Eng.* A, 263, 281-288(1999)
4. Y-W. Kim, *JOM*, 47(7), 39-41(1995)
5. E.A. Loria, *Intermetallics*, 9, 997-1001(2001)
6. H. Clemens, *Z. Metallkd.*, 86, 814-22(1995)
7. Y-W. Kim and D.M. Dimiduk, *JOM*, 43(8), 40-47(1991)
8. G.X. Wang and M. Dahms, *JOM*, 45(5), 52-56(1993)
9. P.J. Maziasz and C.T. Liu, *Met. Trans. A*, 29A(1), 105-17(1998)
10. R.M. Imayev, V.M. Imayev and G.A. Salishchev, *Journal of Mater. Sci.*,27,4465-71(1992)
11. J.M. Larsen, B.D. Worth, S.J. Balsone and J.W. Jones, *Gamma Titanium Aluminides* (TMS Warrendale PA) 821-834(1995)
12. C. Suryanarayana and F.H. Froes, *Advanced Materials*, 5(2), 96(1993)
13. Wegmann G., Gerling R., and Schimansky F., *Acta Mater.*, 51(3) 741-752(2003)
14. T.S. Srivatsan, M. Strangwood and W.O. Soboyejo, *Journal of Mater. Sci.*, 31, 2193-98(1996)
15. V. Recina, J. Ahlstrom and B. Karlsson, *Materials Characterization*, 38,287-300 (1997)
16. J. Kumpfert, Y-W. Kim and D.M. Dimiduk, *Mater. Sci. & Eng.* A, 192-193,465-73(1995)
17. P.J. Maziasz, R.V. Ramanujan, C.T. Liu and J.L. Wright, *Intermetallics*, 5, 83-95(1997)
18. M.M. Keller, P.E. Jones, W.J. Porter III and D. Eylon, *JOM*, 49(5), 42-44(1997)
19. P. Bartolotta, J. Barrett, T. Kelly and R. Smashey, *JOM*, 49(5), 48-76(1997)
20. H. Umehara, T. Suzuki, and R. Hayashi, *J. Japan Inst. Metals*, 58(9), 1050-54(1994)
21. D.E. Alman and C.P. Dogan, *Metall. Trans. A*, 26A(10), 2759-63(1995)
22. A. Jakob and M.O. Speidel, *Materials Science and Engineering*, A189, 129-36(1994)
23. T.J. Jewett and M. Dahms, *Journal of Materials Synthesis and Processing*, 3(5), 339-51(1995)
24. Neves F., Marcelo T., and Carcalho M.H., *Key Engineering Materials*, 188,27-36(2000)
25. J.-G. Luo and V.L. Acoff, *Proceedings of the 79th Annual AWS Convention* (AWS Miami, FL), 311(1998)
26. J.-G. Luo and V.L. Acoff, *Proceedings of the International Symposium on Gamma Titanium Alumindes* (TMS Warrendale, PA), 331 (1999)
27. J.-G. Luo and V.L. Acoff, *Joining of Advanced and Specialty Materials II* (ASM International Materials Park, OH) 97-102(1999)
28. J.-G. Luo and V.L. Acoff, *Welding Journal*, 79(9),239-s-243-s(2000)
29. D.J. Goda, N.L. Richards, W.F. Caley and M.C. Chaturvedi, *Mater. Sci. & Eng A*, 334(1-2),280-90(2001)
30. A. Sagel, H. Sieber, H.-j. Fecht and J. H. Perepezko, *Acta Mater.*, 46(12), 4233-41(1998)
31. F. J. J. van Loo and G. D. Rieck, *Acta Metallurgica*, 21, 73-84(1973)

Proceedings from Joining of Advanced and Specialty Materials, 13–15 October 2003, Pittsburgh, Pennsylvania, ASM International, 2004

Brazing using 2.45 GHz Microwaves

W.M. Drake, D.D. Kautz, R.U. Vaidya, P.W. Hochanadel, V.R. Dave
Los Alamos National Laboratory
P.O. Box 1663, MS E506
Los Alamos, NM 87545

Presented at
ASM Materials Solutions
Conference and Exposition
David L. Lawrence Convention Center
Pittsburgh, PA
October 13-15, 2003

Abstract

Brazing using microwave technology to heat the filler metal possesses many advantages over conventional heating methods. Microwaves are an efficient source of heat, reducing the power needed for brazing due to the localization of the heating. This technology will also allow multiple assemblies to be brazed during a single brazing furnace cycle. Microwaves can easily be configured for various geometries. Applications where different atmospheres are required can all be accommodated by this technology. Several joints can be brazed during one cycle with an incremental increase in the power applied to specimens. Microwave heating reduces the need for water cooling of furnace components. Several joints have been fabricated using a microwave heat source. The metallurgical, as well as, mechanical properties of the joints have been determined and measured. This research has shown that joints produced by microwave heating are similar in structural integrity to those produced by conventional methods, while still providing the process improvements and features listed previously.

Introduction

Microwave technology has been used since the late 1930's when a German refugee living in England, Kassner, patented a "process for altering permanently as well as temporarily the energy content of dipolar substances by exposing them to rapidly oscillating electromagnetic fields [1]." However, "substantial industrial use did not start until after World War II [2]."

Conventional thermal processing utilizes convection, conduction, and radiation to transfer energy to the surfaces of materials to be heated. Microwaves do not transfer heat as these other processes do. Rather, thermal energy is converted from electromagnetic energy via molecular interaction or coupling. Microwaves can penetrate into materials and release energy throughout the volume. In this manner, microwaves generate heat inside materials rather than just at the surface. Since heat isn't diffused into the material from the surface, microwaves provide a quick and uniform heating of thick materials. This provides a means for reduction in conventional process time, and therefore improved efficiency [3].

Microwaves will interact with materials in one of three general ways which are determined by the material's dielectric loss factor. Materials with low dielectric loss factors will be transparent, while materials with high dielectric loss factors (conductors) will be opaque to the microwaves and reflect them. Between these two ends there exists a region where microwaves will be absorbed depending on the dielectric loss factor. A material's dielectric loss factor is also dependent on temperature. Alumina is transparent to microwaves until a temperature of approximately 1000°C is reached at which time it will begin to absorb microwave energy [4].

Silicon carbide's dielectric loss factor exhibits a strong dependence on temperature. When exposed to a 2.45 GHz microwave field the loss factor is 1.71 at room temperature. However, when the temperature is raised to 695°C the loss factor increases to 27.99. This temperature dependency stems from the increase in mobility of dipoles within the silicon carbide [3, 5]. The preferred defects in the silicon carbide structure are vacancies in the silicon or carbon and a substitutional atom in either lattice. These defects preserve the tetrahedral symmetry. There exists a static attraction between defects of opposite charge that form the dipoles [6].

This work explores the use of microwave energy for brazing. Since the metals used for this work reflect microwave energy, a silicon carbide susceptor is used to convert the energy into heat. The susceptor then transfers the heat to the braze joint by radiation.

The purpose of this work is to show that microwave energy is an economical choice for process improvement. The optimum brazing temperature and soak time are discussed. Also, multiple joints can be brazed simultaneously with a minimal increase in power input. The literature reviewed for this work is not inclusive; rather readers are encouraged to refer to the reference section for further study.

Experimental Procedure

Microwave Furnace - This study used a variable power, fixed frequency (2.45 GHz) microwave furnace built by Micramics Inc. This frequency is denoted as one of the internationally recognized Industrial, Scientific, and Medical (ISM) bands. In order to maintain a uniform microwave field, the furnace is designed to produce multiple resonant modes inside the chamber. The furnace chamber is entirely made of aluminum because it reflects microwaves. Also, the size of the chamber plays a key role in ensuring a uniform field with regards to the number of modes and each mode's resonant frequency. The number of possible modes (N) within a volume is given by [7]:

$$N = \frac{8\pi}{3}\frac{V}{\lambda^3} \qquad \text{Eq. 1}$$

where:

V = chamber volume
λ = wavelength

According to Eq. 1, the number of possible modes in the chamber is proportional to its volume. For a rectangular chamber, the resonant frequency for each mode is given by [3, 8]:

$$f_{nml} = c\left[\left(\frac{l}{2d}\right)^2 + \left(\frac{m}{2b}\right)^2 + \left(\frac{n}{2a}\right)^2\right]^{\frac{1}{2}} \qquad \text{Eq. 2}$$

where:

f_{nml} = mode's resonant frequency
c = speed of light
n, m, l = the number of half sinusoid variations in the standing wave pattern along the x, y, and z-axes
a, b, d = dimensions of the chamber in the x, y, and z directions

Eq. 2 illustrates the degree in which a mode's resonant frequency increases with the number of half sinusoid variations. The Micramics Inc. microwave furnace is cylindrical with a radius of 24.8 cm (9.75 in) and a height of 57.15 cm (22.5 in). The door is not circular and the depth of the chamber is 37.5 cm (14.75 in) therefore the chamber volume is approximately 0.08 m^3 (3 ft^3).

Multiple microwave inputs will also assist in uniformity [3, 9]. Therefore this chamber has two waveguide inputs that are orthogonal to each other. The power is supplied by two switching power generators supplying the magnetrons with a capacity of 3 kW forward power each, giving a maximum of 6 kW forward power to the chamber. According to Eq. 1, the number of possible microwave modes per waveguide is approximately 400.

Reduction of the reflected power that could damage the magnetrons must be mitigated. Therefore, these waveguides use a three stud tuner which helps the user to match the impedance of the microwave source to the load.

Materials - The braze materials used in this study consisted of a 1.524 mm (16 GA), 304 stainless steel base metal joined with 0.0762 mm (0.003 in) thick filler metal, AWS designation BVAg-32. The nominal composition of this filler metal is 54Ag/21Cu/25Pd. This combination of brazing alloy for stainless steel exhibits a high degree of wettability. The wettability index number for these materials is 0.226 at 1000°C. Wettability index "values greater than 0.10 are indicative of excellent performance during brazing [10, 11]." Also, a high temperature flux (AWS designation FB3D) was brushed onto both sides of the joint. This flux is active from 760 to 1205°C.

A silicon carbide ceramic was used as a susceptor due to its optimal dielectric properties previously discussed. The exact processing and bonding of the susceptor is proprietary information.

Ceramic casketing – The braze metals are surrounded by the ceramic susceptor which radiates heat to the metal. This type of design is referred to as a casket. The metal to be brazed is placed on top of a silicon carbide plate with a mass of 731.6 g (1.6 lb). Then another silicon carbide plate weighing 62.3g (2.2 oz) is placed on top of the joint. The top ceramic serves a dual purpose:

- to heat the metal
- to serve as a weight to ensure intimate braze metal to base metal contact

The casket is then finished when the susceptor is surrounded by a ceramic layer of fibrous alumina. This microwave transparent layer assists in reducing the thermal gradient due to heat loss at the surface of the silicon carbide [4].

Temperature measurement – A Mikron M190R 2-color infrared thermometer was used to measure the temperature of the ceramic casketing. The unit was standing on a tripod and focused on a mirror allowing clear sight of the top ceramic plate through a hole in the alumina insulation.

Joint design – A braze that is easy to assemble as well as test was needed. Therefore, a lap joint that corresponds to ASTM D 1002-01, (Standard Test Method for Apparent Shear Strength of Single-Lap-Joint Adhesively Bonded Metal Specimens by Tension Loading (Metal-to-Metal)), was chosen. Figure 1 shows the ASTM standard specimen dimensions [12]. This standard method calls for a 2.54 cm (1 in) wide braze and the stainless coupons were 7.62 cm (3 in) long. However, different over lapping (bonding) lengths were used during the experiments. In order to optimize the braze parameters of temperature and soak time, a bond length of 2.54 cm (1 in) was fabricated. Next, to show that two joints could be brazed simultaneously, a bond length of 12.7 cm (0.5 in) was fabricated. Lastly, in order to determine the shear strength, a bond length of 4.6 mm (0.18 in) was used.

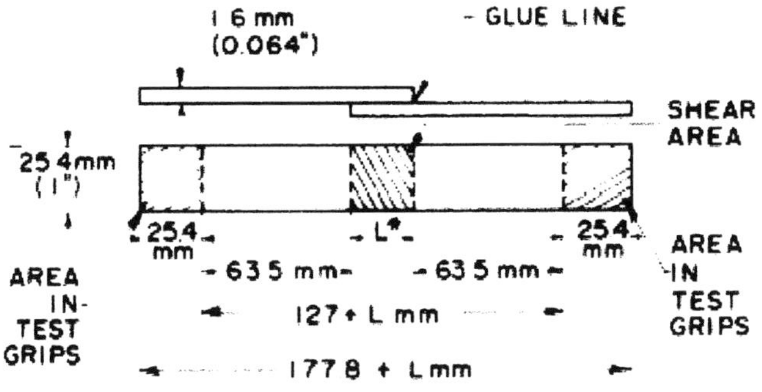

Figure 1. ASTM lap joint standard specimen dimensions.

Destructive testing – Braze joints were tested for shear strength using an Instron 5500R tensile testing machine. Data was collected at a rate of 10 Hz and a cross-head speed of 0.508 mm/min (0.2 in/min). Also, a 30 kN load cell was utilized.

Scanning electron microscopy – Selected braze joints were cross sectioned and metallographically prepared for analysis in a Jeol 6300 scanning electron microscope (SEM) at 20-25kV. Energy dispersive X-ray analysis was conducted across the cross section using a Princeton Gamma Tech system.

Results

Experimental matrix – In order to determine the optimum brazing temperature and soak time for this microwave, a series of five joints were soaked for five minutes from 960 to 1100°C. Another series of joints were soaked for ten minutes at the same temperatures. During shear testing of this first group of joints it was obvious that the braze area was too large for the 16 GA stainless steel. The base metal necked and the braze slowly peeled away during testing, making it difficult to quantify the shear strength. All of these specimens failed between 22 and 27 kN (4900 and 6000 lbf).

Visual inspection of this group of ten specimens' fracture surfaces showed many to have large areas where the braze material didn't completely wet the base metal. The best fracture surface belonged to the braze (braze #18) held at 1106°C for 5 minutes. Braze #18 fractured in the middle of the braze material showing a ductile fracture surface. Figure 2 is an SEM micrograph of this fracture surface.

During SEM analysis the joint shown in figure 2 (braze #18) proved to have few voids, no cracks, and few inclusions. However it is important to characterize the fusion of the joint.

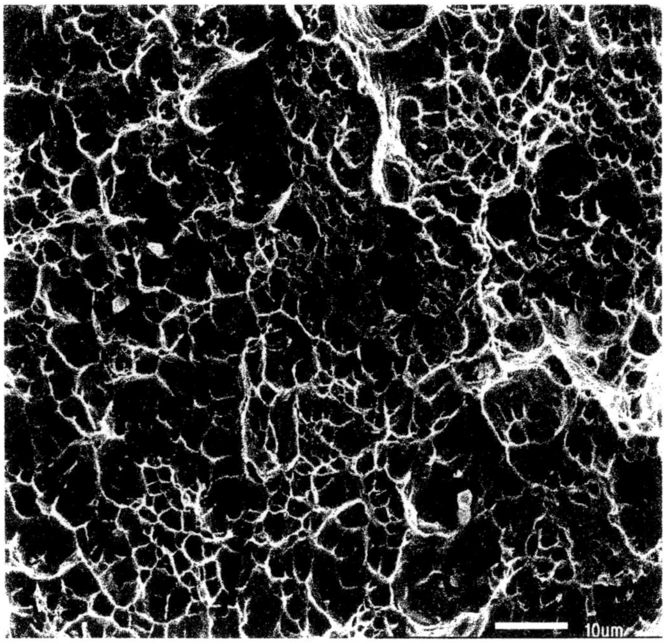

Figure 2. SEM micrograph of the filler metal fracture surface. Note the ductile nature of the fracture.

Another braze joint (braze #7) was held at 1096°C for 5 minutes. This joint was cross sectioned and polished down to 3 microns for SEM analysis.

SEM analysis of braze #7 shows a heterogeneous assembly of at least three different phases across the joint shown in figure 3. Figure 3 is a back scattered electron image, therefore the darker areas correspond to lighter elements. The base metal is the darkest area. The area adjacent to the base metal is a solid solution of mostly Fe and Cu. Silver and Pd form a solid solution, and therefore make up the majority of the lightest areas. Also, closer inspection reveals a eutectic phase, composed of Ag and Cu, in the light areas. Figure 3 shows a void of approximately 1 micron. Also, fusion is present at the braze to base metal interface.

A dispersive X-ray spot analysis using the Imix system was conducted across the cross section of braze #7. Analysis took place at the interface, 10 microns into the base metal, and 30 microns into the braze metal at 5 micron increments. Figure 4 denotes these approximate areas with arrows. Note that in figure 4 there is only one void of approximately 1 micron in diameter.

The results of the semi-quantitative analysis are listed in figure 5. The iron diffused 15 microns into the braze metal, while the nickel and chromium diffusion is inconsequential. Also, copper was the only braze metal to diffuse 10 microns

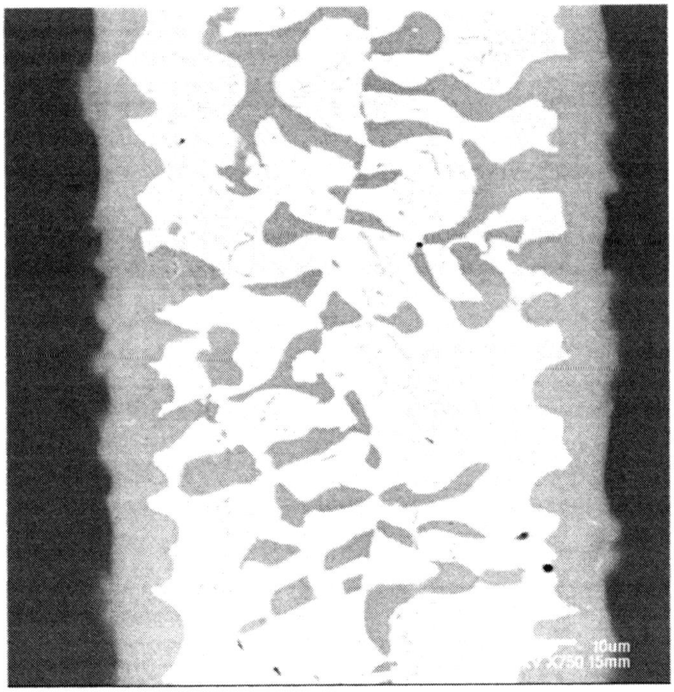

Figure 3. SEM micrograph of braze joint number seven. Note a Ag-Cu eutectic phase in the lightest area.

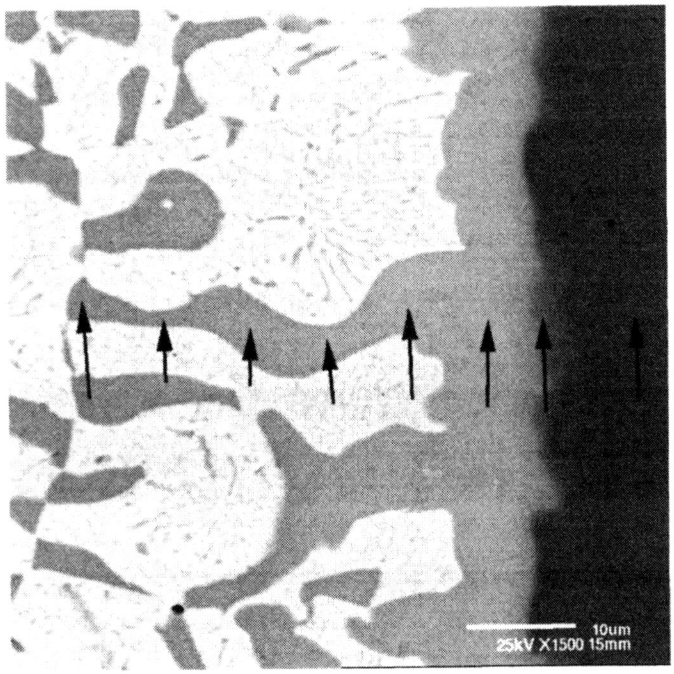

Figure 4. SEM micrograph of half of braze joint #7. The arrows denote where energy dispersive X-ray spot analysis was conducted.

into the base metal. This analysis shows definitively the fusion and bonding that these brazing parameters provide.

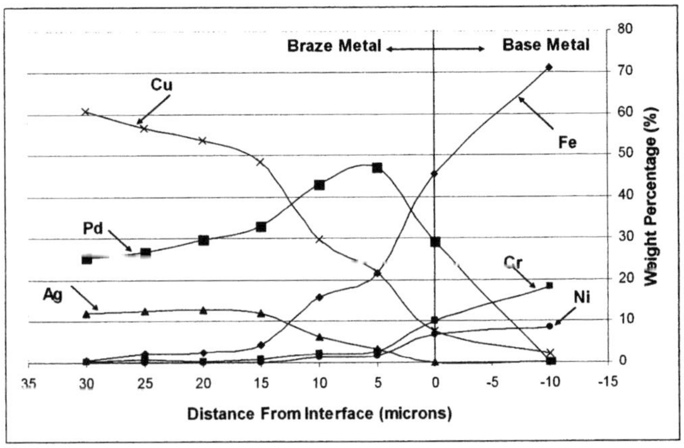

Figure 5. Semi-quantitative analysis showing weight percentage of elements throughout braze #7.

An interesting note, although this analysis is semi-quantitative the nominal Fe, Cr, and Ni compositions for 304 stainless steel are 66.3-70.8, 18-20, and 8-10.5 weight percent, respectively.

Brazing procedure – The AWS Brazing Manual has a standard form for recording brazing procedures [13]. The following is the brazing procedure resulting from these experiments:

- figure 1 shows a sketch of the lap joint, with an overlap of 4.6 mm (3/16 in)
- the base metal is a 16 GA 304 stainless steel the joint area is degreased and thoroughly cleaned with methanol
- apply enough flux (AWS FB3D) to cover the braze area with no excess
- the filler metal is AWS BVAg-32
- joint clearance needs to be approximately .003 in
- place the joint(s) in the ceramic casketing
- purge the microwave furnace of air and backfill with industrial Ar to approximately 1.5 psi
- use 1 kW forward power between both magnetrons for 5 minutes, then increase the power to 2 kW until red heat is reached (approximately 5 minutes)
- increase the power to 2.6 kW between the magnetrons until 1090°C is reached (approximately 9 minutes)
- decrease the power to 2 kW and continue to decrease as needed to keep the temperature at 1100°C for 5 minutes
- the power will need to be decreased to approximately 1.2 kW between the magnetrons
- turn the power off and cool the braze in the furnace at approximately 30°C/min cooling rate
- wait at least 30 minutes before discontinuing the Ar gas flow and opening the chamber

- no postbraze heat treating is necessary
- use a wire brush to clean the joint as needed tensile tests were used to measure the strength of these joints

Simultaneous brazing - Using the aforementioned brazing parameters, joints were created in pairs. However, a lap size of 1.27 cm (1/2 in) was used. Table 2 shows the peak load prior to fracture for each braze.

Braze number	Peak Load (lbf)
27	5134.752
28	5203.441
31	5125.482
32	5224.442

Table 2. Table showing peak loads of joints brazed simultaneously. Braze #27 was brazed during the same cycle as #28, and #31 was simultaneously brazed with #32.

Table 2 shows that brazes made simultaneously exhibit similar strengths, but the lap area was still too large to quantify the joints' shear strength. This is due to the base metal necking, as well as the braze peeling away before fracture.

However, the data can still be used for comparative purposes. The peak loads of the simultaneously brazed joints are within 2% of each other.

Shear testing – From a compilation of articles published by ASM International, the following equation gives a calculation for the desired length of lap area [14]:

$$X = \frac{Y \cdot T \cdot W}{L} \qquad \text{Eq. 2}$$

where:

X = length of lap area
Y = safety factor (1)
T = base metal UTS (75 ksi)
W = base metal thickness (1.524 mm)
L = shear strength of braze metal (25 ksi)

Using a lap length of approximately 4.6 mm (0.19 in), another braze (braze #35) was fabricated with the brazing parameters discussed earlier. When this joint was tensile tested, the lap area was small enough where very little base metal necking occurred. Also, there was minor peeling of the joint prior to fracture. The stress – strain data is shown in figure 5. The shear strength of braze #35, as tested in the Instron 5500R, was measured to be 138 MPa (20 ksi). There were some alignment issues with the testing equipment which caused the strain to rise too quickly in the beginning of the test.

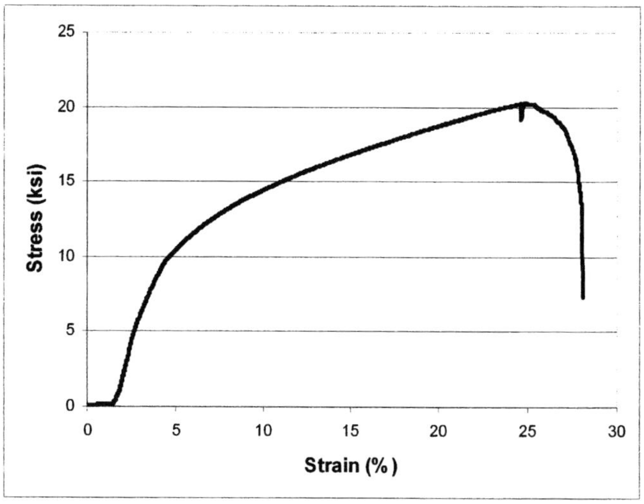

Figure 5. The stress strain data showing the shear strength of braze # 35.

Summary and Conclusions

The results of this work indicate that brazing with microwave energy is an efficient alternative to conventional brazing methods. These joints were fabricated in less than 25 minutes, while some brazing procedures for conventional furnaces require many hours for process time. The inspection methods used in this work clearly indicate sound braze joints. Visual inspection of the fractured joints shows no voids or inclusions. SEM analysis showed very minor voids (< 1 micron) and minimal flux inclusions. Also, the joints were shown to have favorable fusion at the braze to base metal interface.

The ability to braze multiple joints simultaneously was proven in this study. There was no need to increase the power for the extra joint. This was due to the mass of the bottom ceramic susceptor being much greater than the mass of the joints.

The shear strength of the joints was measured to be 138 MPa (20 ksi). This proved to be close to the braze metal's nominal shear strength of approximately 172 MPa (25 ksi). However, "brazing temperature is but one factor among several which influence the strength of the brazed joint [15]." The joint clearance plays a key role in determining the braze strength. Perhaps 0.08 mm (0.003 in) isn't the optimum clearance for this particular braze. The influence of joint clearance will be addressed in future work.

During the brazing procedure, specifically once the maximum temperature is reached, it was necessary to continue to decrease the power to keep the ceramic's temperature from continuously rising. The power was reduced approximately 40 percent during the short 5 minute soak time. This is due to the SiC dielectric loss factor increasing by a factor of 16 [3].

Acknowledgements

This research was funded under DOE contract number W-7405-ENG-36. The authors wish to thank Carlos Dozhier for his machining expertise, Steve Brandenberg for the use of the induction heating facility, Dr. David Olson for his technical review of this manuscript, Mark Richards at CSM and Gary Gladysz for their help with tensile testing, Matt Ruggiero for the metallographic preparation, Rob Dickerson for the SEM images, Stan Morrow for his help with the ceramic casketing, and Hugo Huey of Micramics Inc. for his microwave expertise.

References

1. E. E. W. Kassner, "Process for altering the energy content of dipolar substances," U.S. Patent 2,089,966
2. M. Rothstein, "Dielectric heating," *Encyclopedia of Polymer Science*, vol. 5. New York: Interscience Publishers (1974)
3. E. T. Thostenson, T. W. Chou, "Microwave processing: fundamentals and applications," *Composites: Part A*, 30, 1055-1071 (1999)
4. D. E. Clark, D. C. Folz, J. D. West, "Processing materials with microwave energy," *Materials Science and Engineering*, A287, 153-158 (2000)
5. W. H. Sutton, "Microwave processing of Ceramics-an overview," *Microwave processing of materials III*, Materials Research Society Proceedings, San Francisco, 269. Pittsburgh: Materials Research Society, 3 (1992)
6. J Sun, J, Li, G, Sun, B Zhang, S, Zhang, H. Zhai, "Dielectric and infrared properties of silicon carbide nanopowders," *Ceramics International*, 28, 741-745 (2002)
7. D. A. Copson, *Microwave Heating*, p 124, The AVI Publishing Company, Connecticut (1975)
8. R. E. Collin, *Foundations for microwave engineering.* McGraw-Hill, New York (1966)
9. V. N. Tran, "An applicator design for processing large quantities of dielectric materials," *Microwaves: theory and application in materials processing, ceramic transactions,* Proceedings of the symposium held during the 93rd Annual Meeting of the American Ceramic Society, Cincinnati, 21, 683-692 (1991)
10. ASM Handbook, *Welding, Brazing, and Soldering*, 6, p 116, (2000)
11. W. Feduska, "High-temperature brazing alloy-base metal wetting reactions," *Weld. J.* 38, 122s-130s, (1959)
12. ASTM Designation: D 1002 – 01, "Standard test method for apparent shear strength of single-lap-joint adhesively bonded metal specimens by tension loading (metal-to-metal)," ASM International (2003)
13. *Brazing Manual*, American Welding Society, p 102, (1976)
14. *Source book on brazing and brazing technology*, Editor M. M. Schwartz, ASM International, 8, (1980)
15. N. Bredzs, "Investigation of factors determining the tensile strength of brazed joints," *Weld. J,* 545s- 563s (1954)

Thermal Cyclability of Reactive Air Braze Seals in Anode Supported Solid Oxide Fuel Cells

John S. Hardy*, Jens T. Darsell, Christopher A. Coyle*, Jerome C. Birnbaum*, and K. Scott Weil***
* Pacific Northwest National Laboratory, Richland, WA 99352
** Washington State University, Pullman, WA 99164

Abstract

The popularity of anode-supported solid oxide fuel cells (SOFC) has increased in tandem with the ability to fabricate thinner gas-tight yttrium-stabilized zirconia (YSZ) electrolyte layers, which can now be routinely produced on the order of 7 to 10 μm thick. While this has significantly improved power output and decreased the required fuel cell operating temperatures, the ability to reliably seal fuel cells remains a concern. The seals must be hermetic and be robust enough to retain their hermeticity even under the extreme operating conditions of SOFCs. Perhaps the largest contributor to stresses experienced by the seal is the fact that the SOFC is an assembly of many different materials with different thermal expansion properties. Although every effort is made to minimize thermal expansion mismatches across the seals, the stresses developed during thermal cycling still jeopardize seal integrity. Reactive air brazing (RAB), a method of joining that employs a metallic, and therefore non-brittle, seal material has been used to seal electrolyte/anode bilayers, such as those in anode-supported SOFCs, to Crofer-22 alloy. The results of rupture strength testing will be reported for as-brazed and thermally cycled samples and the effect of thermal cycling on the RAB seal microstructure will be shown.

Introduction

Solid oxide fuel cells function due to an oxygen gradient that drives oxygen ion diffusion across a yttrium-stabilized zirconia (YSZ) electrolyte. In planar SOFCs, the seal bonding the electrolyte to the support frame (see fig. 1) must be hermetic in order to maintain this gradient and maximize the performance of the device. Long-term operation requires a rugged and stable seal material so that hermeticity can be maintained during thermal cycling and/or continuous high temperature operation. Specifically, the seal should be capable of withstanding at least 5,000 hours of operation at 650 - 800°C, including intermittent cooling and heating. Because the fuel cell comprises many different materials, the cooling and reheating cycles will likely produce some of the greatest stresses to be experienced by the seal due to thermal expansion mismatches. Mitigation of these stresses requires either careful thermal expansion matching or the use of a seal material that is compliant to the imposed stresses.

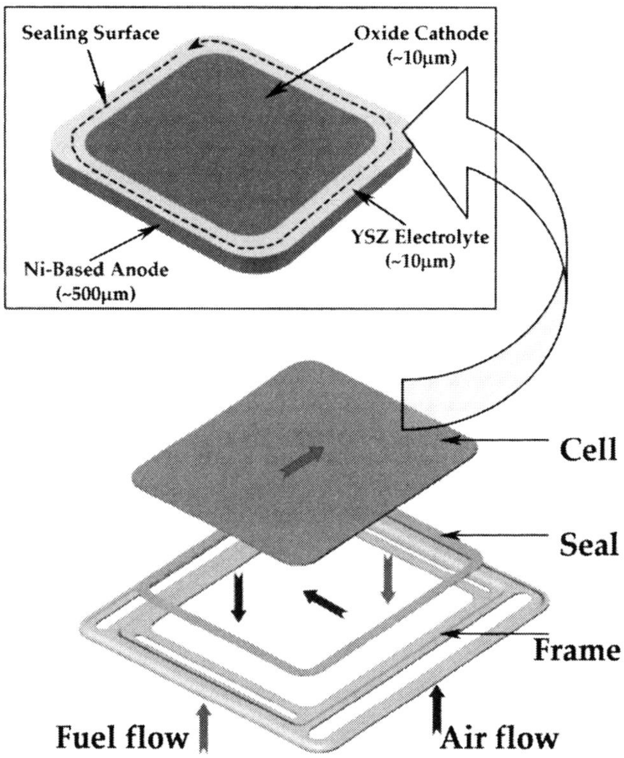

Fig. 1. Diagram illustrating cell-to-frame seal assembly in an anode supported planar SOFC.

Presently, the most prevalent method used for sealing SOFCs utilizes high temperature glass compositions. Glass seals provide hermeticity, moderate strength, and can be formulated to closely match the thermal expansion characteristics of adjacent components. However, fuel cells utilizing glass seals are limited to temperatures below the softening point of the glass. In some cases, the sealing glasses cause accelerated corrosion of metal sealing surfaces[1]. Additionally, the brittle nature of glass calls into question its ability to retain hermeticity when exposed to the stresses associated with long-term operation. Studies have shown that glass devitrifies during prolonged high temperature operation, often leading to increased brittleness and a deviation from the carefully engineered thermomechanical properties of the glass[2]. For example, depending on the choice of alloy for the metallic support frame, crystalline $BaCrO_3$ or $BaAl_2Si_2O_8$ has been found to develop at the glass-to-alloy interface within several hours of operation when chromia- or alumina-forming alloy frames, respectively, are used[3,4]. The thermal expansion coefficients of these unwanted phases are vastly different from both the adjacent alloy and the surrounding glass matrix. During thermal cycling, the presence of interfacial phases can significantly increase the internal stresses at the seal interface, which is inherently a weak point in the seal.

Alternative high temperature joining techniques include active metal brazing and compressive mica seals. Active metal brazing must be carried out in a reducing atmosphere to keep the active species, typically titanium, from pre-oxidizing. However, in addition to being more expensive than air-firing, firing under the required high vacuum or reducing-gas conditions is also intolerable for many oxide SOFC components. Active metal brazes also tend to attack the grain boundaries of, and thereby weaken, ceramic sealing surfaces. Moreover, recent studies on the oxidation behavior of active metal brazes indicate that they are unreliable at temperatures above 500°C in air, at which point they eventually oxidize completely, conferring little or no strength to the joint[5,6].

Mica seals afford compliance to shear stresses such as those caused by thermal expansion mismatches between the materials being sealed together. This occurs through slippage of the laminar shear planes characteristic of mica materials. However, mica seals are not completely hermetic[7] due to interlaminar gas permeation through the mica and leaks at the interface between the mica and the sealing surfaces caused by a combination of the innate surface roughness of the mica and the inability of the mica to comply to any roughness that might be present on the sealing surfaces. These leak rates are low and can be further minimized by increasing the compressive load on the seals. However, increasing the load also increases the interlaminar friction, which may inhibit the mica's ability to comply to shear stresses. The other consideration when using mica seals includes the need of a load frame that is compact, stable, and inexpensive, and the impact that a compressive load may have over the lifetime of the device on other SOFC components.

The recently developed RAB technique reactively modifies one or both oxide faying surfaces with an oxide compound that has been at least partially dissolved in a noble metal solvent, e.g. silver, gold, or platinum, such that the newly formed surface is readily wetted by the remaining molten filler material. In many respects, this concept is similar to active metal brazing, except that the joining operation can be conducted in air and the final joint is resistant to oxidation at moderate-to-high temperatures. One material system that is particularly well suited for RAB is Ag-CuO. Studies have shown that 1.4 to 8 mol% CuO additions in Ag result in an optimal balance of braze ductility and wettability on oxide surfaces therefore producing strong joints[8,9]. RAB seals provide the rare combination of high-temperature oxidation resistance, metallic ductility, and oxide surface wettability. In this study, the effects of aggressive thermal cycling on the rupture strength and microstructure of RAB seals consisting of 96 mol% Ag and 4 mol% CuO braze (AgCuO-4) for anode-supported SOFC applications will be reported.

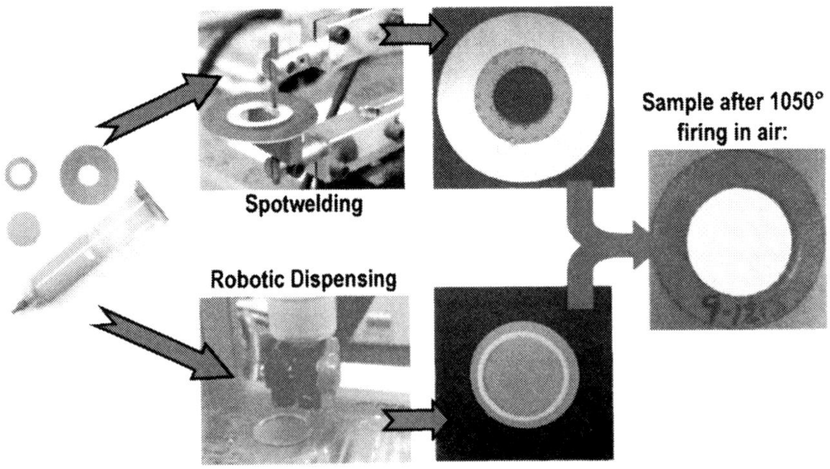

Fig. 2. Flowchart of sample preparation for rupture strength testing of seals.

Experimental

The procedure for assembling samples for thermal cycling and subsequent rupture strength testing is illustrated by the flowchart in fig. 2. Washers were cut from Crofer 22 mesh that had been rolled to a thickness of approximately 150 μm and from 0.5 mm thick Crofer 22 sheet. The inner and outer diameters of the mesh washer were approximately 1.5 and 2.5 cm, respectively, while those of the sheet washers were 1.5 and 4.5 cm. Meanwhile, NiO-YSZ anode / YSZ electrolyte bilayers were fabricated in two thickness, 0.5 mm and 1 mm, through tapecasting and subsequent lamination. The resulting bilayer laminates were laser cut into disks such that the resulting sintered diameter is approximately 2.5 cm. After thorough burnout of the organic tapecasting additives, the bilayers were cosintered at 1350°C for 2 hours.

Crofer mesh washers were then spotwelded concentrically to the Crofer sheet washers with a circle of welds lining both the inner and outer circumferences of the mesh. The space between welds was approximately 3 mm. The mesh is incorporated to act as a spacer and to contain the molten braze during firing. A 2 cm diameter ring of braze paste weighing approximately 0.32 g was applied to the electrolyte face of the bilayer by a robotic dispenser and allowed to dry for 30 min. at 80°C. The ring of braze on the bilayer was then placed concentrically atop the mesh spotwelded to the washer and fired at 1050°C.

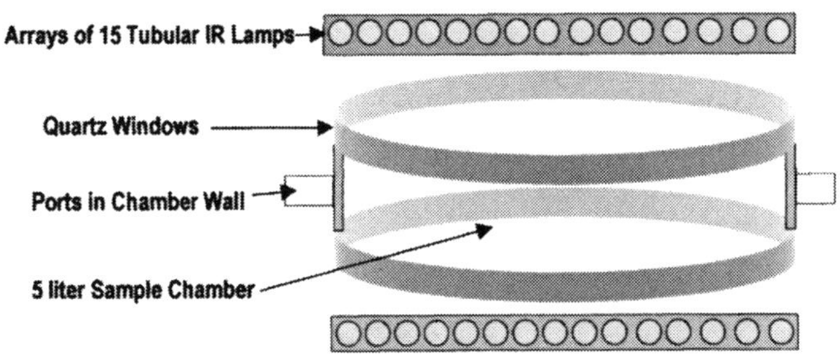

Fig. 3. Schematic of the infrared furnace used for thermal cycling of samples. Thermocouples are fed through ports in the chamber walls and placed in contact with the samples.

Samples were thermally cycled in an infrared furnace such as is illustrated in fig. 3. The infrared furnace consists of a 5-liter sample chamber between two arrays of 15 tubular quartz infrared lamps that irradiate samples through quartz windows placed atop and below the chamber. The infrared furnace is capable of heating to 750°C in less than 30 seconds, and easily follows the temperature profile selected for thermal cycling the popgun samples in order to mimic an aggressive 10 min. SOFC heat up. The samples were heated to 750°C at 75°C/min. where the temperature was held for 10 min. followed by a 75°C/min. cool down. This cooling rate was sustained down to temperatures of less than 400°C, after which, the sample took an additional 25 min. to cool to less than 100°C. This schedule was then repeated for the prescribed number (5, 10, 20, or 50) of thermal cycles.

After thermal cycling, the rupture strength of the samples was evaluated using a semi-quantitative method referred to as pop-gun testing. The apparatus used for these tests is shown in fig. 4 and measures the amount of gas pressure the seal can hold before gas escapes the seal through a leak or rupture. Samples are tightened into the popgun apparatus against an O-ring seal and the gas pressure is increased manually with the regulator. Meanwhile, the gas pressure behind the bilayer-to-crofer seal is measured using a digital pressure gauge that displays the maximum gas pressure exerted on the sample. Thus, when the sample ruptures, the pneumatic pressure at the point of failure is retained by the gauge.

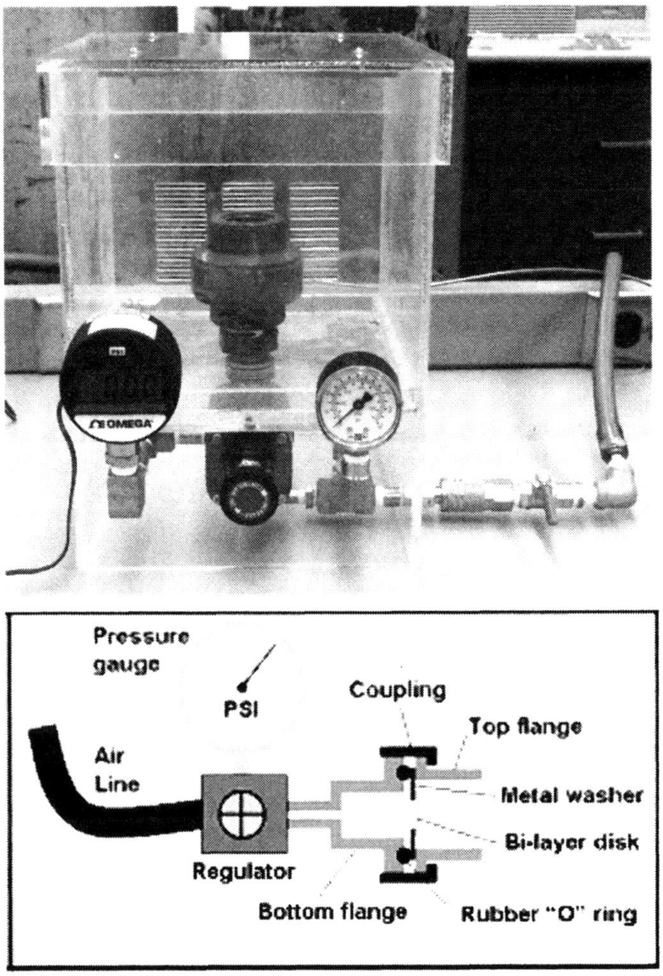

Fig. 4. A picture of the popgun test apparatus for evaluating the rupture strength of sealed samples. Below the picture is a labeled schematic diagram.

Microstructural analysis was performed using a JEOL JSM-5900LV scanning electron microscope (SEM) on samples that had been cross-sectioned and polished after popgun testing. The SEM is equipped with an Oxford energy dispersive X-ray analysis (EDX) system, which employs a windowless detector for quantitative detection of both light and heavy elements. To avoid electrical charging of the samples in the SEM, they were carbon coated and grounded.

Fig. 5. Samples with standard 0.5 mm thick bilayers ruptured the bilayer before breaking the RAB seal bonding it to the stainless steel washer.

Fig. 6. Samples with thicker 1 mm bilayers did not rupture at the maximum pressure limit of the popgun apparatus.

Results and Discussion

Initially, the thermal cycling studies were focused on emulating seals between crofer support frames and bilayers of the prototypical 0.5 mm thickness used in SOFCs. However, all of these samples ruptured in the center of the bilayer (see fig. 5) instead of at the seal, indicating that the RAB seal has greater rupture strength than a standard 0.5 mm thick bilayer, even after 50 aggressive thermal cycles. However, to ensure that the pressure exerted on the seal was not limited by the strength of the bilayers, the bilayers were doubled in thickness to 1 mm and the test was repeated. None of the thicker bilayers ruptured (see fig. 6) but in this case the pressure limits of the popgun apparatus was the limiting factor as none of the samples ruptured at the test limits of approximately 130 psi (0.9 MPa). Figure 7 plots the maximum pneumatic pressure to which the seals were exposed as a function of the number of thermal cycles.

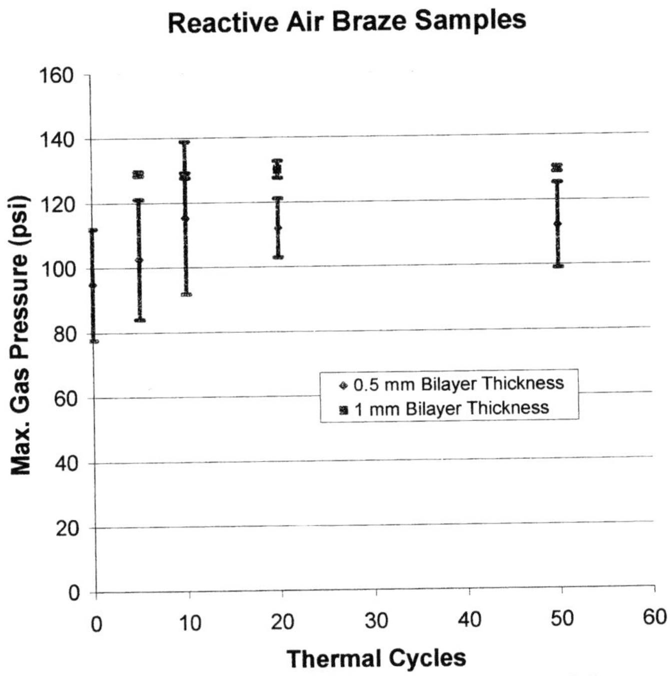

Fig. 7. The results of popgun testing on reactive air braze seals bonding crofer 22 washers to 0.5 mm and 1 mm thick bilayer disks.

Although pop-gun testing is a simple and efficient method of semi-quantitatively comparing different sealing techniques, the results of the pop-gun test cannot easily be converted to tensile and shear strengths without a large margin of error. However, the chart in fig. 8 shows the popgun test results for a number of samples sealed using different sealing techniques.

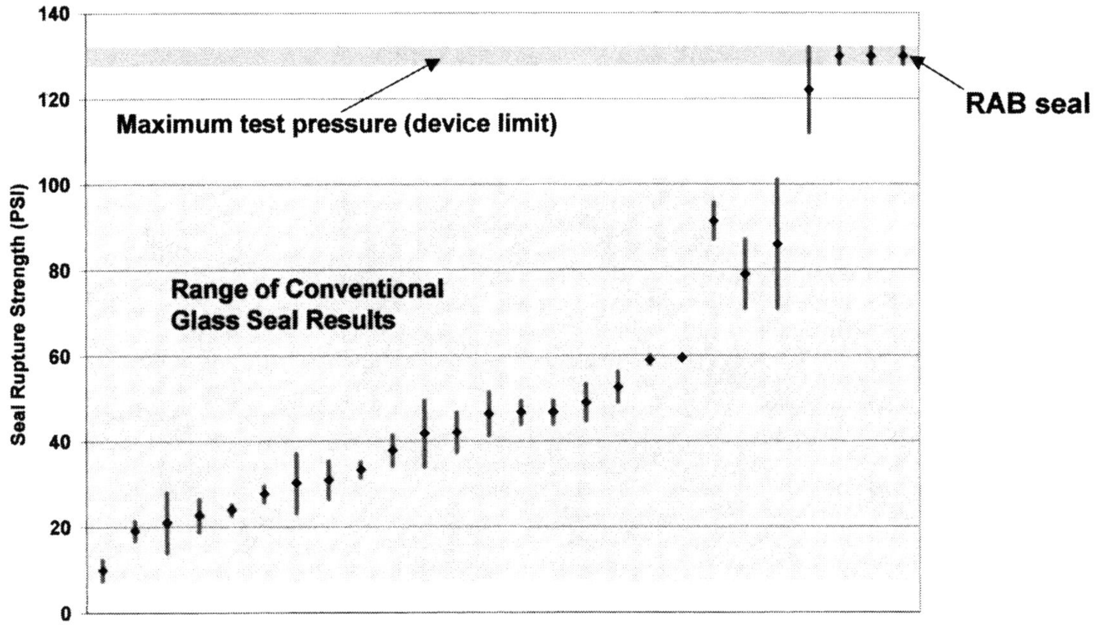

Fig. 8. Popgun results for a number of seal configurations. The range of maximum test pressures and the range of test results for conventional glass seals are highlighted.

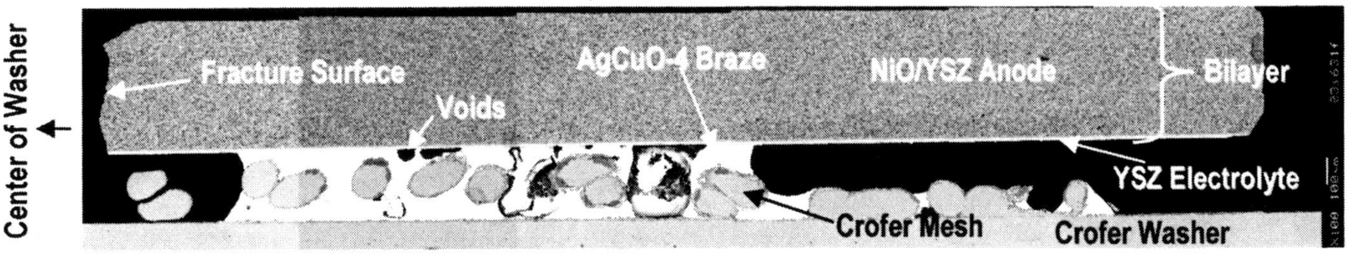

Fig. 9. SEM micrograph of a typical ruptured popgun sample.

Among different high temperature glass seals, which are the current industry standard, the highest rupture strengths measured using the popgun method were approximately 100 psi. On the other hand, after 50 aggressive thermal cycles, the rupture strengths of RAB seals were still above the 130 psi limit of the popgun apparatus.

Figure 9 shows a typical cross-sectional view of RAB seals as seen in SEM evaluation. At the top of the sample is the anode/electrolyte bilayer, which is composed of the porous anode with a thin layer of dense YSZ electrolyte laminated on its underside. At the base of the picture, a portion of the crofer washer is visible. Between the washer and the bilayer is the AgCuO-4 braze which appears as a light-colored phase. There are also a number of oval bodies of the same color as the crofer washer that are present in the braze as well to the right and left of the braze. These are the cross-sectioned wires of the rolled crofer mesh. At the surfaces of the crofer are darker colored bands of oxidation product which consist mainly of chromia. Finally, the darkest inclusions in the braze are voids.

SEM analysis revealed that after 10 thermal cycles, there are areas where delamination of the braze from the YSZ electrolyte is visible (see fig. 10). Although delamination is visible with the SEM in some areas, it does not appear to be continuous across the entire width of the seal and did not have any measurable effect on hermeticity or rupture strength. In samples with 5 thermal cycles or less, micrographs showed that the braze was found to remain well-adhered to the YSZ.

Figure 11 is a micrograph of a sample that had been subjected to 5 thermal cycles. A crack had propagated through the anode, forked into two cracks just before reaching the electrolyte, and penetrated through the electrolyte. Upon reaching the braze, both cracks were deflected to follow the braze/YSZ interface instead of continuing through the braze. Both traveled in opposite directions along the interface until being arrested after a short propagation distance.

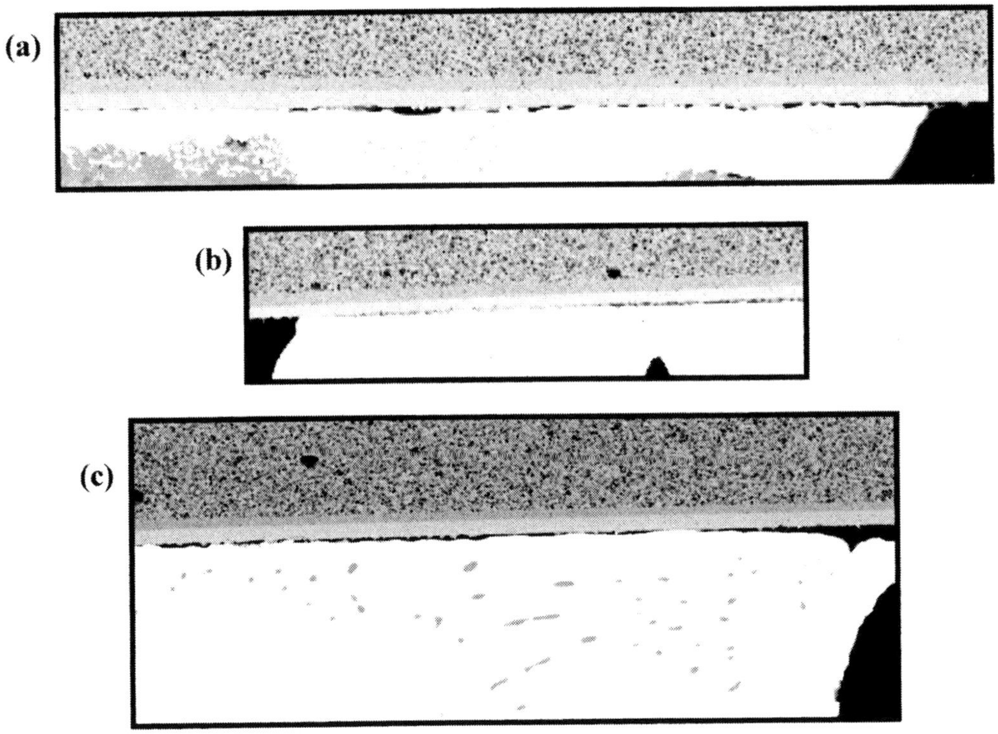

Fig. 10. Micrographs of popgun samples that had been thermally cycled (a) 10, (b) 20, and (c) 50 times in which areas of delamination were observed.

However, in samples that had been thermally cycled more than 10 times and had therefore begun to weaken at the braze/YSZ interface, similar cracks were not as quickly arrested as in the samples with only 5 thermal cycles that showed no signs of delamination. The micrographs in fig. 12 include two such examples (see figs. 12a and 12b) in which only a fracture surface remains in the bilayer because delamination has dislodged the bilayer fragment that was originally on the opposite side of the crack from the braze to which it was once adhered. Thus, even after 20 to 50 aggressive thermal cycles, the deleterious effects of thermomechanical shear stresses on the strength of a RAB seal interface with YSZ only becomes apparent when a crack that originates in a fractured bilayer propagates to the braze/bilayer interface. In such a case, the hermeticity of the cell would have already been compromised. Nonetheless, because the tip of a propagating fracture front is a point of severe stress concentration, the stresses that caused the damage to the braze/YSZ interfaces shown in fig. 12 were probably many times greater than those to which the braze seal would otherwise normally be subjected during popgun testing.

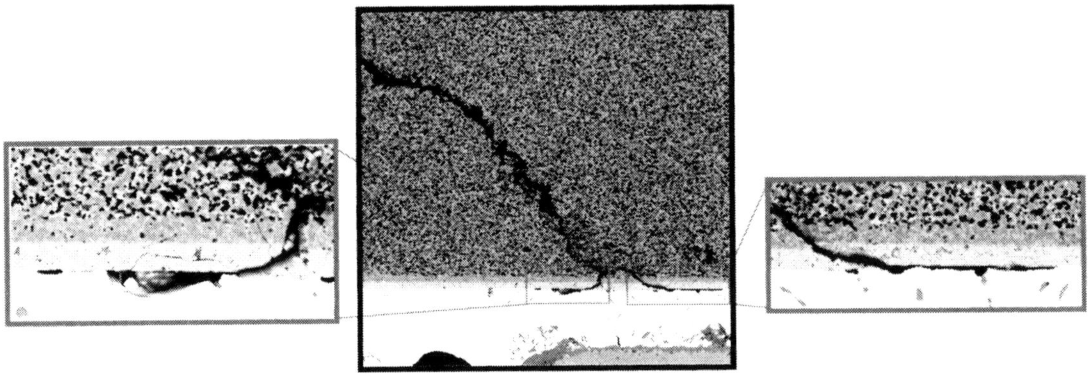

Fig. 11. Micrographs of a popgun specimen that had been thermally cycled 5 times show a crack that was arrested soon after propagating from the bilayer into the braze/bilayer interface.

(a) (b)

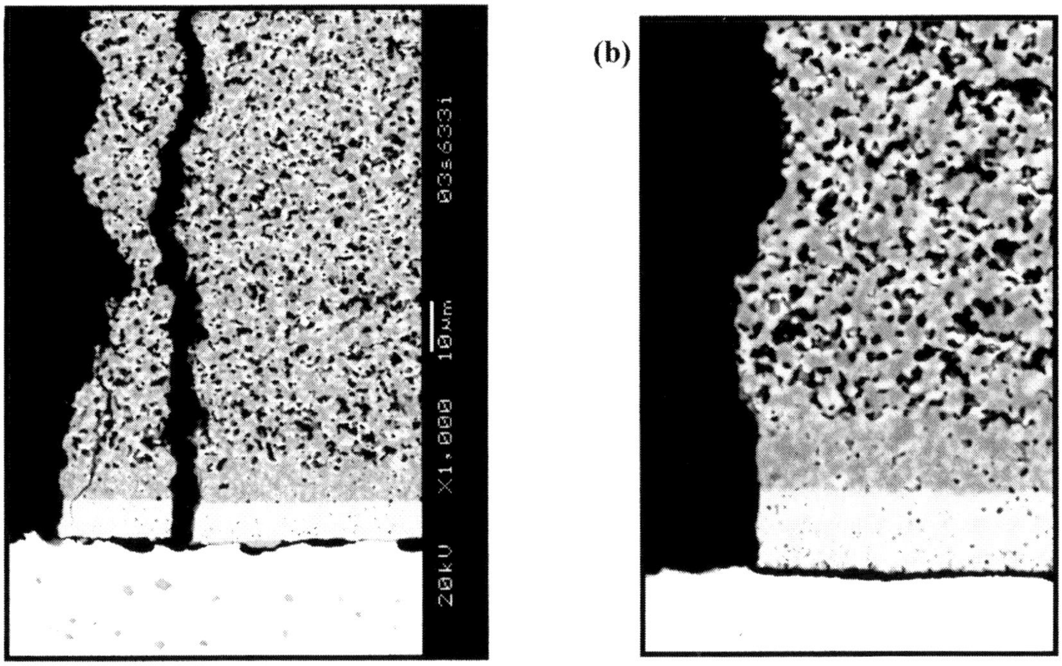

Fig. 12. Examples of crack propagation along the braze/bilayer interface causing bilayer fragments to be completely released from the braze in samples with (a) 20 and (b) 50 thermal cycles.

Summary and Conclusion

After being subjected to 50 aggressive thermal cycles that included heating and cooling rates of 75°C/min., the RAB seals showed greater ruptures strengths than prototypically thick (0.5 mm) SOFC bilayers and greater rupture strengths than the measurement limit of the popgun test apparatus (~ 130 psi) when thicker (1 mm thick) bilayers were used. The rupture strengths of a number of different conventional high temperature glass seals have been measured using the popgun method and the highest rupture strengths measured among them have been approximately 100 psi. Thus, RAB seals, even after thermal cycling, exhibit much higher rupture strengths than conventional glass seals.

Microscopic evaluation revealed areas where delamination at the braze interface with the bilayer had occurred in samples that had been exposed to 10 or more thermal cycles, while the braze remained well adhered to the bilayer in those that had undergone 5 thermal cycles or less. The deleterious effect of interfacial delamination on the strength of this interface was not significant enough to decrease the rupture strength of the RAB seal to a level measurable by the popgun test. The only indication of weakening in the braze/bilayer interface due to thermomechanical stresses was longer crack propagation distances observed in samples with more than 10 thermal cycles. In samples with 20 or more thermal cycles, in which delamination was visible with the SEM, cracks reaching the braze/bilayer interface propagated the entire length of the interface. Meanwhile, in a sample with 5 thermal cycles, two cracks propagating in opposite directions along the interface were both arrested very soon after reaching the interface. Although the effects of aggressive thermal cycling were visible in the form of microscopic areas of delamination and crack propagation, after 50 thermal cycles the RAB seals still held approximately 130 psi of pneumatic pressure without leaking.

Acknowledgments

The authors would like to thank Nat Saenz, Shelly Carlson, and Jim Coleman for their assistance in sectioning and polishing the popgun samples and conducting the metallographic and SEM analysis work. This work was supported by the U. S. Department of Energy, Office of Fossil Energy, Advanced Research and Technology Development Program. The Pacific Northwest National Laboratory is operated by Battelle Memorial Institute for the United States Department of Energy (U.S. DOE) under Contract DE-AC06-76RLO 1830.

References

1. Z. G. Yang, K. S. Weil, D. M. Paxton, and J. W. Stevenson, *J. Electrochem. Soc.*, 150, A1188 (2003)
2. K. Eichler, G. Solow, P. Otschik, and W. Schaffrath, *Proceedings of the Fourth European Solid Oxide Fuel Cell Forum*, pp. 899-906, European Solid Oxide Fuel Cell Forum, Oberrohrdorf, Switzerland (2000)

3. Z. G. Yang, K. S. Weil, K. D. Meinhardt, J. W. Stevenson, D. M. Paxton, G-G. Xia, and D-S. Kim, *Joining of Advanced and Specialty Materials V*, ASM International, Materials Park, OH (2002).
4. Z. G. Yang, K. S. Weil, K. D. Meinhardt, D. M. Paxton, and J. W. Stevenson, *Joining of Advanced and Specialty Materials V*, ASM International, Materials Park, OH (2002)
5. J-H Kim and Y-C Yoo, *Journal of Materials Science Letters*, 16, 1212-15 (1997)
6. J.P. Rice, D.M. Paxton, and K.S. Weil, *Proceedings of the 26th Annual Conference on Composites, Advanced Ceramics, Materials, and Structures*, in publication.
7. S. P. Simner and J. W. Stevenson, *J. Power Sources*, 102, 310 (2001)
8. J. S. Hardy, J. Y. Kim, and K. S. Weil, submitted to *J. Electrochem. Soc.*
9. J. Y. Kim, J. S. Hardy, and K. S. Weil, submitted to *J. Am. Ceram. Soc.*

Developments in Friction Stir Welding

By W M Thomas, and Gil Sylva MTS
TWI Ltd, Gt Abington, Cambridge, CB1 6AL, UK
MTS, 14000 Technology Drive, Eden Praire, MN 55344 2290, USA

Keywords: Friction stir welding, Solid-phase, Welding, Rapid Prototyping, Aluminium, Lap welds.

Abstract

Friction stir welding has gained increasing interest since its invention at TWI some 12 years ago. The continuing development of the process is described. Particular attention is paid to, tool motions, process variants and processing methods. A key factor in the successful application of the FSW process is the ratio between the swept volume of the probe of the tool and the volume of the probe itself. Furthermore, details of a self reacting pin tool developed at MTS are described.

Introduction

Invented at TWI in 1991[1], substantial development of Friction stir welding (FSW) has been conducted subsequently. It allows metals, including aluminium, lead, magnesium, steel, titanium, zinc, copper, and metal matrix composites to be welded continuously [1-14]. Many alloys, which are regarded as difficult to weld by fusion processes, may be welded by FSW. The basic principle of the FSW process is shown in Fig 1.

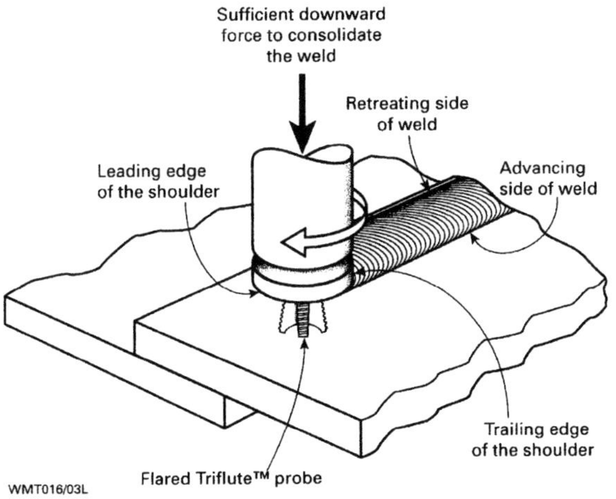

Fig. 1 Principle of friction stir welding *(Copyright © 2003, TWI Ltd)*

The present paper describes recent developments in FSW tools and process variants.

Reversal Stir Welding - Re-stir™

The following describes preliminary studies being carried out on Re-stir™ welding at TWI. The salient features of the Re-stir™ welding technique are illustrated in fig 2. This illustration applies to both angular reciprocating, where reversal is imposed within one revolution, and rotary reversal, where reversal is imposed after one or more revolutions.

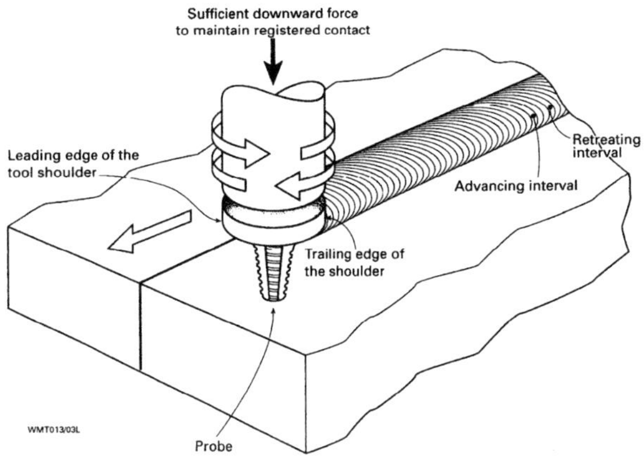

Fig 2. The basic principle of Re-stir™, showing the reversal technique *(Copyright © 2003, TWI Ltd)*

The use of the Re-stir™ welding technique provides a cyclic and essentially symmetrical welding and processing treatment [1& 5]. Most problems associated with the inherent asymmetry of conventional rotary FSW are avoided.

Figure 3a shows the appearance of the weld surface that is formed beneath the tool shoulder produced at a welding travel speed of 1.6 mm/sec (96 mm/min), using 5 revolutions per interval.

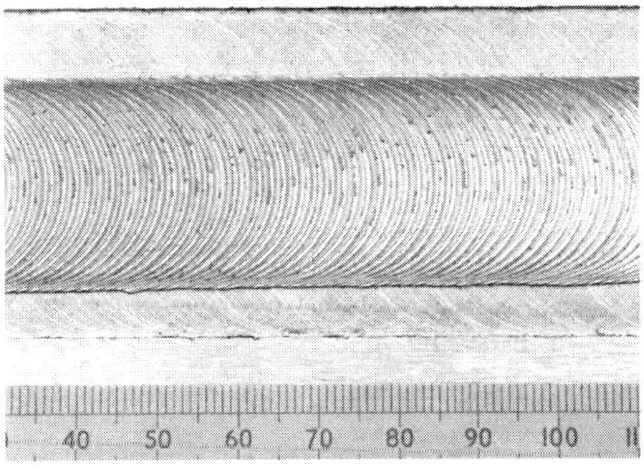

Fig. 3 Surface appearance of Re-stir™ welds made in 6 mm thick 5083-O condition aluminium alloy at a welding travel speed of 1.6 mm/sec (96 mm/min), using 10 revolutions per reversal interval. *(Copyright © 2003, TWI Ltd)*

Fig 4 shows the detail of the surface of a weld made at 4 mm/sec (240 mm/min) travel speed, using 10 revolutions per interval. The fine surface ripples reveal the number of rotations and the extent of the interval, while the less frequent, coarser and wider surface ripples reveal the position of the change in rotation direction. For Re-stir™, the distance and time between each interval depends on the combination of rotational speed and the travel speed used.

Fig. 4 Close up of Re-stir™ weld surface formed beneath the tool shoulder showing surface rippling and reversal interval. Produced at 4 mm/sec (240 mm/min) welding travel speed), using 10 revolutions per interval. *(Copyright © 2003, TWI Ltd)*

Macrosections of a lap weld made by Re-stir™ are shown in fig 5 a b & c. This weld was made in 5083-O condition aluminium alloy, using a Flared-Triflute™ type probe designed for rotary stir, at a travel speed of 3.3 mm/sec (198 mm/min) using 10 revolutions per interval. Figure 5a shows an essentially symmetrical weld. It should be noted that this weld shows detrimental plate thinning/hooking owing to the non-optimisation of welding parameters but does serve to illustrate the symmetrical nature of the weld produced by the Re-stir™ technique.

The longitudinal section shown in fig 5b is taken at a position at the edge of the weld region and shows the effect of the change in the direction of rotation. The plan view in fig 4c reveals a patterned weld region surrounded by a HAZ. There is some evidence that during the reversal stage some of the 'Third-body' plasticised material close to the probe is 're-stirred' back in the opposite direction.

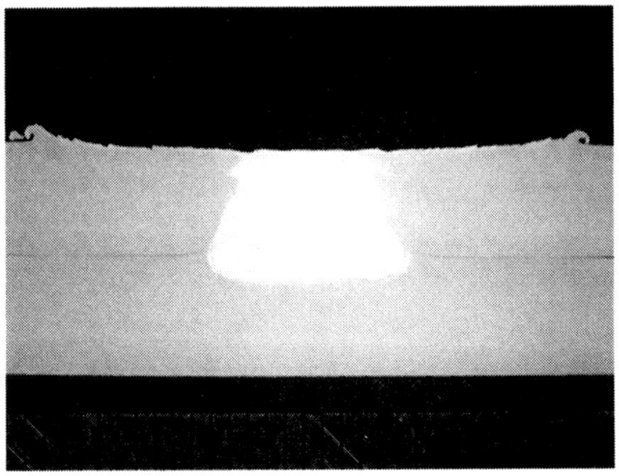

a)

b)

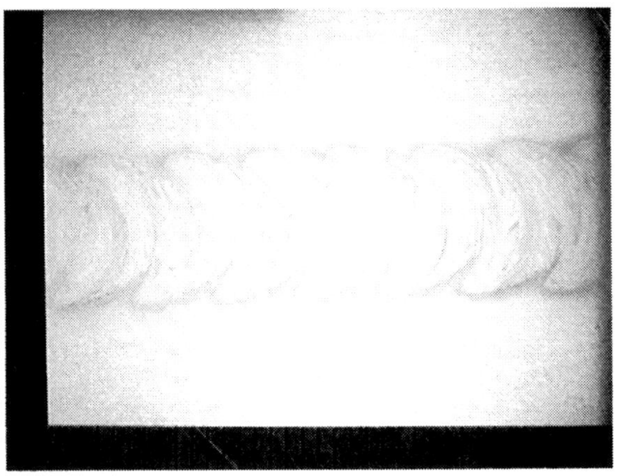

c)

Fig. 5 Metallurgical sections showing the effect of the Re-stir™ technique on the weld shape, produced at a welding speed of 3.3 mm/sec (198 mm/min), using 10 revolutions per interval. *(Copyright © 2003, TWI Ltd)*

a). Macrosection shows an essentially symmetrical dovetail shaped weld with a similar amount of upturn of the plate interface each side of the weld.

b). Longitudinal macrosection showing regular patterns caused by rotation reversal.

c). Plan macrosection taken mid thickness showing the effect of reversal motion.

The Re-stir™ process requires further optimisation to achieve welds of reproducibly high quality and freedom from defects but early trials suggest benefits in terms of weld symmetry. Initial work using an A-skew™ probe also suggests that it may be possible to achieve a slight down turn in the overlapping plate/weld interface at the outer regions of the weld which may be beneficial in particular structures and loading situations. Figure 6 a b & c illustrates this effect in an overlap weld in 5083-O condition aluminium alloy.

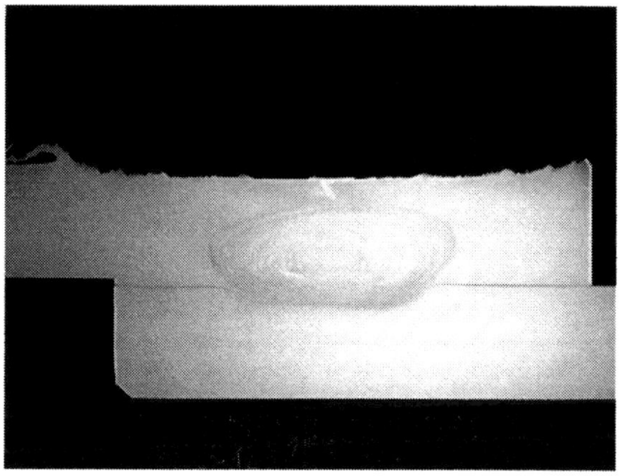

a)

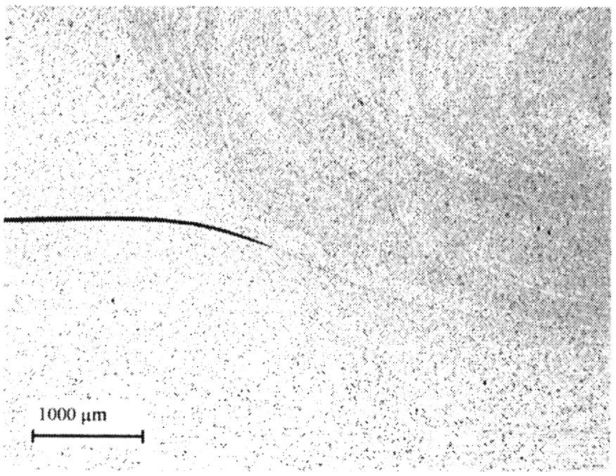

b)

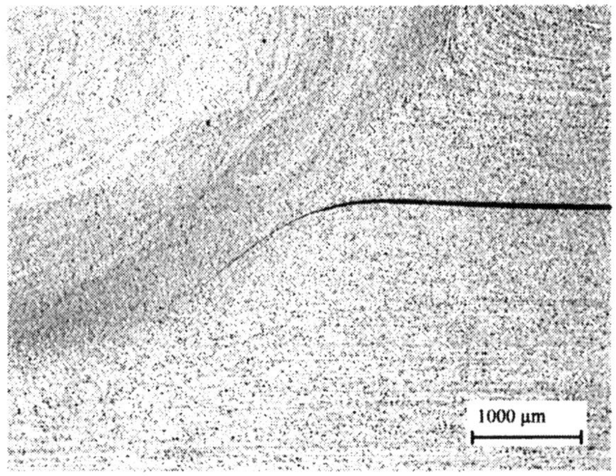

c)

Fig. 6 Detail of the outer regions of a Re-stir weld made with an A-skew™ probe in combination with a skew motion, at a travel speed of 1.6 mm/sec (96 mm/min), using 8 revolutions per reversal interval. *(Copyright © 2003, TWI Ltd)*

a) Macrosection
b) detail of notch (that would formerly have been at the retreating side with conventional rotary FSW)
c) detail of notch (that would formerly have been at the advancing side with conventional rotary FSW)

For conventional FSW a pair of hammer 'S' bends would be used to establish weld quality on both sides of the lap weld. This arrangement effectively tests the lap weld for plate thinning and for any notch effect (hook) by subjecting both sides of the asymmetric weld to the same test criteria. For Re-Stir™, providing the width of the bend specimen is more than two-reversal intervals in length, one hammer 'S' bend will suffice, see fig 7. (Any asymmetric type defect such as plate thinning or notch effect formed will be tested on this one specimen).

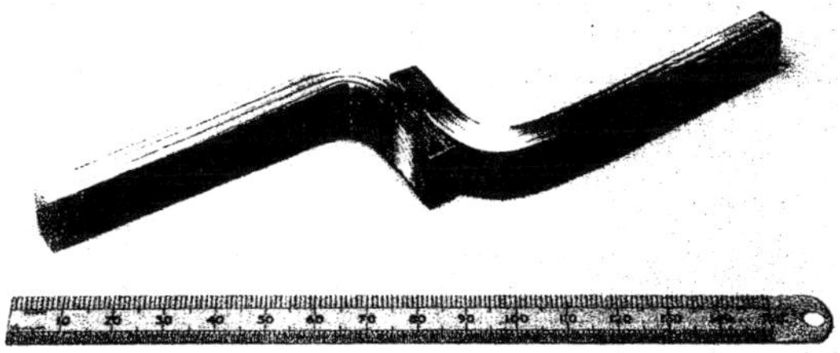

Fig. 7 Hammer 'S' bend of a Re-stir™ lap weld made in 5083-O condition aluminium alloy material showing goodweld integrity *(Copyright © 2003, TWI Ltd)*

Compound Motion Stir Welding –Com-stir™

A new friction stir welding (FSW) variant has been developed – Com-stir™ - which involves the application of a rotary motion in combination with an orbital motion, see Fig.8. This technique has the following potential advantages over conventional friction stir welding:

- It is capable of forming wider welds than obtained with conventional FSW, which is particularly advantageous when welding lap and Tee joints and when processing material (material conditioning).
- It should give higher quality welds than conventional FSW as surface oxide fragmentation is more efficient and because heat generation is more uniform across the weld.
- Less workpiece jigging should be necessary than needed for conventional FSW because of the reduced resultant torque applied during welding. A similar benefit is also achieved when compound motion is applied during machining operations.
- It is speculated that welding of dissimilar materials and the use of the technique for spot welding will also prove advantageous.

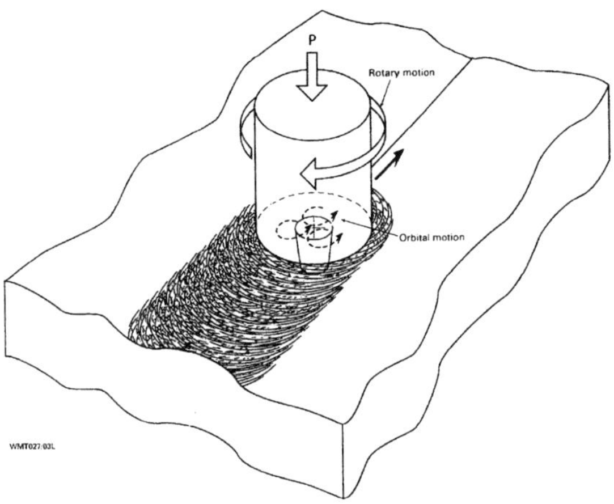

Fig. 8 Principle of Com-stir™ *(Copyright © 2003, TWI Ltd)*

In conventional rotary FSW the relative velocity of the tool increases from zero at the probe centre to maximum velocity at the outer diameter of the shoulder. The combined orbital/rotary Com-stir™ technique can significantly modify the differential in velocity between the probe centre and the shoulder diameter. This technique provides a range of options between an almost a complete orbital motion to an almost a complete rotary motion by selectively changing the relative rotational speed. This feature can be pre-selected, or varied automatically by in-process control to suit the conditions. Although opposed rotations are shown in Fig. 8, rotations that are in the same direction could also be used.

Essentially, orbital motion results in a relative interfacial velocity of constant magnitude but of continuously varying direction [15-18]. Reported work for bar-to-bar orbital friction welding indicates that higher heating rates and increased m etal flow can be expected [15]. The use of compound rotary and orbital motion for friction surfacing has previously been demonstrated [18].

The Com-stir™ welding technique which also uses both rotational and orbital motions benefits from the higher heating rate and greater metal flow.

Surface appearance of Com-stir™

Figure 9 shows the appearance of the weld surface that is formed beneath the tool shoulder after Com-stir™ welding. An additional surface ripple can be seen superimposed on that which is obtained from conventional FSW.

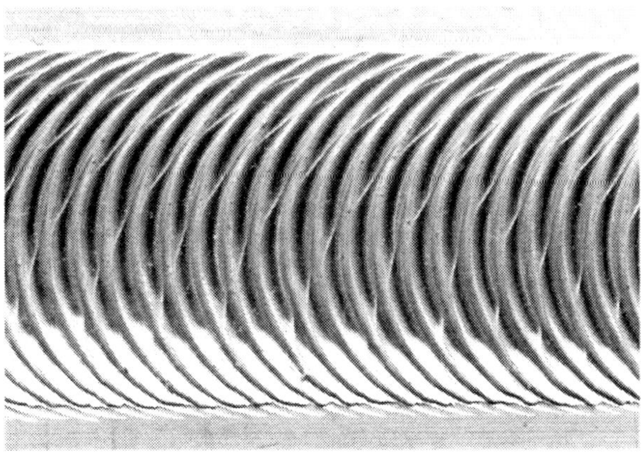

Fig. 9 Surface appearance of Com-stir™ weld made in 6mm thick 6082 T6 condition aluminium alloy at a weld travel speed of 3.6 mm/sec (220 mm/min), using an orbital speed of 0.12 of that of the rotary speed. (Weld track approximately 25 mm wide). *(Copyright © 2003, TWI Ltd)*

When Com-stir™ is used for material processing, the increase in the width of the processed zone (and hence the volume of material processed) is particularly advantageous. Fig 3 shows a wide weld in a lap joint made by Com-stir™. In this example the weld width measures 230% of the top plate thickness (cf. 110% for conventional FSW). Much wider weld regions are believed to be possible and this is being further investigated.

Fig.10 Macrosection showing a wide weld region produce by Com-stir™. *(Copyright © 2003, TWI Ltd)*

The exit hole left at the end of the weld reveals multi-faceted highly reflective internal features that characterise the Com-stir™ technique, see fig 11. This illustrates greater working than is obtained with conventional FSW and hence the greater efficiency of surface oxide fragmentation.

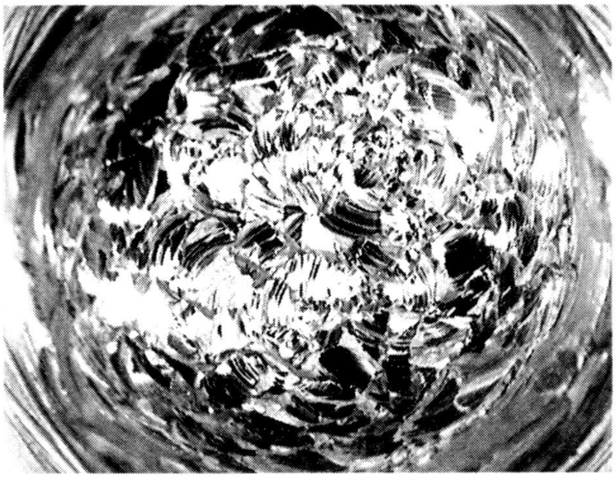

Fig. 11 Exit hole of a Com-stir™ weld (Approximately 14 mm in diameter). *(Copyright © 2003, TWI Ltd)*

Near-Net Shaped Manufacture by FSW - Pro-stir™

A novel near-net shaped prototyping technique is under development at TWI. Rapid prototyping is the most common name given to a host of related additive technologies that are used to fabricate physical objects directly from sheet, or powder material. These methods are unique in that they add and bond materials in layers to form objects. Near net shaped additive technologies offer advantages in many applications compared to classical subtractive fabrication methods such as milling or turning:

- Near-net shape manufacturing systems reduce the construction of complex objects to a manageable, straightforward, and relatively fast process.
- Objects can be formed that have geometric complexity or intricacy without the need for elaborate machine set up.
- Environmental benefits, reduced machined waste, energy, and waste dispersal.

Many welding techniques have been adapted and developed for rapid prototyping and near-net shaped manufacture. Figs. 12a & b show the concept of using the Pro-stir™ method to manufacture near-net shaped components.

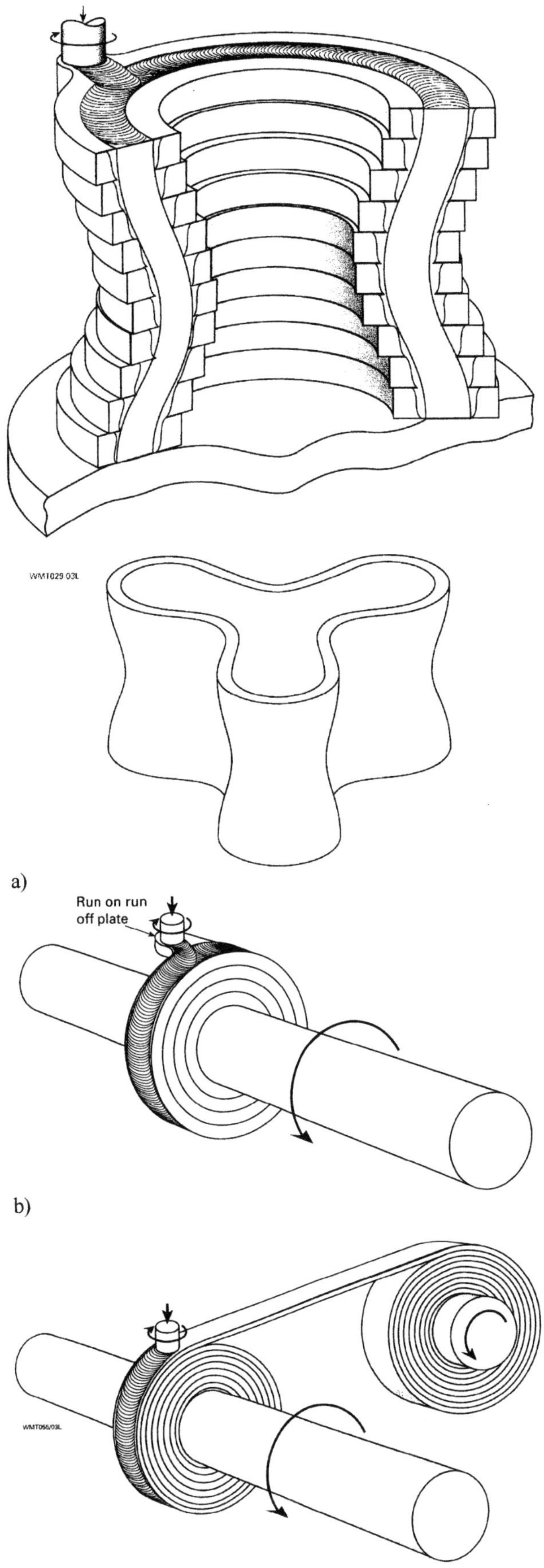

Fig. 12 Principle of Near-net shaped manufacture by Pro-stir™
a) Flat surface - profiled plates
b) Curved surface - annular rings
c) Continuous strip welding (with continuous surface dressing not shown) *(Copyright © 2003, TWI Ltd)*

Following the development of Flared-Triflute™, probes and the alternative stir welding motions such as Skew-stir™, Re-stir™, and Com-stir™, significant improvement in the quality of the lap weld has been achieved. The use of FSW for near net shape manufacture by a layering system is becoming increasingly practical. Fig 13 shows a small test trial using 6 mm thick 5083 aluminium alloy. The same principle would apply for 25 mm thick plate.

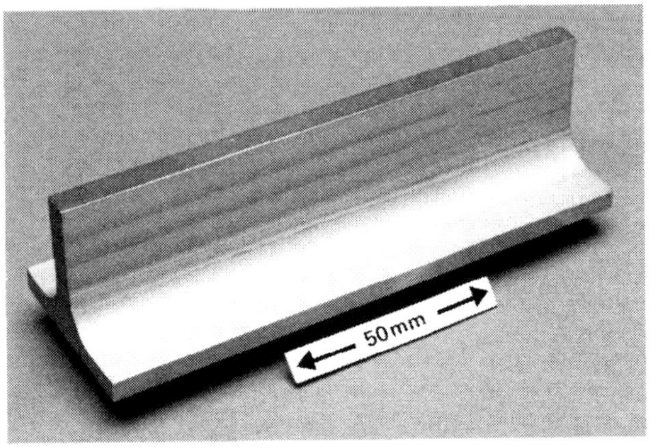

Fig. 13 Near-net shape test sample, - Six 6 mm thick plates welded one on top of each other *(Copyright © 2003, TWI Ltd)*

The advantage that FSW near net shape, prototype-processing techniques has can be summarised as follows:

- High deposition rate when used with comparatively thick plate.
- Able to deal with comparatively thin sheet and thick plate
- Low distortion
- Three-dimensional processing technique
- Strategic regions of the component can be tailored with material to provide different properties.
- The product comprising processed hot forged material.
- It is solid-phase technique not subject to gravity (most rapid prototyping systems are gravity restricted) This means we can 'grow' additional parts in situ if necessary

A Self Reacting FSW Approach on Thin Section Aluminium

In many instances FSW has been shown to be, effective in welding aluminium of approximately 1mm thickness using a single side fixed pin approach. While this has been successfully applied across many different product forms there are also many hardware configurations for which a single sided approach is not possible. For these configurations, such as hollow extrusions, it would be desirable to use a Self Reacting Pin Tool (SRPT) or "Bobbin Tool" approach thereby eliminating the need for internal tooling to react the loads imparted on the work piece.

To date little work has been reported on the bobbin type method. A feasibility study was therefore initiated by MTS to determine what the machine capabilities and SRPT configurations need to be in order to successfully weld thin sheet material. From a practical perspective, a target thickness of 2mm and a target weld speed of at least 1 meter per minute were desired in order to improve on the processing times of the incumbent fusion weld processes and minimize weld processing cycle time in production.

The objective of the feasibility study is to ascertain what practical limits there may be in the FSW approach to thin section aluminium welding of Al 6061-T6 using a self-reacting tool. In this case the thin section is defined as a range between 2mm and 4mm with weld travel speeds starting at 1 meter per minute. Machine performance data was to be collected and reviewed to establish performance envelopes critical to the process.

Approach

New tools were designed for this thickness and alloy (Fig. 14) shown next to a tool successfully used on thicknesses greater than 6mm. A weld system was used incorporating an AdAPT weld head in order to generate the sensitivity needed to perform the welds.

Fig. 14 Self-reacting pin tools (SRPT) *(Copyright © 2003, MTS)*

The tool shown on the left-hand side is suitable for 2 -4 mm thick plate. The tool shown on the right-hand side is suitable for thickness greater than 6 mm. The scroll features on the tool shoulders are also shown.

Weld trials showed that the required loads were to be approximately 300 lbs with rotation speeds of >1000rpm and travel speeds of >1M/min. While this weld system with a maximum load capacity of 20,000 lbs was not designed for these specific loads, consistent performance with little variation between command and actual values was observed (Fig. 15).

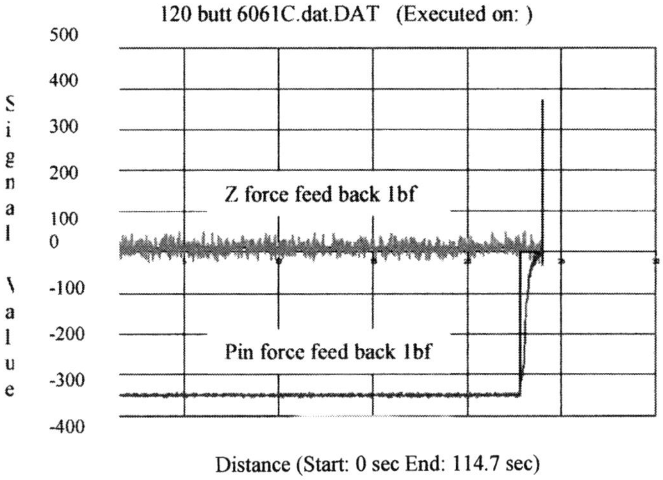

Fig.15 Instrumentation chart showing force feed back *(Copyright © 2003, MTS)*

Several trials were carried out on 3mm thick material successfully with weld speeds greater than 1M/min with no detrimental effects seen in the machine performance, tooling or material. At that point test panels of 1.9mm thickness were fit into the tooling and a weld made with the same parameters as used on the 3mm thick material. A maximum weld travel speed of 1.2 M/min (Fig. 16) was reached using the smaller diameter weld tool and the weld trials were ended.

Fig. 16. Surface appearance - spindle side *(Copyright © 2003, MTS)*

A successful result in these trials would indicate that the lower limit of material thickness and upper limit of travel speed have not been fully appreciated for this alloy. It would seem reasonable to expect that 1mm thick material and travel speeds approaching 3M/min could be anticipated in the near term. Additionally, this study would indicate that other alloy systems and product forms should be considered well within the range of expectation. Further testing in the near future should bear this out.

Summary Remarks

This paper has described recent developments in FSW when using reversal motion (Re-stir™) and combined motion (Com-stir™) techniques. Both Re-stir™ and Com-stir™ motion variants gave an improvement in lap weld quality in that upper plate thinning was reduced as compared with conventional pin type probes using conventional rotary motion. Re-stir™ offers benefit in generating essentially symmetrical welds and hence has the potential to overcome some of the problems associated with the asymmetry inherent in conventional friction stir welds. While Com-stir™ provides for a much-improved path for metal to flow around the tool during friction stir welding and results in a good quality wider weld zone. These wider welds are more suitable for lap welds and material processing.

Preliminary trials using a FSW method of near-net shape manufacture and material processing is also described.

Furthermore, progress and the ongoing development of a self-reacting pin type tool (SRPT) or 'Bobbin tool' for thin section aluminium is described. Typical travel speeds of up to 1.2 m/min for welding 2 - 3 mm thick 6061-T6 condition aluminium alloy are reported.

Acknowledgements

The authors wish to thank R E Dolby (Retired), C S Wiesner, D G Staines, I M Norris, E D Nicholas, S D Smith, E Watts and P Evans for their much valued help and support.

References

1. Thomas W M, Nicholas E D, Needham J C, Murch M G, Temple-Smith P and Dawes C J. European Patent Specification 0 615 480 B1.
2. Mahoney M W, Rhodes C G, Flintoff J G, Spurling R A, and Bingel W H, 'Properties of Friction stir welded 7075 T651 Aluminium' Metallurgical and Materials Transactions A, Volume 29A July 1998.
3. Christner B, 'Development and testing of friction stir welding (FSW) as a joining method for primary aircraft structure', ICAS Congress 2002.
4. Thomas W M, Staines D G, Norris I M, and de Frias R, 'Friction stir welding –tool and developments' FSW Seminar Instituto de Soldadura e Qualidade, Porto, Portugal, December 3, 2002.
5. Nicholas E D, Thomas W M, and Needham J C. 'Improvements relating to friction welding'. International Publication Number WO 96/38256. (Published but not maintained)
6. dos Santos J F, Cam G, von Strombeck A, Ventzke V, and Kocak M, 'A comparison between microstructure, properties and toughness behaviour of power beam and friction stir welds in Al-alloys', TMS 2001, 130th Annual Meeting and Exhibition, New Orleans Louisiana, USA, 11-15 February 2001.

7. Waldron D J, Roberts R W, Dawes, C J, Tubby P J, 'Friction stir welding characterisation for aerospace structures', Trends in Welding Research, Proceedings 5th International Conference USA.
8. Boon T, Thomas W M and Temple-Smith P: 'Friction welding Apparatus and Methods', Patent No US 6,325 273 B1 (priority date December 6, 1996).
9. Ramirez A J, and Juhas M C: 'Microstructural evolution in Ti-6Al-4V friction stir welds', Materials Science Forum Vol. 426-432 (2003) pp 2999-3004, online at **http://www.scientific.net** Trans Tech Publications, Switzerland.
10. Andersson C G, Andres R E, Dance B G I, Russell M J, Olden E J, and Sander R M, 'A comparison of copper canister fabrication by the electron beam and friction stir processes'. Swedish Nuclear Fuel and Waste Management Co (SKB) and TWI. 1st FSW Symposium, 14-16 June 1999, Thousand Oaks, California, USA.
11. Colligan K J, Fisher J J, Gover E J, & Pickens J R, 'Friction stir welding in the advanced amphibious assault', Advanced Materials and Processes, September 2002.
12. Lienert T T, Stellwag W L, Grimmett B B, & Arke R W, 'Friction Stir Welding studies on Mild steel', Welding Journal, January 2003, pp 1-5 - 9-5.
13. Karlsson L, and Berqvist E L, 'Application of Friction Stir Welding to Dissimilar welding IIW DOC. IX2015 – 01.
14. Arbegast W J, Hartley P J, 'Friction Stir Weld Technology Development at Lockheed Martin Michoud Space System-An Overview' Trends In Welding Research, 1-5 June 1998, Pine Mountain, Georgia, USA.
15. Cousens. A. K., and Camping. M. J., 'An initial investigation of the orbital friction welding of mild steel'. TWI Research Report 270, April, 1985.
16. Searle. J., 'Friction welding non-circular components using orbital motion'. Welding and Metal Fabrication, Vol.39, No. 8, August 1971. Pp. 294-297.
17. Dawes. C.J., 'An examination of orbital friction welding using axial offset.' TWI Research Bulletin, Vol 12, No. 6, 1971. Pp. 161-167.
18. Thomas. W. M., and Nicholas E.D., 'Surfacing method'. European patent application number 89303501.4, priority date 11 April 1988, publication number 0 337 691, (published but not maintained).

Proceedings from Joining of Advanced and Specialty Materials, 13–15 October 2003, Pittsburgh, Pennsylvania, ASM International, 2004

Hot-Stage Light Microscopy Studies of the Wettability of Nickel-Base Superalloys during Transient Liquid Phase Bonding

Lewis N. Payton
Auburn University, Auburn, Alabama 36849

Subhadravalli V. Chitti, Dina R. Taarea, Nofrijon I. Sofyan, William F. Gale, Daniel A. Butts, Rajeev Aluru and Robert D. Love
Materials Research and Education Center, Auburn University, Auburn, Alabama 36849

Abstract

The wettability of various polycrystalline and single crystal nickel-base superalloys by a composite interlayer has been examined dynamically, using hot-stage light microscopy on sessile drop specimens (supplemented by post-spreading characterization), in order to explore the origins of differences in wettability. In this paper, the roles of both substrate-oxide composition and substrate microstructure in wettability are discussed.

Introduction

The Ni-base single crystal superalloys possess excellent mechanical properties at elevated temperatures typical of today's modern gas turbines (marine, air and land) [1, 2]. For these reasons, research concerning the brazing characteristics, alloy design, alloy properties, heat treatment conditions and manufacturing techniques continues to be of great interest. The larger land based gas turbines used to generate electricity, in particular, have much longer blades, necessitating the joining of polycrystals to single crystals.

Repair techniques involving molten liquid processes may include brazing, welding and transient liquid phase bonding. Electron beam welding and other fusion welded processes suffer from a number of problems such as stray grain formation and hot-cracking. Satisfactory wettability of the substrates by the interlayer material is prerequisite for more benign processes such as brazing and transient liquid phase bonding.

Transient liquid phase (TLP) brazing is the preferred method for joining in these turbines [1]. The study of wetting at high temperatures has been well documented with many excellent studies of binary and ternary systems [3]. Such studies are extremely fundamental to the TLP brazing technique which has been well studied in the literature [4]. Wettability studies of superalloys, which are by composition an extremely varied substrate, are rarer than binary or ternary systems [5, 1, 6, 7].

Nickel-base superalloy substrates consisting of commercially available polycrystalline alloys (738 and 939) and the commercially available single crystal alloy CMSX-4 (Table 1 details the nominal compositions of the substrates) were investigated. The formed substrate geometry had a diameter of 5 mm, a thickness of 1 mm, and a 6 μm diamond polished finish. Filler-metal disks of 1 mm diameter were mechanically punched from 50 μm thick commercial purity (99 wt% Cu) copper sheet (for temperature calibration) and a proprietary nickel based foil. This consisted of a (relatively) low alloy nickel-base core with boron enriched surfaces (i.e. a boron-rich/nickel-base/boron-rich sandwich). Before testing, the foils were ultrasonically cleaned in acetone and desiccated.

Wettability was characterized using sessile drop testing at a temperature range of 900-1250 Celsius under a vacuum of ~1 mPa. Dynamic observations of the spreading process were made using hot-stage light microscopy together with video recording. These allowed the diameter of the droplet to be measured with a high degree of repeatability.

	738 Alloy	939 Alloy	CMSX-4
Ni	61.0	48.0	60.0
Cr	16.0	22.5	6.5
Co	8.5	19.0	10.0
Mo	1.7	0.0	0.5
W	2.6	2.0	6.5
Ta	1.7	1.4	6.3
Nb	0.9	1.0	3.5
Al	3.4	1.9	5.5
Ti	3.4	3.7	1.0
C	0.11	0.15	0.0
B	0.1	0.009	0.0
Zr	0.05	0.09	0.0
Other	0.54	0.251	0.2

Table 1. Nominal composition of substrates (wt%)

The results of the hot-stage spreading tests were correlated with scanning electron microscopy (SEM) observations and energy dispersive X-ray (EDS) analyses. These analyses were conducted using a JEOL 840 microscope system equipped with an Oxford Instruments ultra-thin window detector and ISIS analyzer.

Initially, pre-oxidized substrate samples were not included in the experiments, but were added afterwards for comparison with the bare substrates. Samples were oxidized for 15, 30 and 45 minutes at 1000 degrees Celsius in an atmospheric furnace. Additionally, in order to compare the possible effects of grain boundaries, the CMSX-4 was reduced in a rolling mill (by ~ 45%), recrystallized and subjected to Ni-flex melts. Also, CMSX-4 was homogenized in a vacuum furnace at 1000 degrees Celsius for 8 hours.

Results and Discussion

Figures 1, 2 and 3 summarize the spreading sequences observed for the 738, 939 and CMSX-4 substrates. The figures represent the cold foil, foil at the onset of melting, molten spreading and onset of isothermal solidification. Post mortem compositional studies were then conducted.

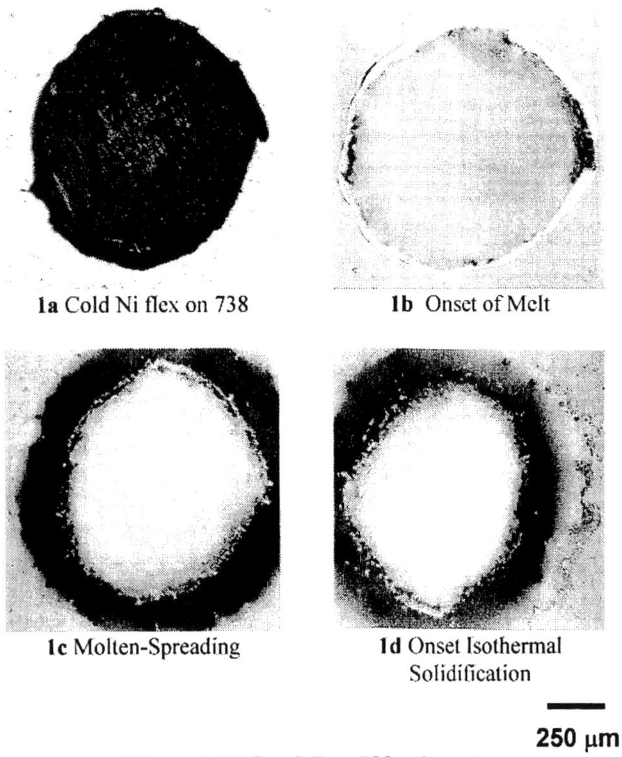

1a Cold Ni flex on 738 **1b** Onset of Melt

1c Molten-Spreading **1d** Onset Isothermal Solidification

250 μm

Figure 1. Ni-flex foil on 738 substrate

There are many factors which might affect the wettability of superalloys. Within the interlayer of this experiment for example, the non-melting core could serve as a sink for boron, leading to early or premature isothermal solidification. The core could also present a physical impediment to flow.

Considering the substrates (CMSX-4, IN-738 and IN-939), the presence of chromia formers in the polycrystals versus alumina formers in the CMSX-4 could produce differences. The lack of a grain boundary structure in the CMSX-4 could modify spreading (hence the recrystallization trial described earlier). The presence of boride formers (such as Cr, Mo and Ti) may also slow wetting. Reactive wetting which depletes boron has also been shown to reduce wettability due to premature isothermal solidification [7].

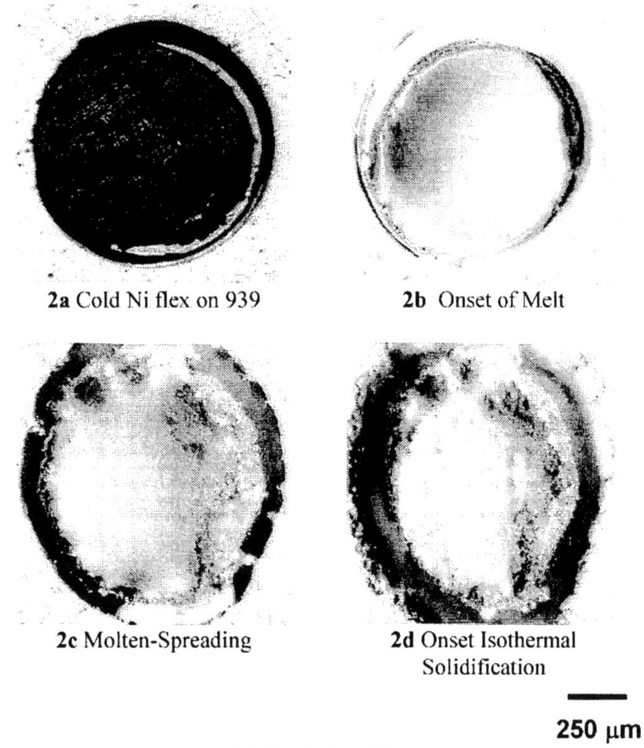

2a Cold Ni flex on 939 **2b** Onset of Melt

2c Molten-Spreading **2d** Onset Isothermal Solidification

250 μm

Figure 2. Ni-flex foil on 939 substrate

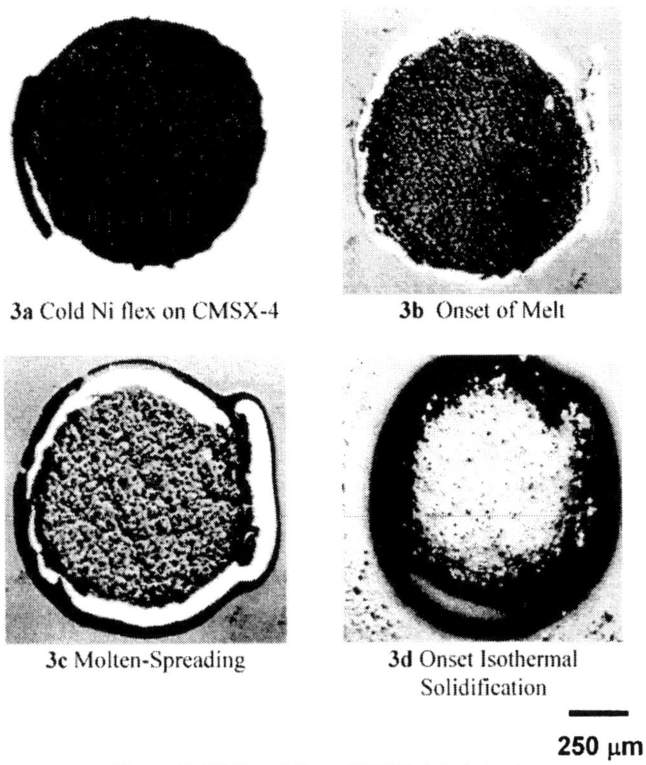

3a Cold Ni flex on CMSX-4 **3b** Onset of Melt

3c Molten-Spreading **3d** Onset Isothermal Solidification

250 μm

Figure 3. Ni-flex foil on CMSX-4 Substrate

Figure 4 graphically compares the typical spreading results. The onset of melting was most consistent for the CMSX-4, followed by the 738 and then 939. The 939 substrate did not spread to as wide a final diameter in all cases. The CMSX-4 and the 939 did not in general exhibit the intermediate plateaus noticeable in the 738 plot. The foil sometimes seemed to de-wet on one side while spreading on the other in the case of the 738 substrate. Summarizing the steps observed in 738 and CMSX-4, the foil first melts and starts spreading from the ends initially. After some time, the material in the center which may be still not be liquid melts due to the increase in temperature and spreads out. In the meanwhile the outer part of the foil that has already melted and spread reaches a condition where it has started resolidifying. The edge of the droplet encounters a "dam" that the liquid must break through. Thus, the liquid from the center flows out over this solid. It is due to this repeated alternating flow and solidification that we observe the steps in the graph (Figure 4). Although the core of the foil should not melt, this is not visible directly in projection. In contrast, the non-melted core could be seen clearly in post-cooling cross sections of the bond.

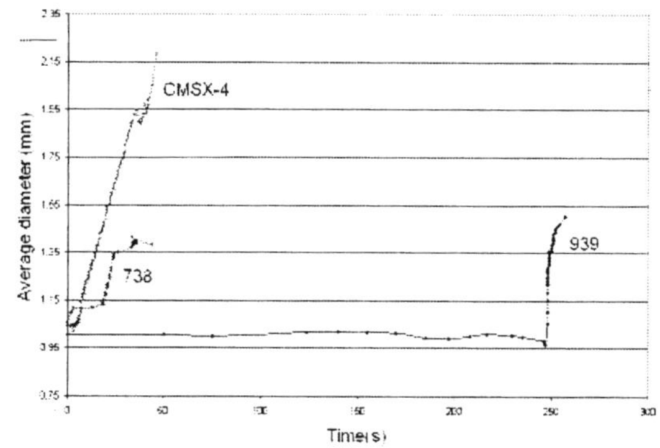

Figure 4. Average diameter versus time (seconds) for substrates

The results of the hot-stage spreading tests were then correlated with scanning electron microscopy (SEM) observations and energy dispersive X-ray (EDS) analyses. Figure 5 shows a typical transit profile conducted across the samples after the foil was allowed to solidify. The Ni-flex/CMSX-4, which generally wet the best, in terms of spreading the most, was greatly enriched in aluminum (>5X) in the dark outer rings of Figure 3d. Chromium tended to concentrate (>3X) in the inner portion of these samples.

The Ni-flex/738 system was not remarkable. There was evidence of a slight tendency of the nickel to decrease and the chromium to increase in the outer rings evident in Figure 1d.

The Ni-flex/939 system showed a very strong decrease in the nickel content of the foil all across the foil (~20+ %). The cobalt concentration generally increased up to 4X throughout the foil as compared to it's "as-received composition.

The spreading and compositional studies were repeated on heavily oxidized samples (15, 30 and 45 minutes in a conventional oven at 800 degrees Celsius). The 738 and 939 substrates tended to de-wet after only 15 minutes of pre-oxidation. This became more noticeable as the period of pre-oxidation increased. The 939 substrate was most reactive, melting completely when pre-oxidized for 45 minutes. The CMSX-4 substrate showed a very strong, consistent de-wetting at all levels of substrate pre-oxidation. Since the de-wetting generally destroyed the foils, or separated them from the substrate, neither sectioned nor plan-view samples were available for analysis. These studies will be repeated with less oxidation.

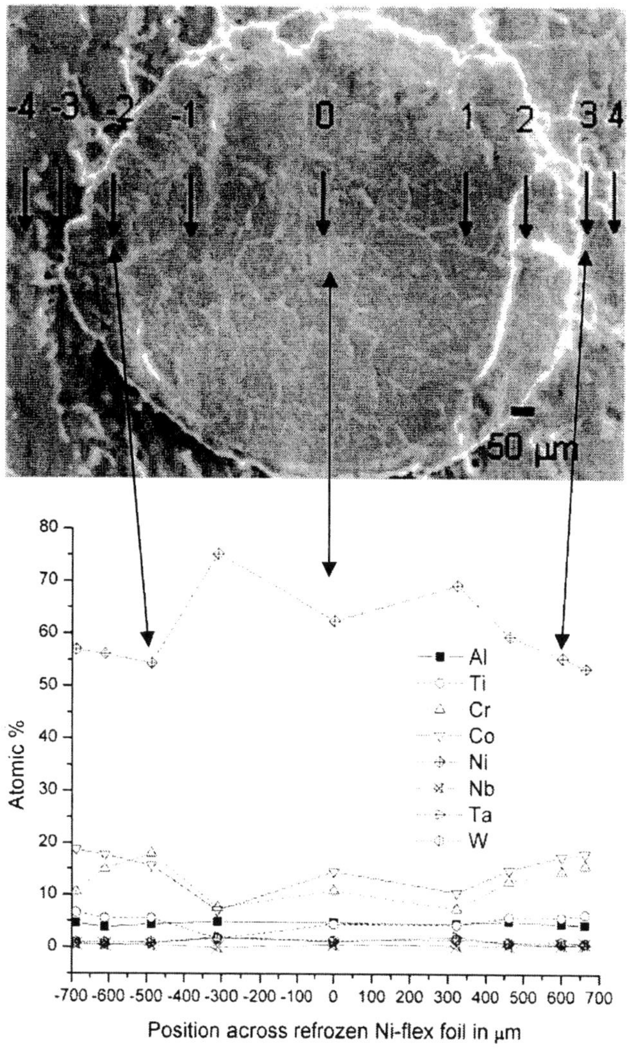

Figure 5. Typical compositional profile across foil after completion of the spreading experiments (Figures 1-4).

CMSX-4 was cold-rolled (45% reduction) until cracking became a problem. Sectioned samples were then placed in a vacuum furnace at a nominal temperature of 1200 degrees Celsius for 8 hours to allow recrystallization and grain growth. The hot stage experiment was then repeated. Again, excellent wetting with liquid spreading uniformly over the entire substrate was observed. The presence of grain boundaries in the single crystal material did change the typical profile for CMSX-4 shown in Figure 4.

The original, as-received CMSX-4 was also homogenized at 1000 deg Celsius over 8 hours for comparison with the recrystallized CMSX-4. These samples exhibited strong de-wetting results. Speculation indicates that perhaps the redistribution of boride formers (such as Cr) may have had a significant effect on wetting; although why this should not have been the case for the recrystallized material is unclear at present.

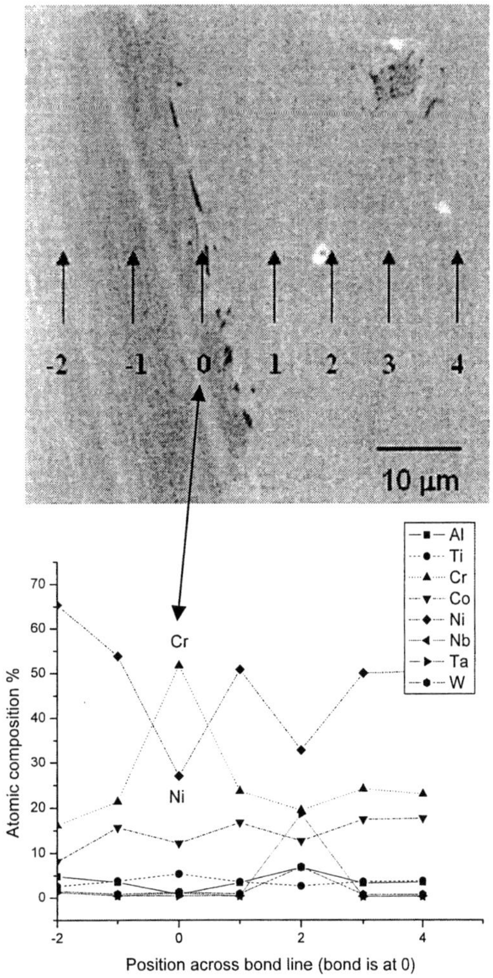

Figure 6. Photomicrograph of 939/Ni-flex cross-section. The there is a large concentration of chromium directly on the bond line. Nickel is increased along the bond in the parent 939 material. Cobalt decreases in the 939 substrate as it evidently migrates to the foil based upon supporting analysis in Figure 5.

Selected samples were then sectioned through the bond line formed between the foil and the substrate for SEM photography and EDS analysis. The 738 and CMSX-4 system bond lines were not very remarkable. Analysis across the bond line in the 939 Ni-flex system was consistent with the notable changes in the earlier compositional studies. Nickel increased in the 939 substrate all along the bond line (at the expense of the foil) where as the cobalt decreased markedly from the nominal "as received" values. This cobalt may have migrated into the Ni-flex foil, accounting for the increases reported in the earlier compositional profiles. There is a marked peak in the chromium concentration at the bond line. Figure 6 provides an SEM photograph of the bond line examined and shows the unevenness of the compositional profile obtained, without detailing the exact composition. Aluminum was consistent from the foil into the 939 material.

Conclusions

Although hot stage experiments do not normally generate a statistically viable population for analysis, this study clearly indicates that the 939/Ni-flex system acts differently than the other combinations studied. Mechanical test results cited in a companion paper [8] are consistent with these observations.

Acknowledgements

Substrate and foil materials were kindly provided by Siemens Westinghouse, Orlando, Florida.

References

1. K. Nishimoto, K. Saida, D. Kim, S. Asai, Y. Furukawa, Y. Nakao, *Welding in the World,* 41, 121-131 (1998).
2. S. Ahmed, *JOM,* 24-26 (1990).
3. N. Eustathopoulos, M. Nicholas, B. Drevet, *Wettability at High Temperatures*, Elsevier Science, Ltd., Amsterdam, Netherlands, 1999, p. 420.
4. W. F. Gale, D. A. Butts, *Adv. Mater. Proc.,* (2003).
5. Y. Shen, W. F. Gale, J. W. Fergus, X. Wen, *Materials Science and Technology,* 17, 1293-1298 (2001).
6. W. F. Gale, R. Wallach, *Journal of Materials Science,* 28, 243-248 (1993).
7. W. F. Gale, Y. Guan, *J. Mater. Sci.,* 34, 1061-1071 (1999).
8. L. N. Payton, R. Aluru, N. I. Sofyan, W. F. Gale, V. G. Krishnardula, D. A. Butts, S. V. Chitti, *International Conference on Joining of Advanced and Specialty Materials VI*, ASM International, Pittsburg, PA, 2003.

Proceedings from Joining of Advanced and Specialty Materials, 13–15 October 2003, Pittsburgh, Pennsylvania, ASM International, 2004

Reactions Between Constituents of Two Molten Nickel-Base Superalloys and Ceramic Materials

Dina R. Taarea, William F. Gale, Jeffrey W. Fergus
Auburn University, Materials Research and Education Center
202 Ross Hall, Auburn University, AL 36849
Tel: (334) 844-3406, Fax: (334) 844-3400
E-mail: taaredr@eng.auburn.edu, wfgale@eng.auburn.edu, jwfergus@eng.auburn.edu

Abstract

The work discussed in this paper grew out of a study of ceramic – metal reactions, during casting of nickel-base superalloys. Wetting promoting reactions involving the melt and such ceramic components as molds, cores and crucibles can be extremely damaging to the quality of precision castings. Conversely, there is the potential to employ these same reactions beneficially in the active metal brazing of structural ceramics.

In this paper, reactions involving doping of the ceramic material, rather than the formation of second phases will be considered. Particular emphasis will be placed on the role of vapor-phase transport.

Introduction

Reactions between the constituents of molten metals and solid ceramics are important in a number of areas, such as active metal brazing of ceramics, metal casting and production of metal matrix composites [1-6]. In the case of metal casting, the observed reactions between melt and ceramic lead to undesired wettability. In contrast, these types of reactions are highly desirable in active metal brazing of ceramics where wettability is required.

Initially, this study focused on metal casting applications, i.e. how to predict and then devise a method to eliminate the unwanted reactions. However, as the study progressed, it became apparent that there seems to be an application in active metal joining of ceramics, since the different applications have potentially similar reaction mechanisms [1-3]. The focus of this project has since shifted to standardized engineering substrates in order to eliminate the variability that exists in foundry ceramics.

Experimental Procedure

The materials used in this study were provided by Rolls-Royce Indianapolis, formerly Allison. Two types of commercial nickel-base superalloys were used; CMSX-4, which is intended for use as a single crystal alloy and CM186LC, a polycrystalline alloy. The ceramic pieces initially used in this study were foundry ceramics (mold, core and crucible) materials used to cast the above alloys. Simple melting procedures were then designed using the ceramics as substrates upon which the nickel-base superalloys were melted.

The compositions of CMSX-4 and CM186LC are shown in Table 1.

Table 1: Compositions of CMSX-4 and CM186LC (weight %)

Alloy	CMSX-4	CM186LC
Cr	6.5	6.0
Co	9.0	9.0
Mo	0.6	0.5
W	6.0	8.0
Re	3.0	3.0
Ta	6.5	3.0
Al	5.6	5.7
Ti	1.0	0.7
Hf	0.1	1.4
C	---	0.07
B	---	0.015
Zr	---	0.005
Ni	Bal.	Bal.

The use of foundry ceramics as substrates yielded variability in the results. In later stages of this research, baseline engineering substrates such as yttria partially stabilized zirconia, high purity alumina and fused silica were used in place of the foundry ceramics to standardize the experiments.

The fused silica used has a nominal melting temperature of 1700°C. However, a large amount of flow is detected at 1450°C, the lowest temperature used for the experiments. This problem necessitated the use of an even lower test temperature of 1250°C. The use of the silica substrates was discontinued due to the lack of reaction between the superalloy and silica at 1250°C.

The bulk of the experiments were done with zirconia substrates since there were less discernable reactions with alumina substrates and lack of adhesion between the superalloy and the alumina.

As can be seen in Table 1, the numerous components of the superalloys make it difficult to determine which components contributed to the occurrence of a 'reaction layer'. A process of elimination was undertaken using simpler alloys containing elements found in the nickel-base superalloys. The binary alloys used were Ni/Cr, Ni/Ti and Ni/Al. Pure Si was also used because of previous work reported in the literature, which suggests that Si plays a major role in the reactions that occur between some ceramics and nickel-base superalloys [7]. A ternary alloy of Ni, Cr and Si, better known as brazing alloy BNi5, was also used since initial observations seem to indicate that Cr is a key element in the reactions that occur. The results of these experiments allow one to better understand the reaction between the nickel base superalloys and ceramic substrates.

To systematically survey the reactions that occur as the alloys melt on the ceramic, two experimental methods were designed. The first method was to melt an alloy on a ceramic substrate using a microscope equipped with a heating stage. This allowed in-situ viewing of the reaction that occurs when the alloy starts to melt. This method is generally useful, but in the present case the reactions involved produce a very small and thin reaction layer on the specimens. It was discontinued since the 'reaction layer' of these specimens yield little information upon analysis.

The second approach was to do the sessile drop method by thermally cycling the ceramic substrate and alloy in a high temperature vacuum furnace.

Two types of furnaces were used for the sessile drop method in this study. Work was initially done using a Lindberg tube furnace, with the specimens ramped up to 1450°C and held at temperature for fifty minutes under pressure of around 10^{-3} Pa. A chamber type Brew vacuum furnace was later on used due to its ability to attain higher temperatures. The Brew furnace has another advantage in that it has a better vacuum system as compared to the tube furnace where leakage through the alumina tube was a concern. The Brew experiments were done at various temperatures, mainly 1450°C and 1600°C, also with fifty minutes holding time at around 10^{-4} Pa pressure.

Results and Discussion

The foundry ceramic materials used in combination with nickel-base superalloys undergo a drastic change in coloring as a result of the stated thermal cycle. The originally white surface ends up with a 'reaction layer' of bright pink or green, or in some cases a combination of the two colors. Other systems involving foundry ceramics and alloys containing Cr, exhibit a similar 'reaction layer' to various degrees. However, no reaction is detected with alloys which contain no Cr.

When zirconia is substituted for the foundry ceramics as the substrate, the thermal cycle causes it to react with the nickel-base superalloys and leads to discoloration and darkening of the substrate.

To characterize the 'reaction layer', the specimens were examined using a scanning electron microscope (SEM). The energy dispersive x-ray spectroscopy (EDS) on the SEM detects Cr on the ceramic surface of specimens that have gone through the cycle, when none can be detected on the substrates in their original state. However, the EDS is unable to detect any other elements not present prior to the thermal cycle. This led to the hypothesis that Cr plays a major role in the reactions that occur between the ceramic and the alloy, while Al, Si and Ti did not appear to play a direct role in the reactions.

There is only a limited amount of literature data available pertaining to this type of reaction, but one paper suggests that Si suboxides play a major role in the reactions seen [7]. In the current case, no reactions were seen with pure Si. However, the presence of both Si and Cr seems to yield a more intense reaction than with Cr alone.

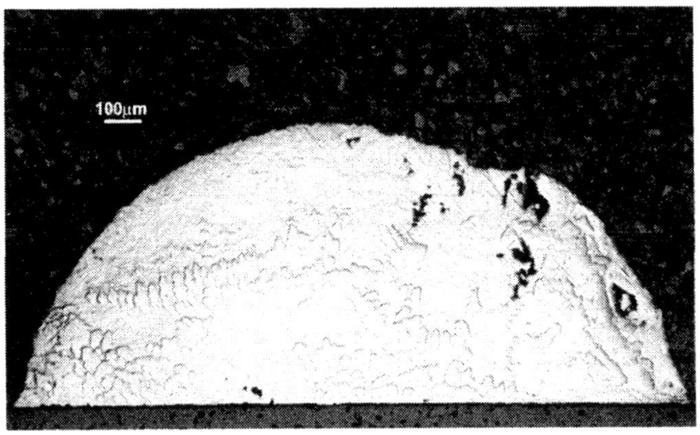

Figure 1a: Light microscope micrograph showing the cross section of resolidified molten CMSX-4 well adhered to zirconia.

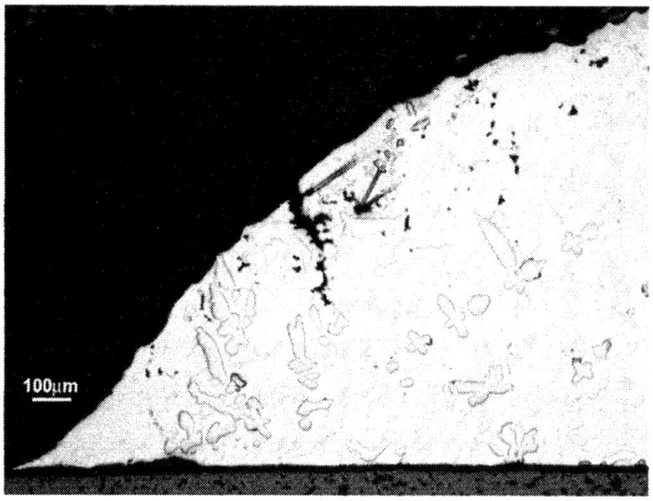

Figure 1b: Light microscope micrograph showing an edge of cross section of resolidified molten CM186LC well adhered to zirconia.

Figure 1a and b were taken with a light microscope. They show significant dendritic growth within the resolidified superalloys. The dendrites seem to mostly congregate near the interfacial area between the superalloy and zirconia. Figure 2a and b were taken using an SEM on the same samples, but at a much higher magnification.

In the SEM micrographs, the dendrites are shown to be much brighter than its surrounding, even in secondary electron images. EDS analysis of the dendrites reveal that the dendrites are composed entirely of tungsten, whereas the dark areas indicate intermetallic phases composed of the other components of the nickel-base superalloy. The overall composition and microstructure of the superalloy droplets was changed radically by the reaction with the substrate, even though a distinct reaction layer was not observed under the droplets.

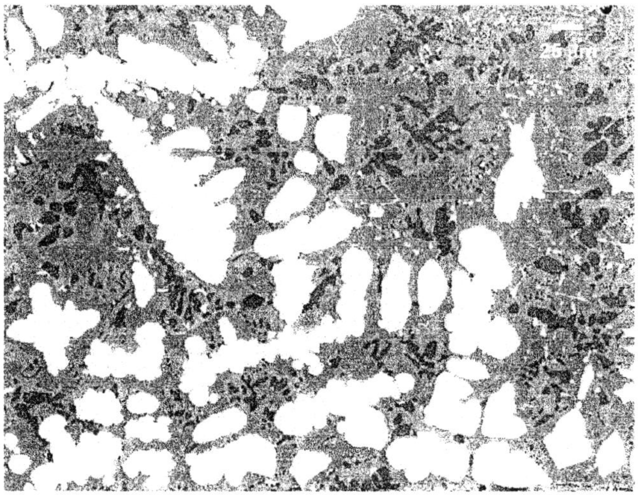

Figure 2a: SEM micrograph of resolidified molten CMSX-4 after a 50 minute thermal cycle at 1600°C.

Figure 2b: SEM micrograph of resolidified molten CM186LC after a 50 minute thermal cycle at 1600°C.

X-Ray diffraction (XRD) was utilized to characterize the phases that may exist on the formed 'reaction layer'. Due to the variability and the complexity of the composition of the commercial foundry ceramics, it is difficult to draw consistent conclusions of existing phases using XRD on those specimens. However, XRD done on zirconia substrates that had undergone the same cycle show a lack of second phase formation, leading to the hypothesis that the visible 'reaction layer' is merely a doping effect.

With the foundry ceramics, wetting is observed, but in general wetting is rather poor and the contact angle is quite large. The use of zirconia as substrates results in very good wetting and adhesion, with a relatively low contact angle.

Hypotheses

Cr content in the alloy seems to have a great impact on the ensuing reactions. The reactions observed started to occur before the melting process began and no reaction occurred in the area where metal and ceramic were in direct contact. These observations led to several hypotheses.

Chiefly, the reactions do not involve reduction of the substrate phases by Cr, since it is thermodynamically impossible in some cases and no evidence of second phase formation was found even when this was possible. Instead, the transport of Cr is through an oxide vapor phase, by which Cr is then incorporated into solid ceramic substrate. The colors of the 'reaction layer' are due to doping from the vapor phase transport of Cr onto substrates, enabling wetting in some cases.

Thermodynamic Calculations, Models and Supporting Experiments

Thermodynamic calculations shown in Figure 3 indicate that the partial pressures of CrO_x vapor species are significant in the expected range of oxygen partial pressure. The oxygen partial pressure in the furnace was calculated to be approximately $2x10^{-5}$ Pa due to residual gas in the vacuum of about 10^{-4} Pa, and is indicated by the dash and dot line in the figure. At this oxygen partial pressure, a few different species exist in such a way that slight atmospheric variations can change the predominant species, which may account for the variability seen in the results.

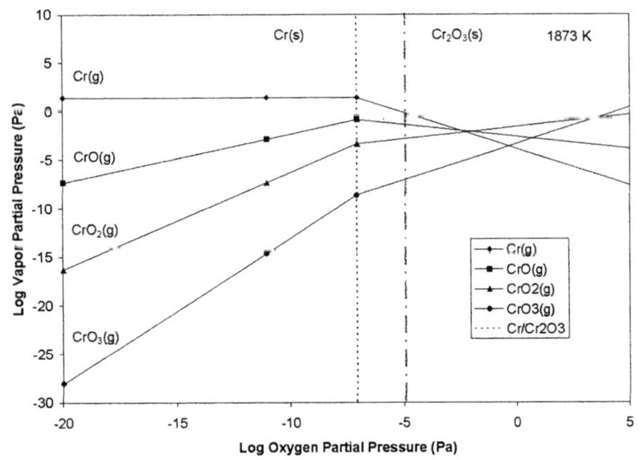

Figure 3: Equilibrium vapor pressure of CrO_x gases at 1600°C (1873 K). Data obtained from reference 8.

In order to better understand the effects of CrO_x gases, a new set of experiments were designed to use metallic and oxidized Cr in place of the nickel-base superalloys.

Small pieces of Cr were preoxidized at 900°C in air for 48 hours. They were then placed on ceramic substrates and put through the same thermal cycle as the nickel-base superalloys, 1600°C (below the melting point of Cr which is 1907°C) at approximately 10^{-4} Pa pressure for fifty minutes. The results were then compared with metallic Cr that had not been preoxidized, but had undergone an identical cycle. With oxidized Cr, an increased and localized amount of reaction was seen as compared to the samples with metallic Cr, supporting the hypothesis that CrO_x vapor phases, rather than just Cr vapors are involved in reaction.

Reactions seen on experiments in the present work are not simple evaporation and condensation, since deposition did not appear in the coldest locations.

Previously, all of the samples had been placed in an open alumina container. A new set of experiments was devised in which two containers were placed on top of each other to create a closed box which restricted but did not eliminate flow. This method decreases the rate of removal of vapor-phase species by decreasing the effective pumping speed of the vacuum system. The specimens produced with closed box method shows an increased rate of reaction with metallic Cr while the combination of oxidized Cr and closed box yield a very intense reaction, both of which support the hypothesis of vapor phase transport through a CrO_x phase.

From the thermodynamic modeling work, it appears that the oxygen partial pressure plays a significant role in determining which reactions can occur. The experiments done took place close to crossover between different dominant vapor species in Cr-O system. Adding a substance that could change the oxygen partial pressure in a preexisting system should conceivably cause a difference in the resulting reaction. Reducing pO_2 in the system atmosphere would favor the following as the dominant species: Cr (g) > CrO (g) > CrO_2 (g) > CrO_3 (g), which results in enhanced reactions. Conversely, increasing pO_2 in the system atmosphere, has the opposite effect in which the reaction is decreased.

Adding Ti into the systems in which Cr is present decreases the oxygen pressure and influences the stability of suboxides. Using thermodynamic data found in reference 8, the calculated equilibrium oxygen partial pressure of Ti is approximately 10^{-15} Pa. Previous experiments have indicated that Ti is not directly involved in the reactions. Ti sponge was chosen due to its large surface area to volume ratio, and was placed inside the closed box experiments of metallic Cr, oxidized Cr, CMSX-4 or CM186LC, placed on ceramic substrates. Ti is known to have a melting temperature of 1668°C, so the samples could be put through the same thermal cycle as has been previously done without melting the Ti. In all cases, the addition of Ti sponge markedly increased the extent of reaction, indicating that at lower oxygen partial pressure the amount of reactions increases as was predicted by the thermodynamic model.

To increase the oxygen pressure, Nickel oxide (NiO) was added to the system. The oxygen supply increases in the container atmosphere through the reduction of NiO. Since Ni has a relatively low melting temperature of 1455°C, the experiments had to be run at a lower temperature of 1400°C. Thermodynamic calculations done with data obtained in reference 9 indicate that the equilibrium oxygen partial pressure of Ni is approximately 4.8×10^1 Pa at 1400°C, higher than that of the vacuum pressure within the furnace chamber. Due to the different temperature employed, the data obtained from this set of experiments cannot be conclusively compared with the result of the experiments run at 1600°C. However, addition of NiO noticeably reduced the extent of reaction in all systems, which suggests that the vapor species contributing to the reactions is one whose partial pressure in equilibrium with Cr_2O_3 increase with increasing pO_2 such as Cr (g) or CrO (g).

Implications for Ceramic Brazing

This study shows that some chromium-containing systems promote wetting between the molten alloy and the ceramic. With zirconia substrates, the superalloy strongly adhered to the ceramic substrate. The lack of second phase formation is an added bonus. In conventional active metal brazing of ceramics, reaction between constituents of the metal alloy and the ceramic leads to the formation of a distinct second phase. This second phase can have a detrimental effect on the observed mechanical properties of the joint due to poor mechanical properties of the second phase and/or coefficient of thermal expansion mismatch. The mechanism observed in this study seems promising for improving active metal brazing of some ceramics.

Summary and Conclusions

The presence of Cr is important to the occurrence of reactions seen in this study between molten nickel-base superalloys and ceramic substrates, while the other elements in the superalloys seem to have little or no direct role in the reactions observed. The Cr from the nickel-base superalloys is transported through an oxide vapor phase. The reaction itself involves doping of the ceramic by Cr, rather than reduction of the ceramic.

Minor changes in oxygen pressure have significant effects on the final result of experiments with reactions most intense and localized at low oxygen pressure in an atmosphere with a limited pumping rate.

Further research needs to be done, but the reactions seen in this study can perhaps lead to a better method of brazing ceramics at relatively high temperatures.

Acknowledgements

This research is supported by NASA Solidification Design Center at Auburn University. The author wishes to thank Dr. R. A. Overfelt, Dr. Yuhui Shen for their invaluable assistance in this research study, Dr. Randolph C. Helmink of Rolls-Royce Indianapolis for supplying the materials, as well as Jim Messick and the other members of the Auburn University Physical Metallurgy & Materials Joining Group.

References

1. W.B. Hanson, K.I. Ironside and J.A. Fernie, *Acta Materialia*, 48, 4673-4676 (2000)
2. Chia-Chin Lin, Rui-Bin Chen, Ren-Kae Shiue, *Journal of Materials Science*, 36, 2145-2150 (2001)
3. Y.N. Liang, M.I. Osendi and P. Miranzo, *Journal of the European Ceramic Society*, 22, 1-7 (2002)
4. J. Hashim, L. Looney and M.S.J.Hashmi, *Journal of Materials Processing Technology*, 119 (1-3), 324-328 (2001)
5. Kun-Fung Lin and Chien-Cheng Lin, *Journal of Materials Science*, 34, 5899-5906 (1999)
6. C.S. Lim and A.J. Clegg, *Journal of Materials Processing Technology*, 67, 13-18 (1997)
7. X.Y. Virieux, J. Desmaison, J.C. Labbe and A. Gabriel, *Key Engineering Materials*, 132-136, 1657-1660 (1997)
8. M.W. Chase, Jr. *NIST-JANAF Thermochemical Tables* (1998)
9. O. Kubaschewski and C.B. Alcock, *Metallurgical Thermochemistry*, Pergamon Press, Oxford, New York (1979)

Proceedings from Joining of Advanced and Specialty Materials, 13–15 October 2003, Pittsburgh, Pennsylvania, ASM International, 2004

The Influence of Ceramic and Metallic Substrates on the Oxidation Behavior of Gold ABA

K. Scott Weil, and Joseph P. Rice
Pacific Northwest National Laboratory, Richland, WA 99352

Abstract

A commercial ceramic-to-metal braze alloy, Gold ABA, was subjected to high-temperature testing in air to evaluate its oxidation behavior, microstructural stability, and material compatibility for potential application in electrochemical devices. The oxidation studies were performed on as-received braze alloy foils, on wetting couples prepared using yttria-stabilized zirconia (YSZ) substrates, and on Gold ABA brazed YSZ/ferritic stainless joints. The results of our investigations demonstrate that the substrates play a significant role in determining the overall oxidation resistance of the brazed joint.

Introduction

Metal braze alloys containing active elements such as titanium and zirconium have been widely used for over 30 years to join ceramics to metals or to other ceramics. These brazes wet ceramic oxides more effectively than non-active brazes because the active metal species partially reduces the ceramic faying surface, forming a sub-stoichiometric oxide interlayer that is readily wetted by the rest of the bulk braze [1,2]. An emerging application for ceramic-to-metal brazing is the hermetic sealing of high-temperature electrochemical devices. Planar solid oxide fuel cells and oxygen-ion generators, for example, employ a series of electrochemically active multi-layer ceramic cells joined to a ceramic or metallic housing that provides structural support and maintains physical separation between the oxidizing and reducing gas streams.

With the advent of thin, tape cast YSZ electrolyte membranes, the operating temperature of these devices can be decreased markedly without significantly reducing electrochemical performance. Lower operating temperatures make it possible to consider commercial oxidation-resistant alloys in place of structural ceramics for use in the balance of the device, allowing one to substantially reduce the cost of the device, as well as improve its structural integrity and thermal management properties and faciltate its design and manufacture. The leading ceramic-metal materials set for these applications is the standard solid-state electrolyte, YSZ, and ferritic stainless steel. Historically high temperature glasses, such as those from the borate- or phosphate-doped aluminosilicate families, have been used to join these materials in commercial electrochemical devices [3]. However, metal alloy brazes promise better performance with respect to thermal shock, mechanical shock, and thermal cycling, key properties for the use of these devices in military and commercial transportation applications.

An example of where ceramic-metal joining is required in an electrochemical device is illustrated in Figure 1, which displays a schematic drawing of a planar oxygen generator. Shown in the inset is a depiction of how oxygen transport takes place in the device. Also noted are the conditions under which the O_2 generator, and therefore the YSZ/metal joint, is expected to operate. The device consists of a stack of individually sealed, serially connected YSZ membranes, which under an imposed voltage catalytically ionize oxygen from an incoming air stream on one side and transport the resulting ions up a chemcial gradient to the other.

To generate a sufficient rate of ionic transport within the stack, it must be operated at high temperature, normally on the order of 500 – 1000°C. Unfortunately, this temperature is also high enough for thermally activated processes to take place which may be deleterious to device performance, such as interdiffusion, oxidation, and interfacial solid state reaction - particularly over the expected lifetimes of these devices, estimated to be 3000 – 30,000+ hours depending on the specific application. Thus, the joint must be functionally stable and hermetic over long periods of time at operating temperature and be capable of surviving numerous thermal cycles from operating temperature to down to room temperature and back up.

Before a given braze alloy can be considered for high

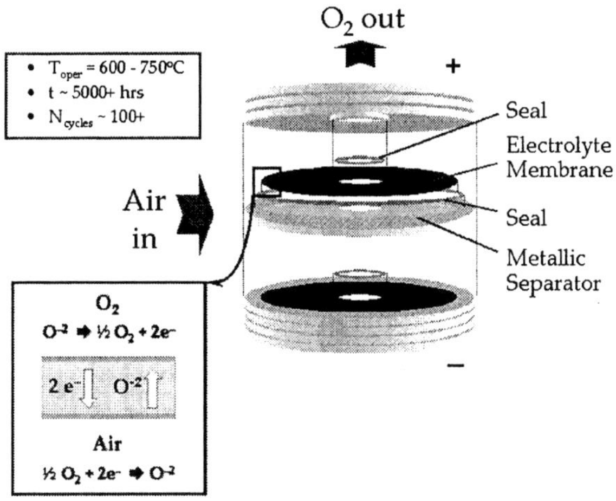

FIGURE 1 A schematic drawing of a planar oxygen generator.

temperature application in air, its oxidation behavior and that of the as-brazed joint must be thoroughly characterized. Previous oxidation studies on active metal brazes have focused primarily on copper-silver based alloys, typically those bearing a small amount of titanium as the active metal species. In general it was found that these particular alloys are not capable of surviving sustained exposure to air at temperatures above 500 - 600°C [4-6]. As part of a broader investigation into the suitability of commercial active metal brazes for electrochemical device application, we have investigated the oxidation properties of the gold-based active metal braze, Gold ABA. This paper presents recent findings from our studies on both the high temperature oxidation behavior of Gold ABA and the effects of the adjoining YSZ and 430 stainless steel substrates on this behavior.

Experimental Procedure

Materials and Sample Preparation. The braze alloy employed in these experiments was Gold ABA, which has a composition of Au-3Ni-0.6Ti. The braze was supplied as 0.150 mm thick foil from Wesgo Metals Inc. Samples measuring ½cm square were used for brazing. 5% yttria stabilized zirconia (5YSZ) and thin gauge 430 stainless steel (430 SS - Fe, 18% Cr, >1% Si) were selected as the model ceramic electrolyte membrane/structural metal system for this study. 5YSZ electrolyte substrates were fabricated by using tape casting and sintering procedure developed at Pacific Northwest National Laboratory. 5YSZ powder (Zirconia Sales, Inc.) was ball milled with a proprietary binder and dispersant system in 2-butanone/ethyl alcohol for 2 days to form a slurry. The slurry was cast onto silicone-coated mylar by the doctor blade technique to form green tapes with an as-dry thickness of approximately 200μm. Thicker tapes were prepared by laminating two of the green tapes at 175°C and 500psi for 2 minutes. Discs measuring 25mm in diameter were cut from this thick tape using a circular hot knife. The green discs were then sintered at 1400°C for 1 hour in air and creep flattened between two ground flat alumina plates at a temperature of 1450°C for two hours. The final diameter and thickness of the YSZ substrate discs were nominally 18mm and 200μm, respectively.

430 stainless steel (430 SS) was chosen as the model representative for the device housing material because: (1) its coefficient of thermal expansion (CTE) most closely matches that of the YSZ, (2) it displays good oxidation properties at high temperature, and (3) it is the most widely produced ferritic stainless product, making it a strong low-cost candidate for commercial high temperature electrochemical devices. The as-received ~½ mm (20 mil) sheet of 430 was sheared into 2 cm x 2 cm square samples and polished on both sides with 1200 grit SiC paper. The samples were flushed with de-ionized water to remove the grit and ultrasonically cleaned in acetone for 10 minutes.

Testing and Characterization. Our objectives in this study were to understand the mechanisms by which oxidation occurs in Gold ABA and what role the substrates might have in modifying this oxidation behavior. As shown in the flow chart in Figure 2, initial oxidation testing of the braze was conducted using a Cahn TG171 thermogravimetric analysis (TGA) unit. The braze alloy was then exposure tested in four different configurations: (1) the braze foil alone, (2) a braze/5YSZ couple, (3) a braze/430 SS couple, and (4) an 5YSZ/Braze/430 SS joint. The oxidation tests were conducted in flowing air at 400°C, 600°C, and 800°C for 5, 10, 20, 50, 100, and 200 hours. Changes in the composition and microstructure of these samples were compared against the baseline microstructure of the alloy in the unoxidized state, as characterized by Scanning Electron Microscopy (SEM) and energy dispersive X-ray analysis (EDX).

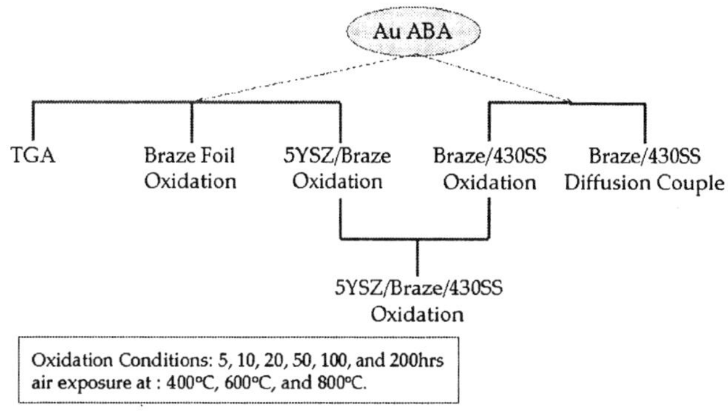

FIGURE 2 Flow chart of the oxidation experiments.

TGA characterization was conducted in slow flowing air for 100 hours at 400°C, 600°C, and 800°C. The weight and surface area of each bare test piece was measured prior to testing. During the test, the braze sample was suspended in the TGA by a quartz wire and the furnace was heated and cooled at 5°C/min. In order to correct for buoyancy effects, an inert sample of alumina with the same weight as the test specimen was heated under the same conditions and the balance fluctuations due to air density changes were recorded and later subtracted from the weight gain data.

Following the TGA experiment, a series of Gold ABA braze coupons were sequentially oxidized in 20cc/min flowing dry air under one of the following time/temperature conditions: 5, 10, 20, 50, 100 or 200 hours at 400°C, 600°C, or 800°C. Heating and cooling rates of 5°C/min were employed in this testing. Similar sets of oxidation experiments were conducted on Gold ABA/YSZ and Gold ABA/430 SS wetting couples and on Gold ABA brazed YSZ/430 joining samples. Prior to joining and testing, all materials were thoroughly cleaned and degreased in acetone and ethanol. The wetting and joining samples were prepared by heat treating under 10^{-6} torr vacuum at 3°C/min to 1080°C, holding at 1080°C for ½ hr, and cooling at 3°C/min. Microstructural analyses of the various oxidation specimens were conducted on as-polished cross-sectioned samples using a JEOL JSM-5900LV SEM equipped with an Oxford energy dispersive X-ray analysis (EDX) system.

Results and Discussion

Oxidation of Au ABAFoil. Shown in Figure 3 are the TGA oxidation curves for Au ABA at 400°C, 600°C, and 800°C. Note that the area specific weight of each sample increases parabolically during the first 30 hours, then stabilizes for the remainder of the test, a behavior that suggests the formation of a nonporous, coherent oxide scale (later determined to be predominantly NiO) that acts as a physical barrier to restrict further rapid oxidation. The growth of this type of oxide scale depends on the ionic diffusion of one or more species through the thickness of the scale [7] which is a thermally activated process, thus explaining the strong temperature dependence of weight gain rate observed in Figure 3. The ragged nature of the three weight gain curves is likely the result of localized spallation of the scale during its growth. Calculations based on the initial weight of the 800°C sample indicate that within experimental error the weight gain attained after 100hrs of oxidation at this temperature corresponds to complete oxidation of the Ni and Ti alloying elements to NiO and TiO_2, respectively.

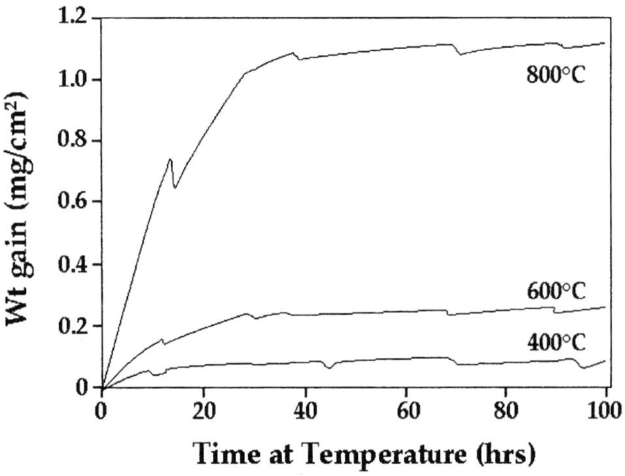

FIGURE 3 Weight gain (mg/cm2) versus exposure time (hrs) for Gold ABA at 400, 600, and 800°C.

Shown in Figures 4(a) - (e) are a series of micrographs that offer a snapshot of the braze microstructure in the as-received state and upon heat treatment in vacuum or air for 100hrs at 400°C, 600°C, or 800°C. As shown in Figure 4(a), the cross-sectional microstructure of the as-received Au ABA foil consists of fine, micron-sized equiaxed Ni/Ti-rich precipitates randomly dispersed in a Au-rich matrix. The precipitates are composed of ~60 a/o gold and 20 a/o each of nickel and titanium, whereas the matrix contains ~90 a/o gold and 8 and >0.5a/o nickel and titanium, respectively. The microstructure near the surface of the

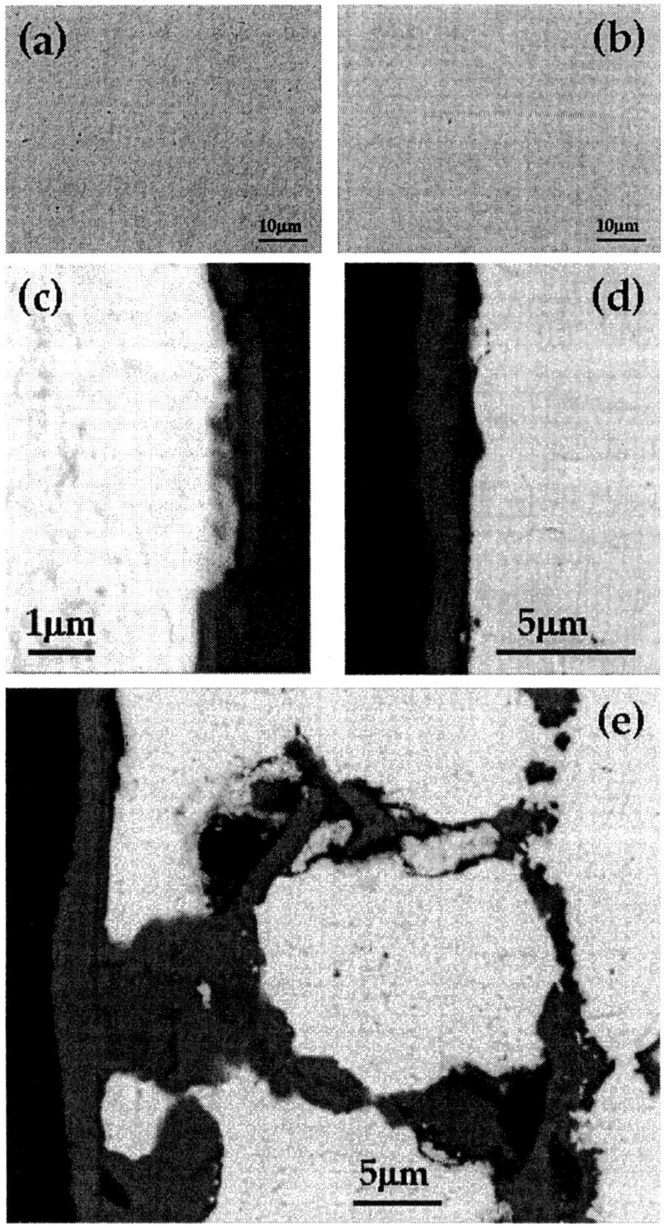

FIGURE 4 Cross-sectional SEM micrographs of Au ABA braze foils in the following conditions: (a) as-received, (b) vacuum heat treated for 100hrs at 600°C, and (c), (d), and (e) oxidized in 20 cc/min flowing air for 100hrs at 400°C, 600°C, and 800°C, respectively.

as-received Au ABA sample was found to be approximately the same as that in the center. We found that precipitate coarsening occurs during long-term vacuum annealing, an example of which is shown in Figure 4(b). This phenomenon becomes increasingly more exaggerated as the annealing temperature increases.

In each of the oxidized specimens there is a precipitate-depletion zone in the braze found beneath the dense scale that forms on the surface. The outer surface of the scales in the 400°C and 600°C specimens is predominantly NiO (dark gray phase). Patches of NiO-doped TiO_2 (black phases) exist along the interface between the outer scale layer and the braze in these samples. Lying just inboard of this interface is a thin 0.1 – 0.5μm zone in the alloy that contains a visible amount of porosity. The voids are likely generated as the result of mass transport of nickel at the braze/scale interface through the oxide layer at a rate which is greater than the inward movement of oxygen; a typical mechanism for NiO scale formation in non-Cr, Si, and Al containing nickel alloys [7]. In the 800°C specimen oxidation is particularly aggressive, with a significantly thicker NiO scale forming at the exposed surface, as well as penetrating along grain boundaries well into the alloy. It was also obvious that this sample was generally devoid of the Au-Ni-Ti precipitates originally found in the as-received braze, indicating that the alloying elements completely oxidized during exposure testing, which confirms our TGA results.

Substrate Effects on the Oxidation of Au ABA. The effects of the 5YSZ substrate on the oxidation behavior of Au ABA at 400°C, 600°C, and 800°C can be seen by comparing the micrographs of the as-wetted and as-oxidized specimens shown

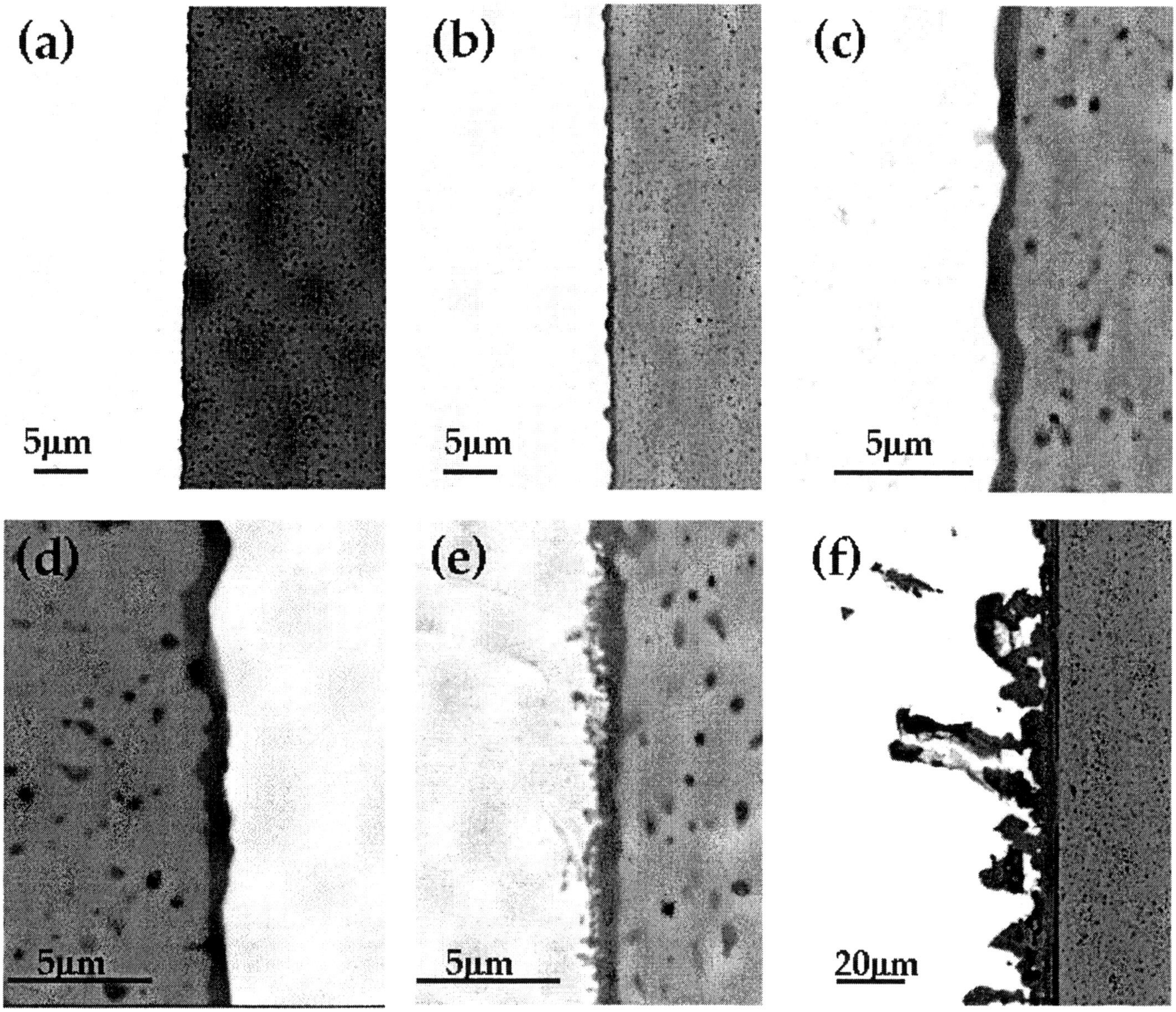

FIGURE 5 Cross-sectional SEM micrographs of Au ABA/5YSZ couples in the following conditions: (a) as-brazed; (b) and (c) vacuum annealed for 100hrs at 400°C and 600°C, respectively; and (d), (e), and (f) oxidized in 20 cc/min flowing air for 100hrs at 400°C, 600°C, and 800°C, respectively.

in Figures 5(a) - (f). Au ABA exhibits good wetting and adherence with 5YSZ. In comparison with the as-received foil, Figure 4(a), the microstructure of the Au ABA in the as-brazed specimen contains virtually no precipitates, appearing as a homogeneous alloy up to the interface with the ceramic substrate. EDX measurements taken at several regions within the bulk of the braze and near the braze/5YSZ interface indicate that the alloy is composed, on average, of ~97 a/o Au, 2.5 a/o Ni, and >0.5 a/o Ti. Along the braze/5YSZ interface is a dense, ½ - 1μm thick reaction zone containing essentially a solid solution of titanium oxide and zirconium oxide. Local chemical analysis suggests that the metal species in this region are fully oxidized. The microstructures of the vacuum annealed Au ABA/5YSZ specimens shown in Figures 5(b) and (c) are nearly identical to that of the as-brazed sample except that reaction zone has thickened slightly. Measurements based on SEM analyses indicate that the reaction zone thickens upon a 100hr soak as follows: $t_{as\text{-}brazed} \sim 0.7\mu m$, $t_{400°C} \sim 1.0\mu m$, $t_{600°C} \sim 1.4\mu m$, and $t_{800°C} \sim 1.5\mu m$. The composition of the reaction zone in each case remains unchanged, a solid solution of titania and zirconia.

The microstructure of the Au ABA/5YSZ specimen oxidized at 400°C, Figure 5(d), is essentially the same as that of the corresponding annealed specimen shown in Figure 5(b). However, when the two 600°C samples are compared, Figures 5(c) and (e), we find that the oxidizing condition has caused the formation of a thin NiO sublayer just beneath the braze-5YSZ reaction zone. Below the NiO is a layer of fine porosity, presumed to originate in a manner similar to that observed in braze foil oxidation, i.e. due a differential in the rates of mass transport between the two reacting species, nickel and oxygen. The formation of the NiO and associated porosity is much more aggressive in the 800°C oxidation specimen, Figure 5(f). Here, the NiO scale is quite thick, on the order of 20μm, and assumes a more convoluted morphology. In addition, the porosity is substantially larger and the couple shows signs of delamination.

The free surfaces on the braze side of the oxidized Au ABA/5YSZ couples are shown in Figures 6(a) – (c). Like the braze foil oxidation samples in Figures 4(c) – (e), a dense, continuous NiO scale forms on the oxidizing surface. In the 400°C and 600°C specimens, this scale is of sub-micron thickness. In the 800°C sample however, the scale is at least 10μm thick in most places and exhibits a significant amount of gross porosity. The morphology of the scale is notably different than that observed in the corresponding braze foil sample in Figure 4(e), displaying a more uniform appearance. It was also noted that unlike the oxidized braze foils, very little TiO_2 was found in or adjacent to the NiO scale on these samples. Presumably most of the titanium has been tied up in the reaction zone at the braze/5YSZ interface.

Shown in Figures 7(a) - (d) on the next page are micrographs of the Au ABA/430 SS couples in the as-brazed and annealed conditions. The microstructure of the braze in the Au ABA/430 SS couple, Figure 7(a), is quite different from that of the as-received Au ABA foil, Figure 4(a), and the braze in the as-wetted Au ABA/5YSZ couple, Figure 5(a). The primary reason for this is the partial dissolution of the underlying stainless steel substrate into the molten braze during joining, which subsequently establishes not only a new microstructure and composition in the braze filler, but also a mixing zone between the braze and the substrate. What is interesting is that the general microstructures of the braze, mixing zone, and adjacent substrate do not change significantly upon annealing for 100hrs at 400 – 800°C. Considering the as-brazed Au ABA/430 SS couple in greater

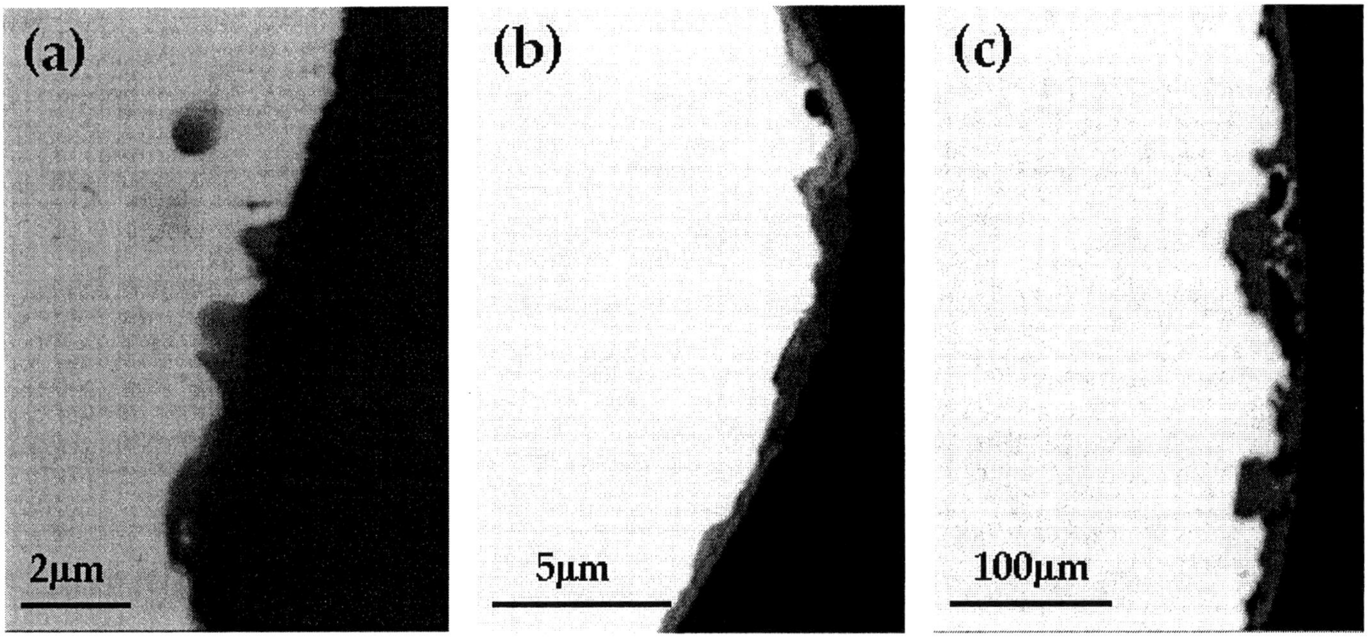

FIGURE 6 Cross-sectional SEM micrographs of the free-surface of the braze in Au ABA/5YSZ couples oxidized at the following temperatures for 100hrs: (a) 400°C, (b) 600°C, and (c) 800°C.

detail, Figure 8, we find the braze consists of Fe-rich precipitates (point 2) in an iron-gold alloy matrix (point 1). The mixing zone found between the braze and substrate (the region surrounding points 3 and 4 in Figure 8) contains submicron-sized Au-rich precipitates in an iron-based matrix.

The microstructures of the Au ABA/430 SS couples oxidized for 100 hours at 400, 600, and 800°C, Figures 9(a) – (c) are strikingly different from the corresponding oxidized braze foil and Au ABA/5YSZ couple specimens as well as from each other. The 400°C sample in Figure 9(a) exhibits a sub-micron thick (~0.6μm) coherent scale on the free surface of the braze that consisists of primarily Fe_3O_4 with measurable amounts of Cr_2O_3 and NiO, apparently in solid solution. The scale on the Au ABA/430 SS specimen oxidized at 600°C is significantly thicker, ~3 – 4μm thick, and is composed of a heterogeneous mixture of Cr_2O_3 (light phase) and Fe_3O_4 (dark phase) in roughly equal

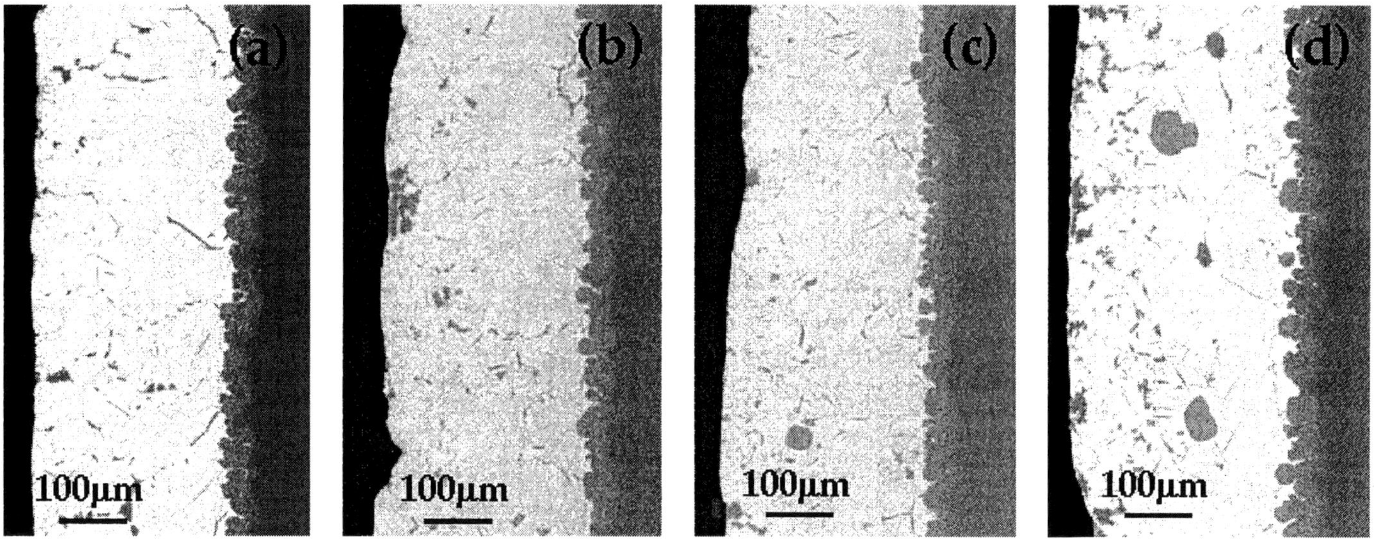

FIGURE 7 Cross-sectional SEM micrographs of Au ABA/430 SS couples in the following conditions: (a) as-brazed, (b) annealed for 100hrs at 400°C, (c) annealed for 100hrs at 600°C, and (d) annealed for 100hrs at 800°C.

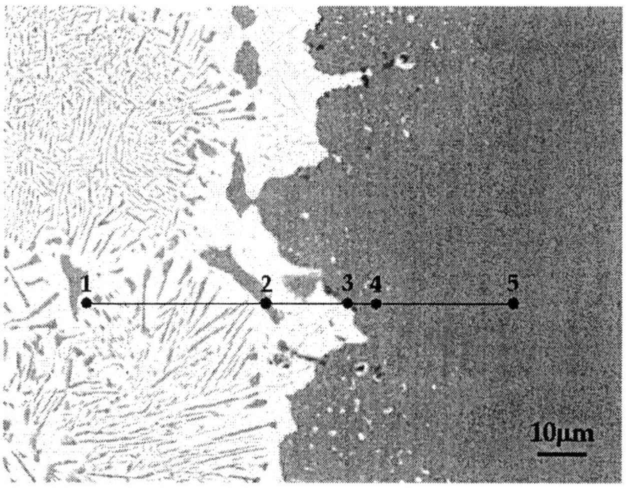

Region	Fe	Ni	Cr	Au	Ti	Si
1	22.14	2.57	5.43	69.86	-	-
2	64.94	12.20	8.37	4.13	0.56	9.80
3	59.40	11.41	6.78	1.63	16.25	4.53
4	75.67	6.06	11.65	5.88	-	0.74
5	81.06	-	18.06	-	-	0.88

FIGURE 8 Local chemical analysis by EDX (in mol%) of a cross-sectioned Au ABA/430 SS couple in the as-brazed condition.

molar amounts. In the 800°C sample, the scale is nearly pure Cr_2O_3, doped with a small amount of TiO_2, that measures 4.4μm thick on average.

EDX analyses of these three specimens suggests that the chemistries of the braze precipitates and matrix in the region adjacent to the scale evolve via diffusion as iron and chromium are "lost" to oxide formation. For example in the 600°C sample, the matrix loses measurable amounts of iron and chromium with an average composition of 18.21 a/o Fe, 6.97 a/o Ni, 3.44 a/o Cr, and 71.38 a/o Au, whereas the precipitates also lose iron and chromium, but to a lesser extent: 63.11 a/o Fe, 13.77 a/o Ni, 7.54 a/o Cr, 5.06 a/o Au, 0.61 a/o Ti, and 9.91 a/o Si. That is, the matrix appears to be the source of iron and chromium in the formation of the mixed oxide scale on this sample. Similar results are found with the 400°C and 800°C oxidation couples.

Oxidation of Au ABA Brazed 5YSZ/430 SS Joints. The combined effects of the 5YSZ and the 430 SS substrates on Au ABA braze joint oxidation at 400°C, 600°C, and 800°C can be seen by comparing the micrographs of the annealed and as-oxidized specimens shown on the next page in Figures 10 (a) – (f). Note that in the annealed samples, the 5YSZ/braze interface looks quite different from those seen in Figures 5(a) – (c). Because of the influence of the stainless steel substrates, there is no continuous mixed oxide reaction zone adjacent to the 5YSZ faying surface. In fact in all three cases, much of the titanium remains tied up in fine scale, equiaxed TiO_2 particles in the braze matrix [seen as small black regions in the brazes of Figures 10(a) – (c)]. The braze side of the Au ABA/5YSZ interface consists primarily of two phases, iron-nickel rich precipitates in a gold-based matrix. Both phases contain measurable amounts of chromium. In the 400°C annealed sample, EDX analysis indicates that the precipitates at the interface with the 5YSZ contain upwards of 50 a/o oxygen with roughly equal atomic ratios of iron and nickel, corresponding to a solid solution composed primarily of FeO and NiO. Approximately 1 – 2μm away from the interface, the precipitates are composed of slightly less oxygen. Also found in the braze microstructures near the Au ABA/5YSZ interfaces of all three joining specimens [although not obvious in the micrograph of the 600°C sample shown in Figure 10(b)] are Widmanstätten-type iron-titanium precipitates. It is interesting to note from the micrographs in Figures 10(a) – (c) that annealing temperature appears have little effect on the braze microstructure at the interface and in general on the overall microstructure of the joint.

However as shown in Figures 10(d) – (f), under an oxidizing atmosphere temperature plays a critical role. For example, in the joining sample oxidized at 400°C, discrete patches of TiO_2 (black regions) begin to form at the Au ABA/5YSZ interface. While still seen throughout the braze microstructure, the blocky FeO-NiO phase is less apparent directly at the interface with the electrolyte. At 600°C, Figure 10(e), the TiO_2 is even more obvious, again appearing as a black phase along the 5YSZ faying surface. Adjacent to the titania is a semi-continuous layer of Cr_2O_3. At 800°C, Figure 10(f), this chromia layer is fully continuous and measures ~ 2 – 3μm in thickness. Note also in Figure 10 (f) that the braze microstructure has also evolved. While the blocky iron oxide-nickel oxide phase remains, the oxygen content has increased, corresponding to a mixed composition of Fe_2O_3 and NiO. The elongated iron-titanium precipitates have shrunken in size considerably, although they remain needle-

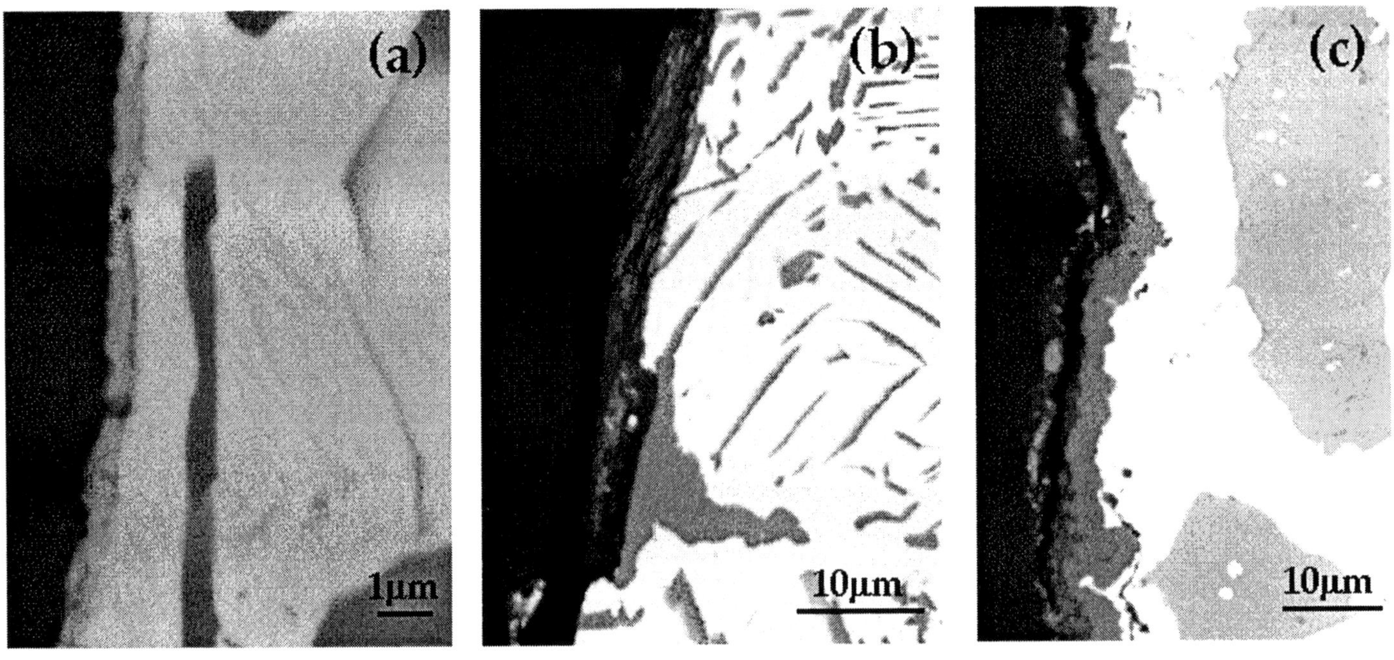

FIGURE 9 Cross-sectional SEM micrographs of the free-surface of the braze in Au ABA/430 SS couples oxidized at the following temperatures for 100hrs: (a) 400°C, (b) 600°C, and (c) 800°C.

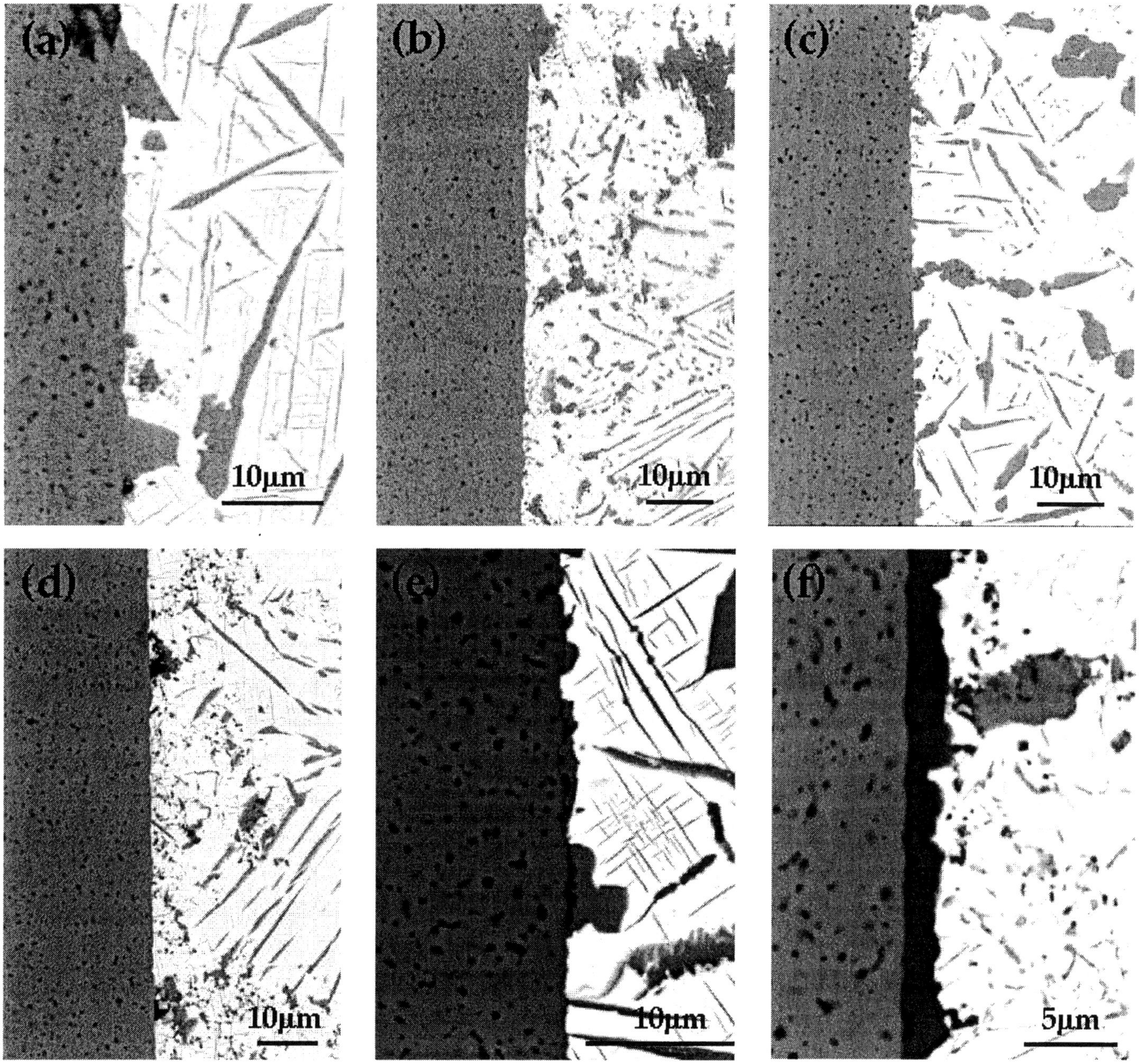

FIGURE 10 Cross-sectional SEM micrographs of Au ABA brazed 5YSZ/430 SS joints in the following conditions: (a), (b), and (c) vacuum annealing for 100hrs at 400°C, 600°C, and 800°C, respectively and (d), (e), and (f) oxidized in 20 cc/min flowing air for 100hrs at 400°C, 600°C, and 800°C, respectively.

shaped. In addition a new precipitate has appeared, also with an elongated morphology but a lower c to a ratio, that forms in random directions as opposed to perpendicular crystallographic directions of the Widmanstätten precipitates. EDX analysis of the precipitates suggests that they are composed of iron, nickel, and oxygen although a quantitative ratio of these elements could not be obtained because of their size. The titania and chromia layers adjacent to the interface with the 5YSZ appear to protect the rest of the braze from the type of oxidation observed in the Au ABA/5YSZ oxidation couple of Figure 5(f). In this way, the stainless steel substrate significantly influences the oxidation behavior of the overall joint.

Conclusions

The oxidation behaviors of the commercial ceramic-to-metal braze Au ABA was characterized at 400 - 800°C using metal braze foil samples, braze/YSZ and braze/430 SS wetting specimens, and brazed YSZ/430 stainless steel joining specimens. Our observations from these tests can be summarized as follows:

(1) When tested for 100 hrs in 400°C, 600°C, and 800°C air, the Au ABA alloy exhibits parabolic growth of a dense, coherent NiO scale. At the two lower temperatures, this scale is protective. However at 800°C, oxidation is extremely aggressive and the oxidation products penetrate along the grain boundaries well into the alloy. Scale formation in this manner was found to be accommodated by internal porosity and it was observed that the nickel and titanium alloying agents were completely oxidized.

(2) In the Au ABA/5YSZ couples, the braze forms a TiO_2/ZrO_2 reaction zone along the interface with the electrolyte. Subsequent long-term annealing of these specimens has little affect on the interfacial microstructure other than to cause the reaction zone to thicken slightly. Oxidation of the couple at 400°C for 100hrs also does not appear to alter the interface significantly. However when air exposure testing is conducted at 600°C, there is evidence of a thin NiO sub-layer forming just beneath the braze-5YSZ reaction zone. At the 800°C oxidation condition, NiO formation is much more aggressive and the couple exhibited signs of delamination.

(3) Joining of the braze to stainless steel gives rise to substantial iron and chromium alloying in the braze due to partial dissolution of the metal substrate. Consequently, the oxidation behavior of the braze is substantially changed. In 400°C air, a thin scale composed primarily of Fe_3O_4 forms on the braze. At higher exposure temperatures, Cr_2O_3 begins to appear in the scale, in roughly equal portions with the Fe_3O_4 at 600°C and occurring as the dominant phase at 800°C.

(4) The presence of the stainless steel in the Au ABA joining specimens dramatically alters the microstructures and compositions of the brazes at the 5YSZ/braze interface. In the annealed specimens, the TiO_2 interfacial layer previously found in the 5YSZ wetting specimens does not exist. Instead, two phases are observed at the interface, a mixed iron-nickel oxide precipitate in a gold-based matrix; both contain measurable amounts of chromium. The titanium appears to be tied up in two separate phases not found directly at the interface with the 5YSZ, microscale TiO_2 and Widmanstätten-type iron-titanium precipitates. Upon 100 hours of oxidation at 400°C, discrete patches of TiO_2 form at the Au ABA/5YSZ interface. At 600°C, the interfacial TiO_2 phase is more continuous and is accommodated by an adjacent semi-continuous layer of Cr_2O_3. At 800°C, the chromia is fully developed, measuring ~2 – 3μm thick, and appears to protect the rest of the brazed joint from the type of aggressive oxidation seen in the Au ABA/5YSZ coupon.

Acknowledgments

The authors would like to thank Jim Coleman, Nat Saenz, and Shelly Carlson for their assistance in metallographic and SEM sample preparation and analysis, Kerry Meinhardt for his assistance in fabricating the YSZ discs, and Dean Paxton for his help in fabricating the braze specimens. This work was supported by the U.S. Department of Energy, Office of Fossil Energy, Advanced Research and Technology Development Program. The Pacific Northwest National Laboratory is operated by Battelle Memorial Institute for the United States Department of Energy under Contract DE-AC06-76RLO 1830.

References

1. C. W. Fox and G. M. Slaughter, *Weld. J.*, 43, 591 (1964).
2. O. M. Akselsen, *J. Mater. Sci.*, 27, 1989 (1992).
3. K. D. Meinhardt, J. D. Vienna, T. R. Armstrong, and L. R. Peterson, PCT Int. Appl., WO01909059 (2001).
4. R. R. Kapoor and T.W. Eagar, *J. Am. Ceamr. Soc.*, 72, 448 (1989)
5. A. J. Moorehead and H. Kim, *J. of Mater. Sci.*, 26, 4067 (1991)
6. D. B. Lee, J.H. Woo, and S.W. Park, *Mater. Sci. Eng.*, A268, 202 (1999)
7. O. Kubashewski and B. E. Hopkins, *Oxidation of Metals and Alloys, 2nd ed.*, Butterworth, London, (1962)

Proceedings from Joining of Advanced and Specialty Materials, 13–15 October 2003, Pittsburgh, Pennsylvania, ASM International, 2004

Microstructural Development and Relationship to Mechanical Properties in Polycrystalline to Single Crystal Nickel-Base Superalloy Transient Liquid Phase (TLP) Bonds

Lewis N. Payton
Auburn University, Auburn, Alabama 36849

Rajeev Aluru, William F. Gale, Nofrijon I. Sofyan, Venu G. Krishnardula
Daniel A. Butts, Subhadravalli V. Chitti and Tao Zhou
Materials Research and Education Center, Auburn University, Auburn, Alabama 36849

Abstract

The microstructures developed in dissimilar superalloy transient liquid phase (TLP) bonds are more complex than those between similar materials. This paper investigates the microstructure and mechanical properties of single crystal to polycrystalline nickel-base superalloy TLP bonds in an attempt to understand phenomena governing microstructural development and structure - property relationships.

Introduction

Brazing, welding, and diffusion bonding have been the main methods commonly used in industry to join superalloys [1, 2]. However, transient liquid phase (TLP) brazing is the preferred joining method for nickel-base superalloys [3-5]. It has the advantage of not needing the high pressures typical of solid state diffusion bonding and is generally considered to be tolerant to the presence of a faying surface oxide layer [1].

This paper examines the TLP bonding of two commercially available forged nickel-base superalloys (IN 738 (738) and IN 939 (939)) to a commercially available cast single crystal superalloy (CMSX-4) using a proprietary foil as the interlayer. Mechanical tests of the bonds were conducted using a shear test rig which minimized the required sample size.

Shear Testing Method and Specimen Preparation

Superalloy substrates consisted of commercially available polycrystalline alloys (forged 738 and forged 939) along with the commercially available single crystal alloy CMSX-4. The brazing foil consisted of a (relatively) low alloy nickel-base core with boron enriched surfaces (i.e. a boron-rich/nickel-base/boron-rich sandwich).

Testing methods for bonds have included shear, bend and peel tests using standard geometries. However, such tests may require a large amount of material in order to fabricate the samples. An alternative method to the standard tests is to use a shear rig of the type shown in Figure 1 [1, 6]. Figure 1 shows a simplified schematic and details the geometry of the specimens that go into the shear rig. This is economical in terms of material used and ease of operator usage.

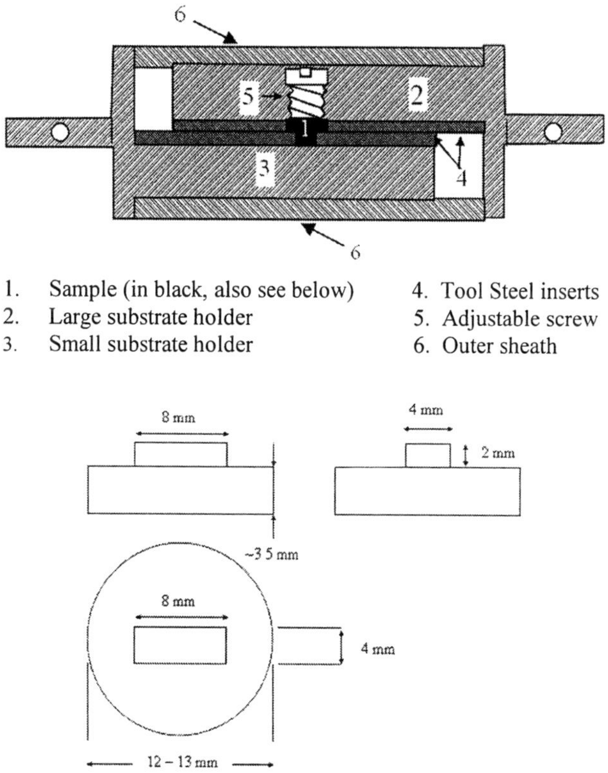

Figure 1. Shear Test Rig and specimen geometry (not drawn to scale)

The nominal composition of the substrates is detailed in Table 1. Polycrystalline alloys (738 and 939) were always used as the larger, round substrate of Figure 1. CMSX-4 was in all cases used as the smaller of the two substrates (fitting precisely into the slot of Figure 1). The adjustable thumbscrew was then used to bring the shoulder of the polycrystalline substrate into contact with the tool steel shoulder of the slotted CMSX-4 holder. When pulled in tension, the bond plane was always in the shear plane formed by the two counter-sliding tool steel faces (producing Mode II loading).

	738 Alloy	939 Alloy	CMSX-4
Ni	61.0	48.0	60.0
Cr	16.0	22.5	6.5
Co	8.5	19.0	10.0
Mo	1.7	0.0	0.5
W	2.6	2.0	6.5
Ta	1.7	1.4	6.3
Nb	0.9	1.0	3.5
Al	3.4	1.9	5.5
Ti	3.4	3.7	1.0
C	0.11	0.15	0.0
B	0.1	0.009	0.0
Zr	0.05	0.09	0.0
Other	0.54	0.251	0.2

Table 1. Nominal composition of substrates (wt%)

The substrates were rough shaped from parent material through the use of Electrostatic Discharge Machining (EDM) and a precision surface grinder. Substrates were then manually finished to the final dimensions of Figure 1 using 1000 grit SiC paper. All surfaces were then ultrasonically cleaned in an acetone bath and desiccated. Immediately prior to bonding, the Ni-flex interlayer was placed between the rectangular CMSX-4 substrate and the round polycrystalline substrates. The whole assembly was wrapped with tantalum wire for fixturing and inserted into a Lindberg Type 54233 vacuum tube furnace. Upon reaching a minimum vacuum of at least 1 mPa, samples were bonded at the temperatures indicated in the experimental design. The <001> axis of the CMSX-4 was always placed perpendicular to the bond line when forming the bonds. The as received forged and cast materials had not been heat treated.

Experimental Design and Procedure

Since a single standard test for shear strength does not exist, and the test used was unconventional, it was necessary to establish a standard by which to compare the shear test results achieved by using the shear rig and geometry of Figure 1. An excellent bond is generally no stronger than the weakest substrate being joined. Therefore, non-bonded specimens were shaped from the polycrystalline 738 and 939 nickel-base alloys using EDM and manual polishing techniques. These "solid" samples were then sheared in the rig depicted in Figure 1. Using a simple ratio, one establishes the following formula for the Quality Characteristic (QCH):

$$QCH = \frac{Shear\ Strength\ of\ bond}{Substrate\ Shear\ Strength} * 100 \qquad (1)$$

Using this formula, the perfect bond would score 100%. Failure would then occur in the base material.

This created a simple 2x2 factorial experiment, with two polycrystalline substrates (738 and 939) and bonded versus solid geometries. Since prolonged exposure to heat is known to affect bond strength during in-service usage, a third factor level, heat treated, was added to the design to examine microstructure instability. This created a 2x2x2 full factorial design for statistical analysis with the following factors:

Factor	Level 1	Level 2
A (Polycrystal)	738	939
B (Shape)	Bonded	Solid
C (Heat treated)	No	Yes

Table 2 Factor Level Combinations

As a matter of clarification, two types of post-bond heat treatment were used in this work. The first (referred to as "heat treated" in Table 2 of the statistical experiment) was used to investigate microstructural stability as detailed above. The second heat treatment was used as an aging treatment for γ' precipitation (referred to as "precipitation heat").

Bonding was accomplished by placing the samples to be bonded and their equivalent solid standards in the vacuum furnace at 1160 degrees Celsius for 240 minutes (4 hours). Ramp time to temperature was approximately 140 minutes. At the completion of the bonding time, samples were removed from the furnace and fan cooled to ambient temperature. The samples subjected to post bond heat aging were then placed in an atmospheric furnace for 7 days at 1000 degrees Celsius.

In addition to the samples produced for shear testing, a series of interrupted bonding times was performed to study bond development in the 738-CMSX-4 and 939-CMSX-4 systems. These bonds were then sectioned and polished for optical and scanning electron microscope (SEM) studies including energy dispersive x-ray spectroscopy (EDS) compositional analysis. The bond-lines of several representative specimens were then studied using a Vickers micro-hardness tester. Post-mortem SEM fractography studies were also performed on all shear specimens tested.

Statistical Data Analysis

Shear testing of the solid 738 and 939 polycrystalline samples established a base average (Equation 1) of 790 MPa for the 738 alloy and 780 MPa for the 939 alloy. The two bulk alloys when tested in the shear rig and geometry of Figure 1 were thus essentially equal. In general though, the 738-CMSX-4 bonds achieved a lower value quality characteristic for the same conditions. A one way ANOVA of the factors depicted in Table 2 yields Table 3 below for the Quality Characteristic:

Source	Sum Squares	DF	Mean Square	F Value	Prob > F
Model	3975.87	5	795.17	33.17	<0.0001
A (Poly Base)	271.35	1	271.35	11.32	0.0035
B (Shape)	2901.80	1	2901.80	121.03	<0.0001
C (Heat treat)	380.01	1	380.01	15.85	0.0009
AxB	160.68	1	160.68	6.70	0.0185
BxC	262.02	1	262.02	10.93	0.0039
Residual	431.56	18	23.98		

Table 3. ANOVA for strength analysis (QCH output) (2^3 experiment of Table 2)

The model F-value of 33.17 implies the model is highly significant. There is only a 0.01% chance that a this large of a data signal could occur due to noise. Values of "Prob > F" less than 0.0500 indicate which model terms are significant. In this case A (738 or 939 polycrystals), B (bonded or solid), C (heat treated) along with the interactions AB and BC are significant model terms. There was *no* detectable interaction between crystal type and heat treated (AC interaction). Not surprisingly, solid versus bonded (Factor B) yielded the strongest F value. Effects of the heat Aging and polycrystal type (738 vs 939) were both significant, but of roughly equal F value. The intereaction between shape and heat Aging was noticeably stronger than crystal selection versus heat Aging as illustrated in Figure 2. Heat Aging lowered the QCH in all bonded samples, but more so in the 738 polycrystal. In general terms, the 738 material performed less well in bonded tests, although the two polycrystals had approximately the same nominal QCH value when tested in the bulk.

Metallographic and Fractographic Analysis of the Results

Figure 3 illustrates typical failures in the bulk samples. All of the bulk substrates exhibited a planar shear failure. However, there were distinct differences in the extent of secondary cracking. The 738 bulk substrates failed by shear in the plane of maximum shear stress. The 738 in Figure 3 exhibits "dimpling" [7]. In contrast, the 939 exhibited extensive secondary cracking, out of the plane of maximum shear stress.

Figure 4 provides typical comparative photographs of the polycrystalline side and the CMSX-4 side for the two systems that were sheared after bonding for four hours. All of the bonds exhibit highly planar shear failure, as expected with the test methodology employed (although some secondary cracking was observed as in the bulk substrates). This is regarded as significant validation of the test, with the fractographic area roughly equal to the ideal cross sectional area of the brazing foil.

Vickers micro-hardness testing was conducted across sacrificial samples prepared using the same bonding protocol as the shear specimens. Figure 5 summarizes the results of those examinations. On average, at least 50 percent of the 738-CMSX-4 bonds showed the low hardness value at the bond line. This low hardness is typical of materials that exhibit "dimpling" on the fracture surface of shear tested specimens. The low hardness may be due to a lack of precipitate hardening on the fracture surface. The low point always occurred in, or near the foil bond line. In contrast, the 939 Ni-flex/CMSX-4 bond lines were consistently uniform, even though some discontinuity in the bond line γ' distribution was apparent.

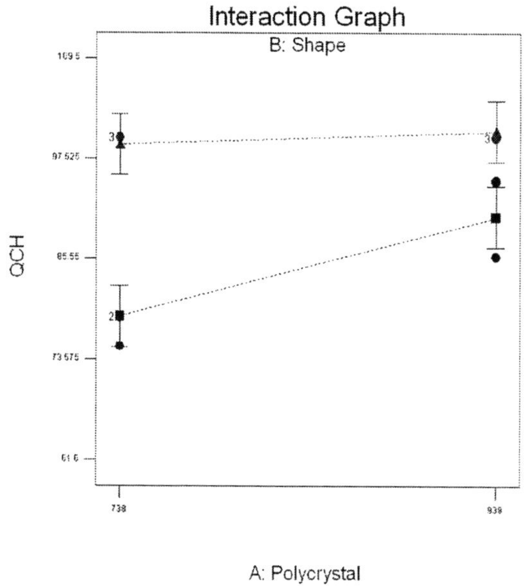

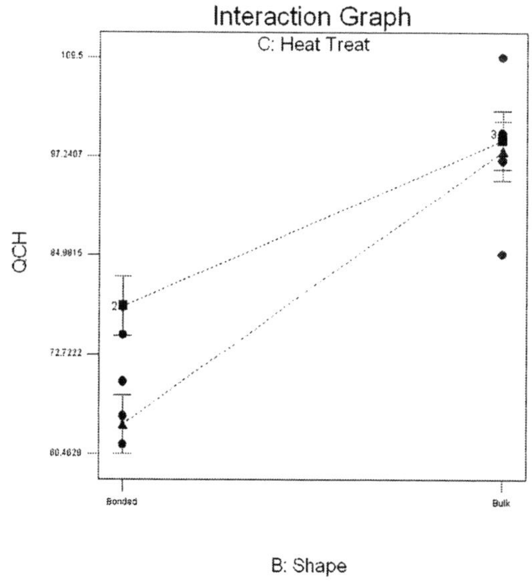

Figure 2. Comparison interaction heat treat (Heat Treat in figure) with shape versus heat treat and polycrystal selection. The much steeper convergence (interaction) between the shape and heat treat, indicating that the bond is greatly affected by the thermal exposure over a long time.

Solid shape nickel-base 939 fractography (500X)

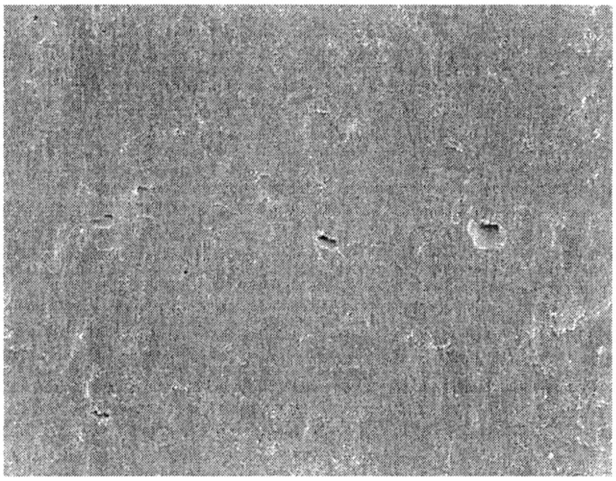

Solid shape nickel-base 738 fractography (500X)

Figure 3. Fractography of solid polycrystals shapes

Additional metallography of a traditionally heat treated bond (referred to earlier as the "precipitation heat": heated to 1093 Celsius, held for 1 hour in vacuum, then held for 4 hours in vacuum at 1000 Celsius, then 4 hours in vacuum at 843 Celsius before holding at vacuum for 16 additional hours as it cooled to ambient temperature) is shown in Figure 6 for the 939/foil/CMSX-4 system using a composite photograph. Notice that no borides remain at the bond line. However the γ' distribution remains somewhat inhomogeneous. The carbide is typical of a number seen in the diffusion affected portion of the polycrystals (738 and 939).

The polycrystalline bond lines formed for SEM analysis were then subjected to energy dispersive X-ray (EDS) analyses. These analyses were conducted across the bond lines using a JEOL 840 microscope system equipped with an Oxford Instruments ultra-thin window detector and ISIS analyzer. Focusing only on the change in atomic percent of nickel and aluminum in the polycrystals, Ni-flex and CMSX-4 yields the graphs in Figure 7. Aluminum increases along the bond line to a higher value in the 939 than the 738. Since a Ti/Al ratio > 1 has been shown to correlate with the precipitation of the γ' phase [8], it should be noted that in the systems studied here, the ratio never exceeds 0.5 in the 738-CMSX-4 system. The Ti/Al ratio never decreases below 1.0 in the 939-CMSX-4 system. In fact, there is a small but statistically valid increase in this ratio (with 90 percent confidence) just inside the 939 base materials where fractography indicates secondary cracking occurs.

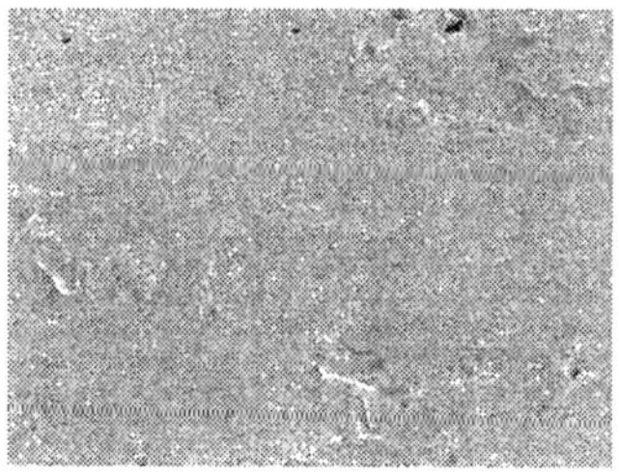

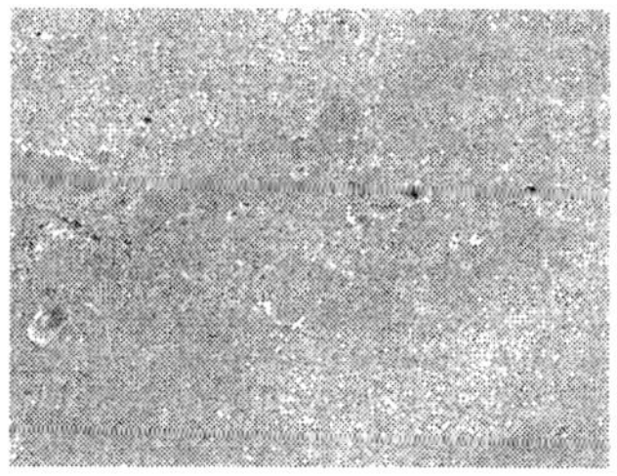

Alloy 738 side of bond (250X) | **CMSX-4 side of bond (250X)**

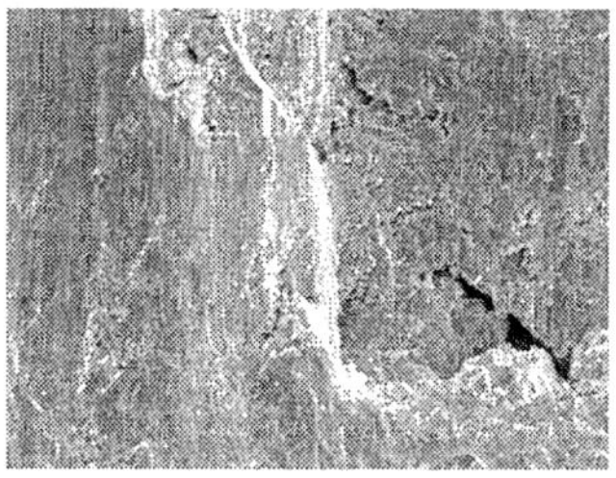

Alloy 939 side of bond (250X) | **CMSX-4 side of bond (250X)**

Figure 4. Fractography of polycrystalline-CMSX-4 bonds. Trace elements indicate that shear occurred on the polycrystalline side of the bond line in all cases or in the Ni-flex foil.

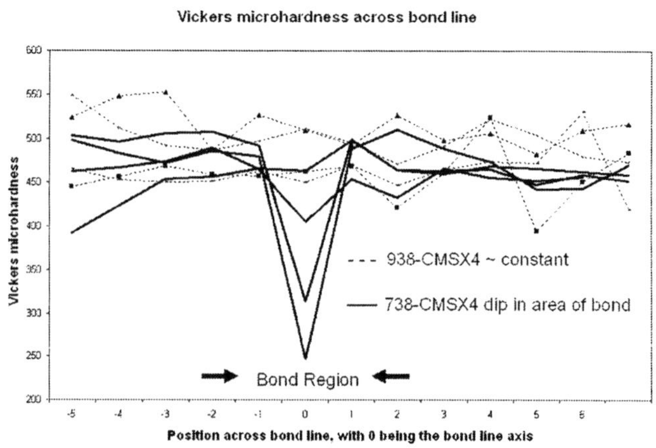

Figure 5. Hardness profile across various 939/CMSX-4 and 738/CMSX-4 bond lines. The low points (see dark lines) are all associated with random 738/CMSX-4 systems at the bond line.

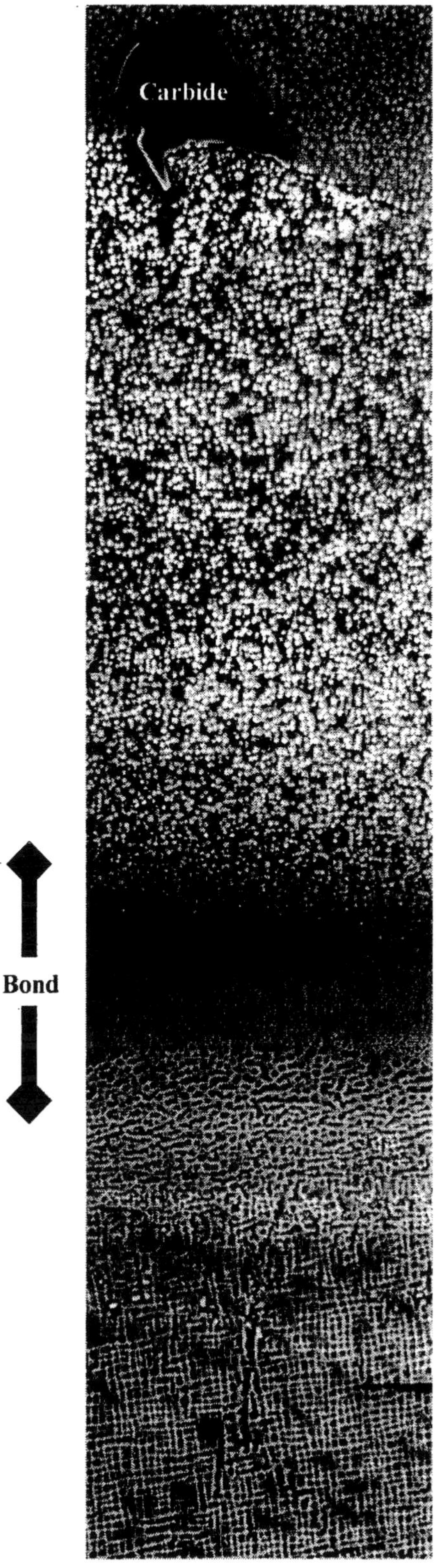

Figure 6. Composite photograph of 939/foil/CMSX-4 bond line observed in this experiment showing γ/γ' and carbide formation (see text for discussion of formation methodology and features).

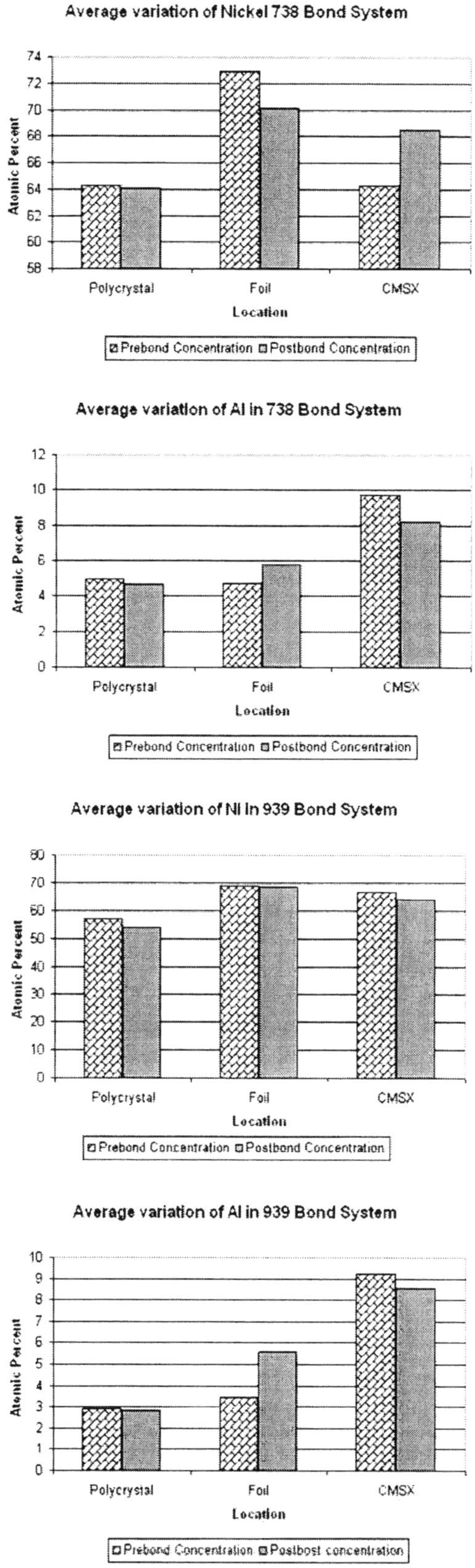

Figure 7. Compositional changes in substrates and interlayer (prior to bonding and after 240 minutes of bonding).

Interrupted bonding indicated penetration of the boron containing outer layer of the foil "sandwich" into the underlying nickel-base core along the grain boundaries of the latter. Figure 8 depicts the interrupted bonding tests showing the formation of a eutectic along the internal grains of the nickel-base. This seems to impede full wetting of the substrates as it wicks the boron in the outer layers of the braze "sandwich" into the internal grains of the inner nickel-base core. This was much less evident in the 939 substrate systems.

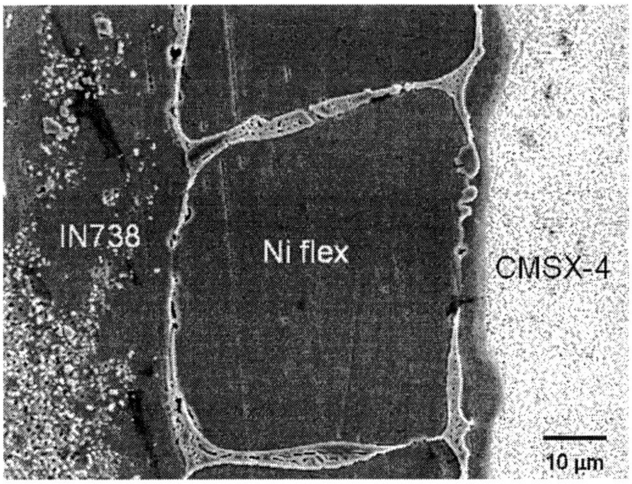

Figure 8. Interrupted bonding (after 1 minute at 1160 Celsius) in the IN 738/Ni-flex/CMSX-4 system. Wetting is occurring into the nickel-base core as evidenced by eutectic visible along the grain boundaries.

Conclusions

The results of this study indicate that the 939 Ni-flex CMSX-4 bond lines are significantly stronger in shear than the 738 Ni-flex CMSX-4 bond lines (with > 99 % confidence). The 939 substrate material shows a slight increase in the Ti/Al ratio after bonding. Precipitation of carbides was evident in both polycrystals, with 939 showing a strong tendency for secondary cracking. The 738 system had trouble wetting the substrates due to penetration of the internal grain boundaries of the Ni-base inner "sandwich" layer of the foil. The amount of γ' in the interlayer appears to be lower in the 738 than in the 939, which tends to promote soft, ductile failure.

Future Work

Auburn University's Materials Research Center will be conducting a detailed microstructural analysis with the Transmitting Electron Microscope (TEM) of the precipitated materials which are so evident in the 939 post bond substrate as well as the 738 system of bonds. Wettability studies of the Ni-flex system upon the 738 and 939 substrates are documented in the proceedings of this conference [9].

Acknowledgements

Substrate materials were kindly provided by Siemens Westinghouse of Orlando, Florida.

References

1. X. Wu, R. S. Chandel, H. Li, *J. Mater. Sci.,* 36, 1539-1546 (2001).
2. C. Mayr, G. Eggeler, G. A. Webster, G. Peter, *Mat. Sci. Eng.,* A199, 121-130 (1995).
3. M. S. Yeh, T. H. Chuang, *Weld. J.,* 76, 517-522 (1997).
4. W. F. Gale, D. A. Butts, *Adv. Mater. Proc.,* (2003).
5. W. F. Gale, Y. Guan, *J. Mater. Sci.,* 34, 1061-1071 (1999).
6. P. Yan, E. R. Wallach, *Intermet,* 1, 83-97 (1993).
7. S. Bhattacharyya, V. E. Johnson, S. Agarwal, M. A. H. Howes, *Iitri Fracture Handbook: Failure Analysis of Metallic Materials by Scanning Electron Microscopy*, IIT Research Institute, Chicago, Illinois, 1979, p. 492.
8. S. Ahmed, *JOM,* 24-26 (1990).
9. L. N. Payton, S. V. Chitti, D. R. Taarea, N. I. Sofyan, W. F. Gale, D. A. Butts, R. Aluru, R. D. Love, *International Conference on Joining of Advanced and Specialty Materials VI*, ASM International, Pittsburg, PA, 2003.

Proceedings from Joining of Advanced and Specialty Materials, 13–15 October 2003, Pittsburgh, Pennsylvania, ASM International, 2004

Nondestructive Evaluation for Characterization of Welding Defects and Development of Welding Technology

Norbert Meyendorf
Center for Materials Diagnostics, University of Dayton
300 College Park
Dayton, Ohio 45469-0121
(937) 229-4830

Abstract

Nondestructive evaluation (NDE) of welds is considered to be the detection of cracks, pores or inclusions in the welding seam and the heat affected zone. Traditional NDE is usually the final step after the production of the welded structure. However, NDE methods are capable of acquiring detailed information about the welding processes such as welding metallurgy and can be used for process monitoring. The following paper will highlight three aspects of the application of NDE in welding technology. First, the automatic inspection of welds, second, the use of multi-spectral information from x-ray radioscopic imaging for analytical purposes and last the use of infrared thermographic techniques to improve welding technology.

Introduction

Testing of welds is the main application of traditional nondestructive evaluation. Radiographic imaging and ultrasonic testing are the standard methods that are available in every laboratory that is responsible for weld testing. Also dye penetrant techniques and magnetic leakage field methods like magnetic particle testing are frequently applied to welded structures. These techniques are described in technical standards and are the topic of courses to train NDE inspectors. However, techniques based on NDE principles have the potential for a much broader application in welding technology. Those applications include completely automated testing, the identification and classification of materials for analytical purposes, monitoring of welding processes and data acquisition for basic metallurgical research to improve welding quality (Figure 1). Examples will be discussed in the following chapters in more detail.

Automated Weld Inspection

Scanners and robotic systems in connection with high efficient data acquisition systems and data image processing allow a completely automated inspection of welds. For example it is possible to detect and classify welding defects with accuracy comparable to an inspection by radioscopic imaging and image processing techniques. The general problem is that the appearance of weld defects is a very rare event in a good production process and that every defect appears differently. Therefore the application of standard image processing tools failed because they are usually designed for the application to microscopic images or other types of images where a-priori knowledge is available. Trained inspectors for x-ray radiographic images have the ability to identify weld defects even if they are barely visible on the film. If an inspector knows he has to expect a defect; he will seldom fail. However, an image processing system can fail.

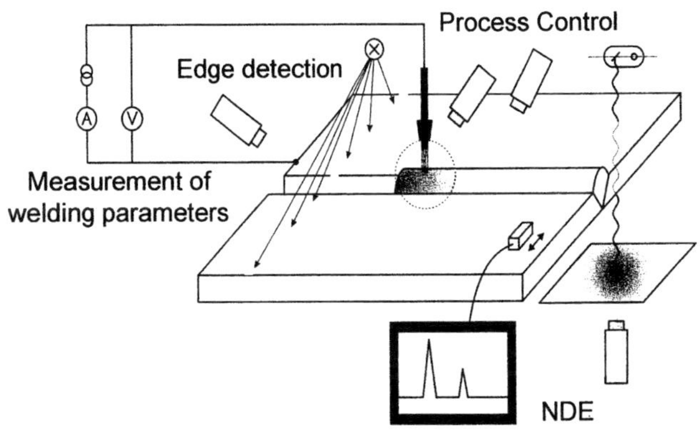

Figure 1: NDE and sensor techniques for quality assurance of welds.

If inspectors work under real production conditions the situation is completely different. Their attention may decrease if they do not have success (finding a defect) for several hours. They may fail even if a very critical defect appears on the screen. The inspector's performance may vary from day to day and throughout the day [1, 2].

The performance of image processing systems and inspectors for detection and classification of weld defects in submerged arc welds was compared in a detailed study [3-5]. Several hundred welds were produced by varying the welding parameters (voltage, current, speed) gap preparation, plate thickness and so on. All these test plates were pre-examined by radiographic inspection (using x-ray film). About 20% of the inspected pieces which really contained some defects were chosen for the detailed analysis. All these selected welding seams were carefully inspected by radiography. Five independent inspectors had unlimited time to identify and classify all welding defects. This result establish a data base that contained more than 2,500 weld defects of all types including cracks, pores, inclusions and root defects. Then all the specimens were imaged by a radioscopic system (image intensifier, TV-camera and image processing system). Image data were stored in a data bank. A magnification factor of 2 was used for radioscopy while the pixel resolution of the imaging system was 512 x 512 pixels. This radioscopic (digital) image data set was now inspected by seven inspectors. They had unlimited time and were allowed to manipulate the brightness and contrast of the images. Because of the lower resolution of the radioscopic image, they did not detect all the defects, which were visible in the radiographic image. The same data set was then analyzed by an image processing system using two different methods for segmentation (detection) and classification of the defects.

Figure 2 summarizes the results of this multi year project. It shows the percentage of cracks, root and bonding defects, and voids and inclusions that were correct, detected and classified by the best and the poorest inspector and by the image processing systems. It is obvious that the performance of image processing and inspectors can be compared. However, the data from image processing are reproducible and reliable. The disadvantage of image processing is that the most dangerous defects, the cracks, are the most difficult to detect.

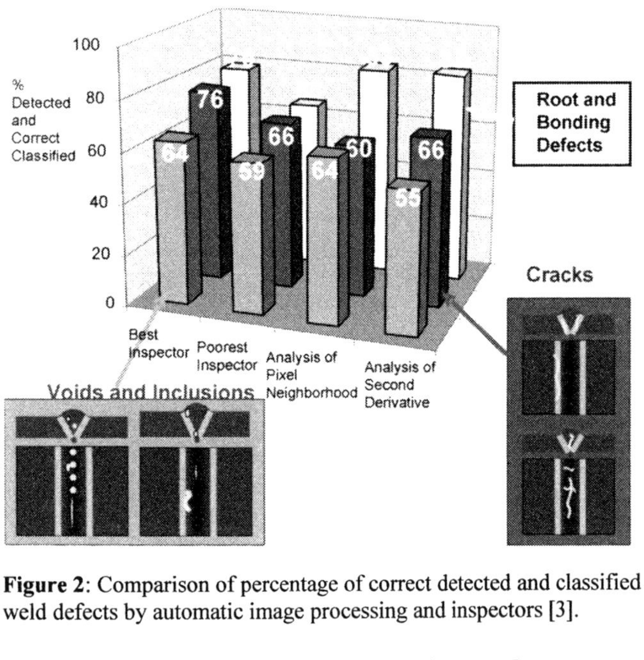

Figure 2: Comparison of percentage of correct detected and classified weld defects by automatic image processing and inspectors [3].

Multi Energy Color Radiography

It might be beneficial to have some additional information about the separate material components and the material composition especially for joining (welding, soldering, brazing) of special materials and material combinations. A tool to achieve this is the multi-energy radioscopic technique. This technique uses the dependence of the variation of the attenuation coefficient with the photon energy upon the atomic number (Z) of the material (Figure 3 bottom right). This effect was successfully used for the identification of different phases or different materials in composite structures by multi-energy computed tomography. Here the reconstructed images allow a direct determination of the attenuation coefficient. However in a conventional radioscopic image, the gray scale (gray values, "G") is mainly determined by thickness variations and different ration contrast for different x-ray energies. One idea to overcome this problem is to calibrate radioscopic images in a way that frames taken at different x-ray energies have a gray scale that varies linearly with the thickness of the material. This can be done if in addition to the materials that have to be inspected a step wedge made from a reference material is imaged.

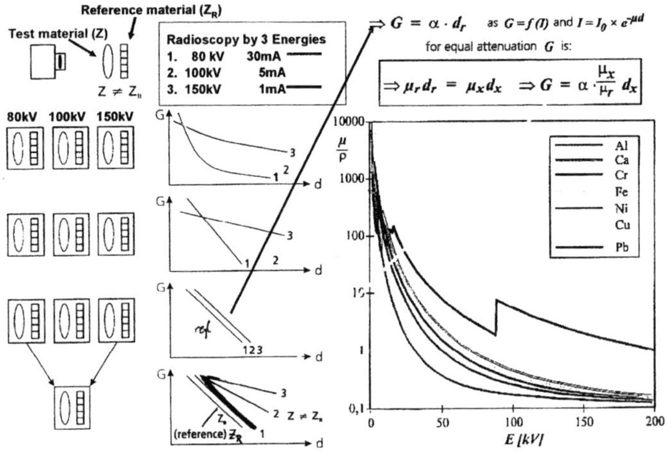

Figure 3: Principle of multi-energy radioscopy, Z: atomic number, G: gray value, d_r: thickness of the reference material, d_x: thickness of the unknown material, μ_r: attenuation coefficient for the reference material, μ_x: attenuation coefficient for the unknown material, α: parameter

In the original images, the gray scale varies with the thickness of the step wedge in a non-linear way. Usually the lower energy generates a higher contrast than the higher energy. If the thicknesses of the different steps of the step wedges are known, this information can be used to linearize the gray scale and then calibrate all three images taken at different energies to the same gray scale/thickness relation. The three images taken at different energies, which are now linearized and calibrated, can be used to generate a color image by feeding them into the red, green and blue channel of a color image processing system. Due to the calibration process, the step wedge of the material will appear as "gray". However, due to the different energy dependence of the attenuation coefficient for other materials, materials with an atomic number different to the reference materials will show up in a characteristic color.

The color image can be described by HSV model: H stands for hue, S for saturation and V for value. By using the above described calibration processes, images of components from one material with one characteristic atomic number will appear with only one color value while the saturation may vary depending on the wall thickness [6-8].

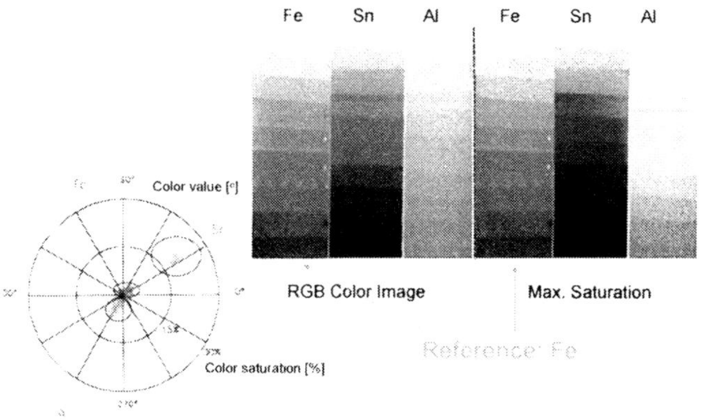

Figure 4: Multi-energy radioscopic color image of step wedges of iron, tin and aluminum and color valued and saturation from the HSV model (lower left corner).

Figure 4 shows color images of three step wedges that are synthesized from three radioscopic images taken at 80kV, 100kV, and 150kV. The three images are synchronized, linearized and calibrated as described in Figure 3. The reference material was iron. Therefore the iron step wedge appears gray in these color images while tin and aluminum show characteristic colors that depend on their atomic number. The real colors are shown in the left frame, the right frame shows the color saturation enhanced. The lower left corner shows the value/saturation plane of the HSV model for color values and saturations measured for varying positions within the step wedges. Values for iron are located at the origin with the scatter due to the quantum noise. Values for aluminum are located at the color value angle of 240°, while values for the tin step wedge are located at a color value angle of 30°. The variations in color saturation are due to the variation of the thickness of the step wedges. This technique was successfully applied to characterize the copper interface layer which was plated on steel with a titanium deposited weld on top [8].

Thermographic Weld Process Monitoring

The mechanical properties of a weld depend on the heating and cooling cycle during the welding process. Thermographic cameras are capable of measuring these temperatures on-line during process. However, the high intensity of the welding arc causes blooming in the camera image and hot molecular gases like water or hydrogen may emit heat radiation in emission bands within the sensitivity range of the camera. The best wavelength range for imaging of weld pools and welding processes is between 1 and 1.2 μm [9, 10]. Using small band filters it was possible to observe a GMA (Gas Metal Arc) welding process even through the arc. By applying gray filters that shadow regions of the image, it was possible to enhance a dynamic range of the camera so that within one thermographic image, the cooling process from solidification to approximately 500°C could be observed (Figure 5).

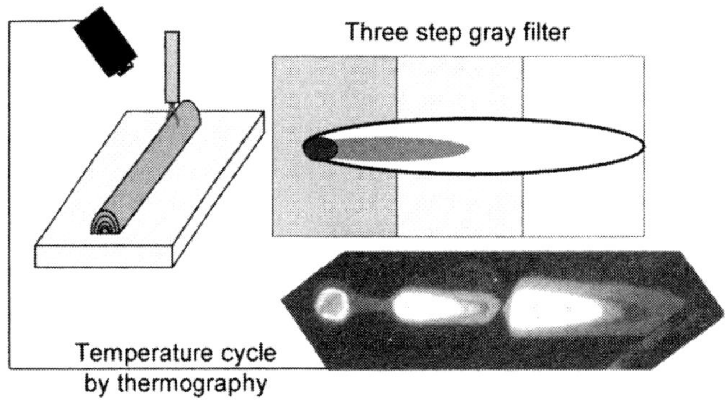

Figure 5: Measurement of thermal cycle for GMA deposition welding by a CCD camera, Sensor element: Si CCD, Spectral filter: 1000 - 1100nm, using a special designed optics with 2 selective gray filters.

Such thermograms allow determination the t $_{8\text{-}5}$ time (the time required to cool the weld down from 800 to 500°C) at the heat-affected zone (HAZ). For given welding speed the hardness at the heat-affected zone can be estimated by applying this data to regression models [11]. This approach should not work for the welding seam. However, a sufficient agreement between the weld metal hardness calculated from the temperature field and the hardness measured on a cross section was found even within the welding material that went through the melting and solidification cycle (Figure 6).

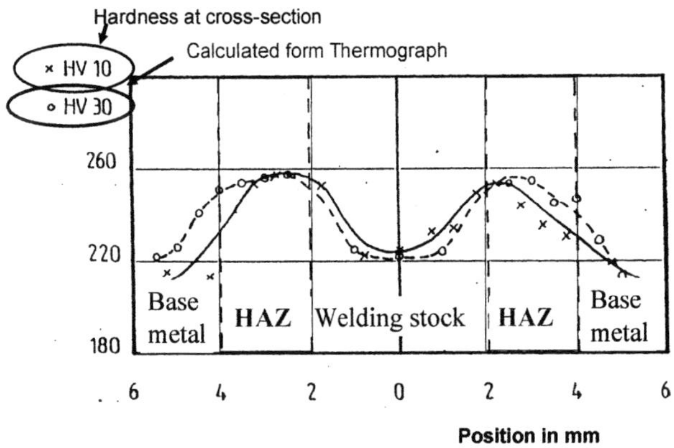

Figure 6: Hardness calculated from a thermographic image (circles) and measured at a cross section (crosses) for a GMA deposition weld. The applied regression formula was developed only for the Heat Affected Zone (HAZ).

Finally, thermal imaging techniques in conjunction with high-speed cameras are capable of imaging the temperature distribution at the welding droplet [10, 12]. This technique was used in conjunction with other techniques to determine the reaction conditions for metallurgical reactions during the droplet transfer for gas metal arc welding of low-alloyed steel materials (Figure 7) [10,14].

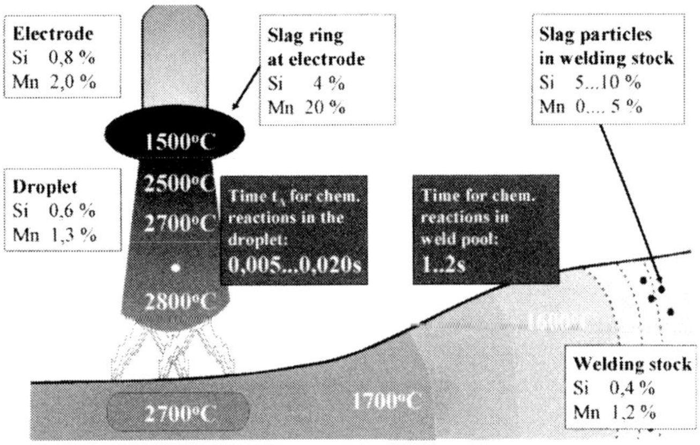

Figure 7: Temperatures and reaction times for GMW welding and an example for manganese and silicon content in the material during the welding cycle [10, 14].

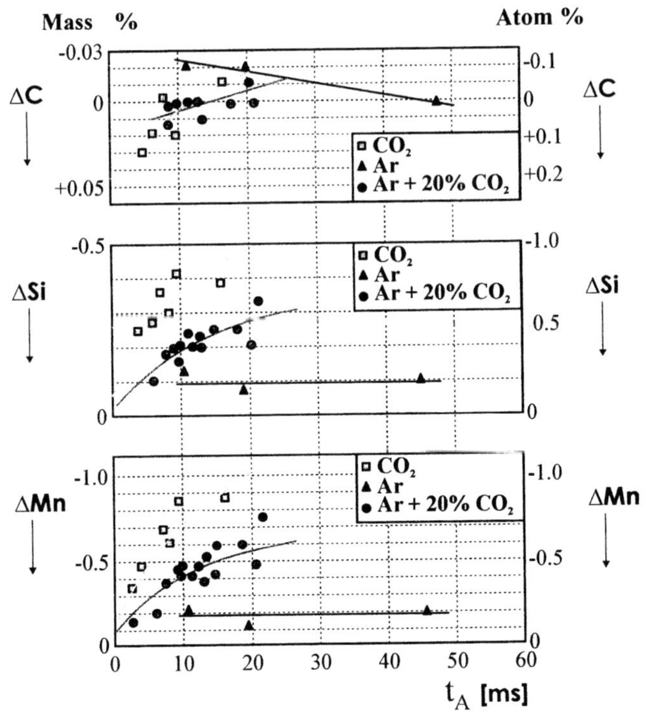

Figure 8: Variation of concentration of chemical elements between wire and welding stock for GMA welding with different gas shields as function of the time between two droplet transitions t_A. Every data point represents one welding process. Parameters varied with current, voltage, pulse rate, and wire diameter [14].

The figure also illustrates the change of alloying elements that can be measured between the welding wire and welding seam. Based on these data it was possible to show that the metallurgical reactions in a Gas Metal Arc (GMA) welding process significantly depends on the arc time, that means the average time that material at the welding tip is exposed to the higher temperatures of the arc (Figure 8) [10, 14]. The conclusion was that it is possible to modify the mechanical properties of a weld simply by modifying the droplet transfer frequency. By using the same welding parameters during a deposition welding process the Charpy notch hardness was significantly different by varying the droplet transfer frequency using pulse techniques [10, 15].

Summary

- Automated welding inspection using radioscopic techniques and automated image processing can achieve a similar probability of detection of defects as the inspection by trained inspectors.
- Nondestructive testing techniques capable of providing the welding engineer with much more information than only detection of cracks or flaws in a post welding inspection. For example multi energy radioscopic techniques can be applied to characterize a material and determine the chemical composition, for example in a layered structure.
- Thermographic techniques are capable of monitoring welding processes and calculating weld properties on-line during the process by determining the 8>5 temperature.
- Thermal imaging techniques have been applied to determine the reaction conditions for metallurgical reactions in the vicinity of the welding tip and the welding pool. These data, combined with other information about the process have been used to develop models for metallurgical processes [10, 14].

References

1. Rose, P., N. Meyendorf. "*Image Processing in NDT*" (German), "*Materialprüfung*", Vol. 35, No. 7-8. München: (1993): p. 231-236.
2. Eckelt, B., N. Meyendorf, W. Morgner, U. Richter. "Use of Automatic Image Processing for Monitoring of Welding Processes and Weld Inspection." Proceedings of the 12th World Conference of NDT (WCNDT), Amsterdam, 1989: p. 37-41.
3. Pohle, R, A. Hilbig, F. Heindoerfer, R. Link, W. Nuding, S. Langrock, H-E Steinecke. "Fully Automated Evaluation of Welding Seam Radioscopy's – Presentation of Results of a Joint Project.", Presented at DGZfP Annual Conference, (German) Proc. Timmendorfer Strand ,(May 9-11, 1994): .
4. Heindorfer, F., R. Pohle, "Feature Extraction and Classification of Welding Defects in Radioscopic Images. " (German), Proc. 3. Kolloquium "*Qualitätssicherung durch Werkstoffprüfung*", Zwickau, (November 23-24, 1993).
5. Pohle, R., A. Hilbig, N. Meyendorf. "Algorithm for the Automatic Defect Detection in Radioscopic Weld Testing", Proceedings of the DGZfP Seminar on Image Processing, Berlin, (November 23, 1992):p. 149-156.

6. Meyendorf, N., A. Hilbig, I. Mikuszeit, Th. Panten: "Real Color Radiographic Images for Materials Characterization" (German), Ingenieurwerkstoffe, Düsseldorf, Vol. 3, No. 11. (1991): p. 59-61.

7. Meyendorf, N., G. Walle, H. Reiter, M. Maisl, H. Bruns, A.Hilbig, F. Heindörfer, R. Pohle, S. Ehlers. "Multi-Energy Radioscopy - An X-Ray Technique for Materials Characterization", *Nondestructive Characterization of Materials VIII*, New York: Plenum Press (editing by R. E. Green) (1998): p. 455-460.

8. Hilbig, A., N. Meyendorf. "Multi-Energy Radioscopy - A Method for Materials Characterization Using X-Ray Transmission Techniques" (German), *Proceedings of Conference, "Werkstoffprüfung 1997"*, Bad Nauheim, (December 1997).

9. Meyendorf, N., R. Nitzsche, V. Schauder. "Visual Sensors for Welding Engineering" (German), *Bild und Ton*, Leipzig, Vol. 41, No. 12. (1988): p. 357-359.

10. Meyendorf, N.: "Reaction-Kinetic Model for the Calculation of the Chemical Composition of Weld Metal in Shielding Gas Arc Welding Processes", *Nondestructive Characterization of Materials VIII*, New York. Plenum Press, (editing by R. E. Green) (1998): p. 257-262.

11. Nitzsche, R., N. Meyendorf. "Computerized Temperature Field Processing for Technological Research and Process Control" (German), ZIS-Mitteilungen, Halle, Vol. 30, No. 1. (1988): p. 100-107.

12. Meyendorf, N., R. Nitzsche, M. Ströfer. "Study of Metal Transfer with High Speed Film in Welding Technology" (German), *Schweißtechnik*, Berlin, Vol. 35, No. 8. (1985): p. 362-363.

13. Probst, R., N. Meyendorf. "Physical-Chemical Processes During Metal Transfer in Arc Welding Processes" (German), *Schweißtechnik*, Berlin, Vol. 38, No. 7. (1988): p. 297-300.

14. Meyendorf, N.: "Physical and Chemical Processes During Material Transfer at Gas Metal Arc Welding Processes" (German), *Doctor of Science Thesis*, Magdeburg, (1986).

15. Meyendorf, N., R. Nitzsche. "Reaction Kinetic During Materials Transfer in MIG and MAG Welding Processes" (German), *Schweißtechnik*, Berlin, Vol. 35, No. 6. (1985): p. 251-254.

Proceedings from Joining of Advanced and Specialty Materials, 13–15 October 2003, Pittsburgh, Pennsylvania, ASM International, 2004

Comprehensive Inspection of Friction Stir Welds using Arrays

André Lamarre **Olivier Dupuis** **Michael Moles**
R/D Tech

Abstract

Phased array ultrasonics can reliably detect all internal volumetric defects in friction stir welds. Spot-focused beams improve detection, and inspection angles can be optimized. Unfortunately kissing bonds (or entrapped oxide defects) can be transparent if very tight. While phased arrays offer some kissing bond detection, a new technique using ultrasonic attenuation measurements shows the presence or absence of conditions for forming kissing bonds. Also, eddy current arrays can be used for surface inspection, and can detect kissing bonds. Using all three approaches, the overall detection capability of kissing bonds is high.

Introduction to friction stir welding

Friction Stir Welding (FSW) is rapidly gaining acceptance in the aerospace and other industries. FSW is a new process, only having been commercialized during the 90's. It is a solid state bonding process, which minimizes contamination. FSW is a very controllable process, and produces a very fine microstructure in the deformed region. This fine microstructure produces a higher tensile strength than other welding techniques, which permits less structural conservatism. FSW is highly repeatable, and offers other advantages like less shrinkage, no porosity, little finishing required, no gas shielding.

FSW is performed using a milling-type tool, which fits into a pre-machined slot. The tool is rotated and pushed along the weld line. The two pieces of metal (usually aluminium, but possibly steel or titanium) are clamped together very firmly with a backing plate. As the milling tool pushes along the weld line, the metal is plasticized and forced around the pin. Once deformed, it rapidly cools and recrystallizes [1].

Inspection of friction stir welds

FSW have some unique features, which makes inspections more challenging. Unlike conventional welding, FSW defects can occur in principle at any orientation and any angle. In practice, most defects apparently occur along the axial and transverse axes. However, the wide range of defect orientations and skews severely complicates any NDE technique; consequently, inspection procedures are typically tailored to the actual inspection process, FSW parameters and expected defects using a Performance Demonstration approach. In general, volumetric defects like worm holes are readily detected.

FSW characteristically produces tight defects, called "kissing bonds" or entrapped oxide defects. These are inherently difficult to detect by any NDE technique. Besides pulse echo ultrasonics, R/D Tech has been working with TWI as part of the Qualistir program to develop alternative inspection approaches for kissing bonds. In practice, attenuation measurements offer significant capability for reliably detecting conditions where kissing bonds may occur, though not the actual kissing bonds themselves. Additional inspections using the eddy current array probe show that kissing bonds may be detectable. However, these results are limited to one manufacturer only.

From a practical inspection aspect, the FSW process generates small "lips" along either side of the OD weld line, where the milling tool deposits excess metal. These lips are typically less than 1 mm high. However, they are sufficient to hinder contact ultrasonic testing, and necessitate some form of immersion like a local water bath. The FSW metal surface finish is good compared with conventional fusion welding processes, but not up to the quality of machined surfaces.

This paper describes a comprehensive approach for detecting all defects, including kissing bonds, using a triple NDE approach:

pulse echo using optimized phased arrays; attenuation measurements (also using phased arrays); and eddy current arrays.

Inspection techniques

Initially, a comprehensive review of NDE techniques was performed. The wide variety of defect orientations essentially precluded any radiographic inspection techniques. Eddy current lacks the penetration to detect defects on the opposite surface, though conductivity measurements have been used to detect poor process control [2]. However, eddy current arrays offer major advantages for surface and near surface inspections, and have the advantage that no couplant is required. Conventional ultrasonics is limited in detection of defects with unusual orientations and skew, though most defects generated to date are axial or transverse. Our primary solution for NDE of FSW was ultrasonic phased arrays, which have the ability to change inspection angles and to skew the beam.

As always with solid state process like diffusion bonding and electric resistance welding, the main concern was tight defects where bonding did not in fact occur. This indicates the need for alternative inspection approaches (in this case attenuation measurements), and the eddy current array.

Phased arrays

Phased arrays use an array of elements to generate an ultrasonic beam, using different time delays. The beams are formed by constructive interference [3], and can be skewed and scanned electronically. Once generated, the ultrasonic beam from a phased array is nominally identical to one generated by conventional ultrasonics.

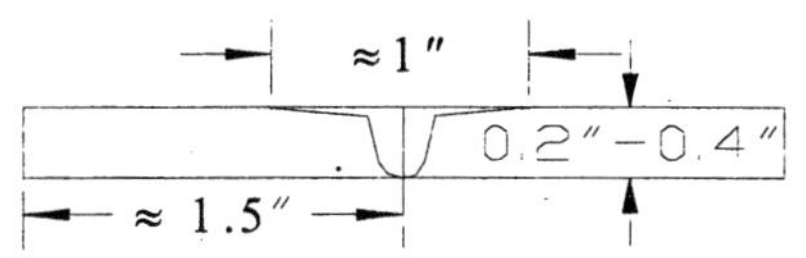

Figure 1: Typical FSW profile and dimensions.

Phased arrays have big advantages over conventional ultrasonics in pulse echo mode: it is possible to change angle every pulse (called sectorial or azimuthal scanning). Electronic (or linear) scanning is possible with linear and matrix arrays, where beams are rapidly scanned in a fixed pattern over a selected area. Focusing can be optimized electronically, and repeated with every set-up. With matrix arrays (or modified linear arrays), lateral scanning is possible to detect skewed defects. Dynamic depth focusing is another capability, wherein the receiver is refocused repeatedly during a single pulse to give the equivalent of multiple conventional transducers. Overall, phased arrays permit complex scans using sectorial, linear, lateral techniques; however, industrial phased arrays are typically customized to the specific application [4,5].

Ultrasonic attenuation measurements for detecting kissing bonds (entrapped oxide defects)

Kissing bonds mainly occur because of low penetration of the tool during the FSW process. This prevents the root region from being properly stirred (see the micrograph in Figure 2). Typically, the weld area has much finer grain size than the parent material due to the plasticizing of this are. Smaller grains mean less ultrasonic attenuation (i.e. less "noise"), and this is clearly visible on ultrasonic B-scans (see Figure 3). The principle of the signal processing is to quantify the attenuation to determine if proper mixing and FSW has occurred. While this approach does not actually detect kissing bonds, it does reliably detect the conditions under which kissing bonds occur [6].

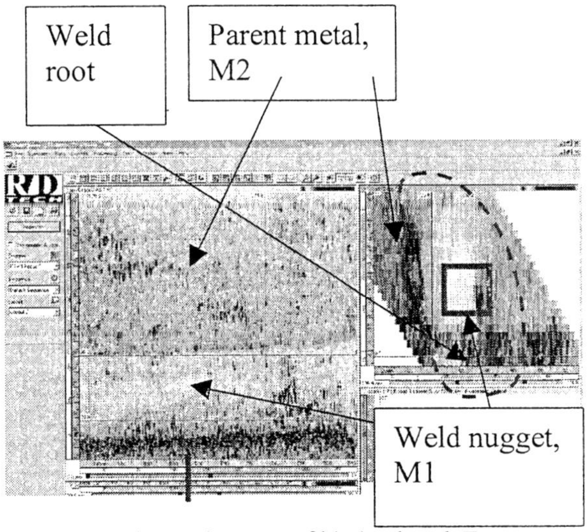

Figure 3: Ultrasonic scans of kissing bond. Top view (C-scan left) and side view (B-scan right) of the phased array inspection results, showing less noise in the weld (M1) than the parent material (M2)

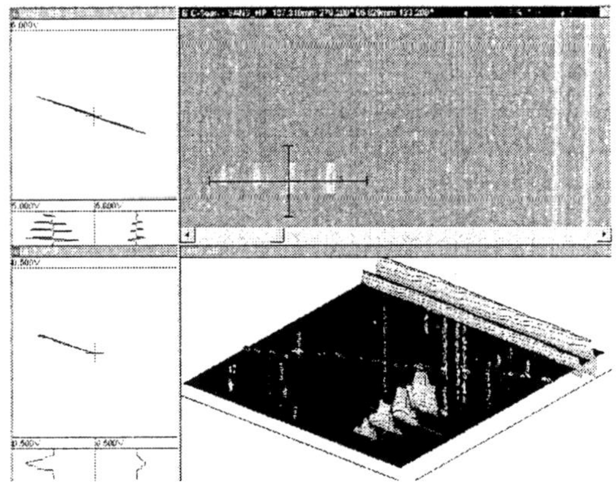

Figure 4: EC array scan results of cal block with notches and holes showing C-scan, isometric and two Lissajous patterns

Eddy current array probe

The eddy current array probe consists of a series of individual eddy current coils, closely packed into a pre-determined array. The coil arrangement typically permits pitch-catch axially and circumferentially, as well as multifrequency, absolute and differential operation. R/D Tech has developed a proprietary multiplexer, which effectively eliminates crosstalk between the coils. As a result, the EC array acts as a multitude of individual coils, but permits unique imaging techniques like C-scans and isometrics. All the data is saved, and individual Lissajous patterns or strip charts can be displayed. Figure 4 a) and b) below shows a typical block and scan result.

The EC array probe has major advantages: good surface and near-surface detection; defect sizing and characterization; axial vs. circumferential discrimination. However, it suffers the same limitations as other electromagnetic techniques, primarily limited penetration.

Equipment and scanning

To determine suitable pulse echo inspection angles, a selection of FSW plates with embedded defects is requested from the customer for inspection with appropriate thickness, welding parameters and defects (see Figure 1). This plate is inspected over a wide range of incident angles (say 35, 40, 45, 50, 55, 60, 65 and 70$^{o)}$ and inspection parameters to optimize detection. Axial defects can be detected using a transverse linear array and sectorial scans if requested. Calibration can be performed on a similar panel containing electro-discharge machined notches in the usual manner.

For the ultrasonic aspects of this study, an R/D Tech FOCUS phased array system is used, with TRAKER manipulator and small water bath. Typical array is a 10 MHz linear array, with 32 active elements on a 0.31-mm pitch to give good beam steering.

Since FSW is a relatively high speed process, suitable scanning speeds must be factored into the inspection process, including number of inspection angles, beam paths, quantity of data collected and display rates.

Suitable samples containing kissing bonds or possible kissing bonds were also inspected using phased array attenuation measurements and the eddy current array.

Typical results

Pulse Echo on Defect Plates

Figure 5 shows a sample scan on a transverse defect. This particular defect was best detected using a lateral scan at 0° skew and 45° refracted angle.

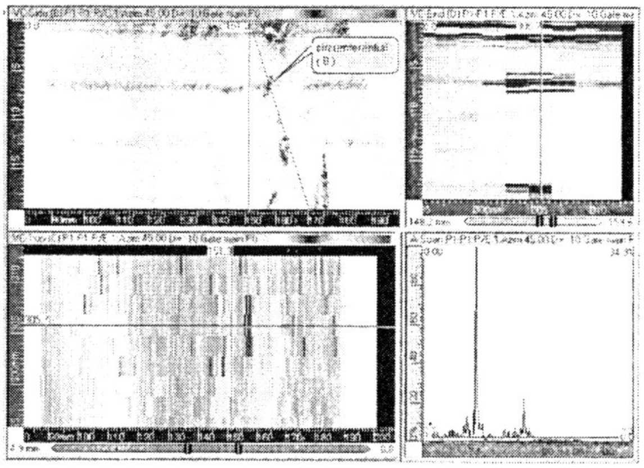

Figure 5. Scan of defect E using lateral scan at 0° skew and 45° refracted angle.

Figure 6 below shows a sample scan on an axial defect, which was a tight defect nominally 0.4 mm deep. This particular defect was optimally detected using 35-40° shear waves, in contrast to other axial root cracks, which were best detected at 65°. The top left B-scan in Figure 6 shows the defect superimposed on the weld profile, which gives a very clear image of the crack.

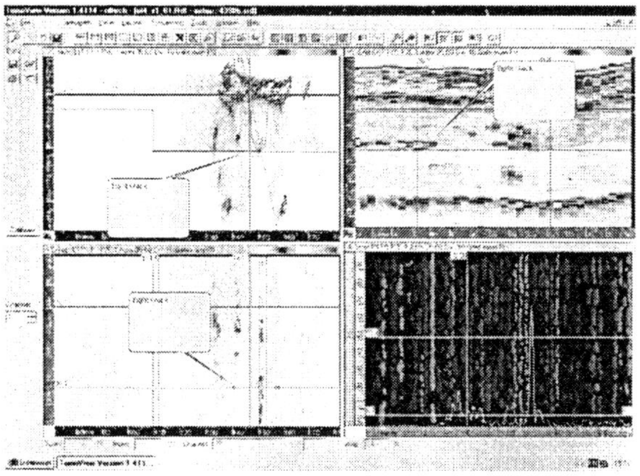

Figure 6. B-, D-, A- and C-scans on axial defect C.

The results from this plate set can be summarized qualitatively as shown in Table 1 below. Each refracted angle is ranked from 0 to 10 for each defect, based on signal-to-noise ratio. The overall results showed that the FSW's are optimally inspected at a combination of angles: 35° for the crown, and both 35° and 65° for the root. For lateral defects, the recommended inspection is a 45° incident angle, with sectorial scanning of ± 30°.

Table 1. Qualitative summary of results on defects A-H

Refracted angle	35	40	45	50	55	60	65	70
Defect A side 1	2	2	5	5	10	10	10	8
Defect A side 2	4	4	4	6	8	10	10	8
Defect B side 1	2	2	2	5	5	8	8	10
Defect B side 2	2	2	2	5	5	8	8	10
Defect C side with lip	10	10	2	2	0	0	0	0
Defect C side opposite to the lip	10	8	5	5	0	0	0	0
Defects D-H for root defects	10	10	10	8	8	8	8	7

Results from Ultrasonic Attenuation Measurements

Results from attenuation measurements of samples with no kissing bonds and with kissing bonds are shown below. Figure 7 shows the difference in ultrasonic noise for a normal bond, with the nugget (red) showing lower noise than the parent material (blue). The nugget has a much narrower profile, indicating a correctly positioned weld.

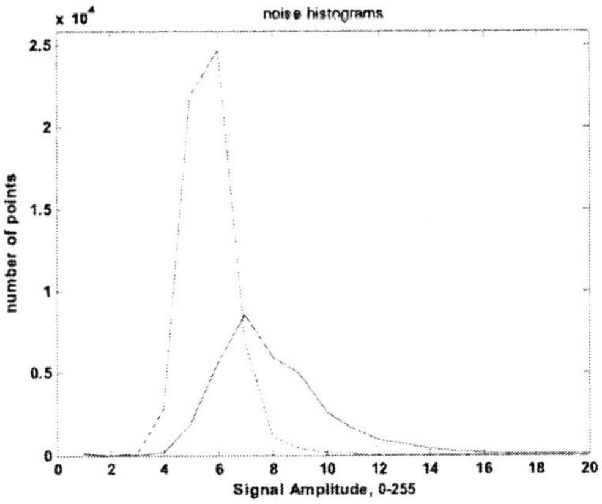

Figure 7: Illustration of the difference in ultrasonic noise: histograms of amplitude level for weld nugget volume (red) and weld root volume (blue).

When the root has been properly positioned, the ultrasonic noise level inside the root should be close to that of the weld nugget because of its small grain size. By comparing the mean level inside the root to that of the weld nugget, the operator has a powerful tool for estimating the pin penetration and therefore the probability of having kissing bonds.

Figure 8 below shows a scan of sample T7 containing kissing bonds (entrapped oxide defects) all along the weld. Only a few indications can be observed by pulse echo technique (red circles).

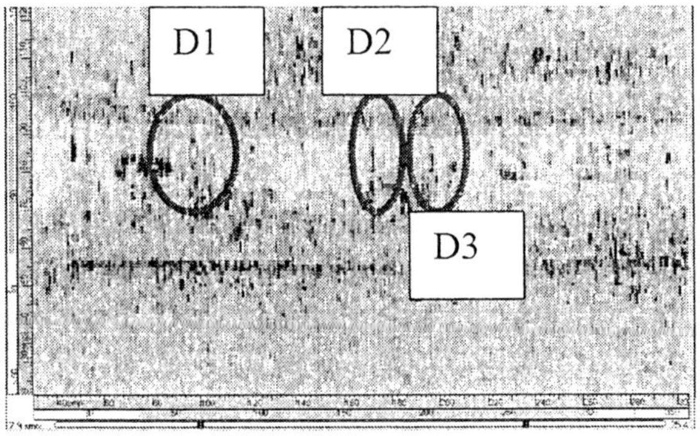

Figure 8: C-Scan image of FSW inspection: the plate T7 contains kissing bonds. Lower attenuation is visible all along, while the kissing bonds only show sporadically from pulse echo.

Though a few indications can be detected at the weld root, it is difficult to say whether there is a kissing bond along the full weld or not. This can be achieved by comparing the ratio M2/M1 of this plate to that of a reference plate containing no defect. The graph in Figure 9 compares two curves: the red curve represents the ratio M2/M1 of T7 sample (containing kissing bonds) and the blue curve represents the same ratio with sample T1 containing no defect.

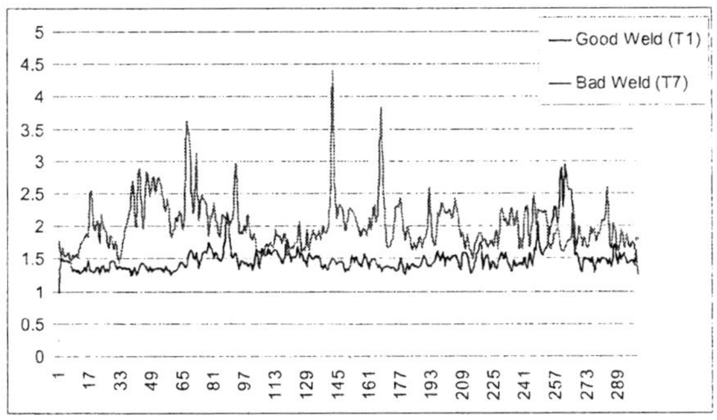

Figure 9: Ratio M2/M1 along with the weld for sample of reference (blue curve) and defective sample (red curve) containing a very tight kissing bond observed by metallography.

The mean value of the ratio M2/M1 is higher when the sample contains kissing bond defects, or potential kissing bonds. The difference in mean value between the weld nugget and the weld root was consistent for most of the samples containing kissing bonds. Figure 10 represents the value M1 and M2 for 13 samples. Sample 1 and 2 have no defects while all the other samples contain kissing bonds. For samples 1 and 2, the attenuation value of the root weld M2 is approximately equal to that of the parent metal. The higher M2 values in the other samples indicates that the root has not been properly stirred (grain size similar to the parent metal), which increases the probability of having kissing bonds.

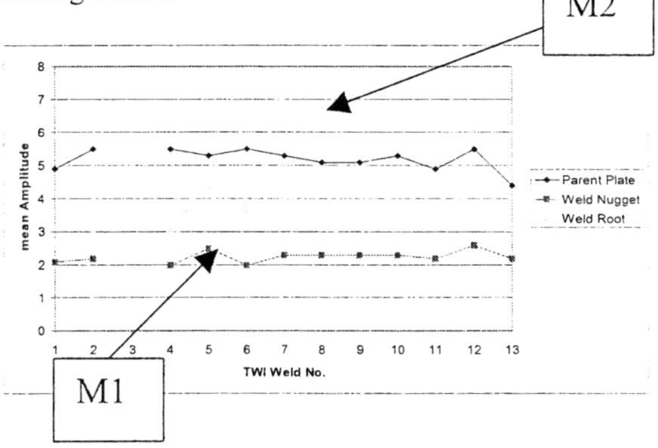

Figure 10: Mean signal inside the nugget and inside the root for different samples

Eddy current inspections of Kissing Bond Panels

Eddy current inspections were done on four 0.8 mm (0.320") thick panels containing no kissing bonds, and kissing bonds of prepared depths of 0.75, 1.0 and 1.5 mm (0.030", 0.040"and 0.060") respectively. The technique used the following parameters: send/receive; absolute; frequency: 800 kHz; scanning resolution: 0.5 x 3 mm; lift off aligned horizontally; median filter.

1. No defect

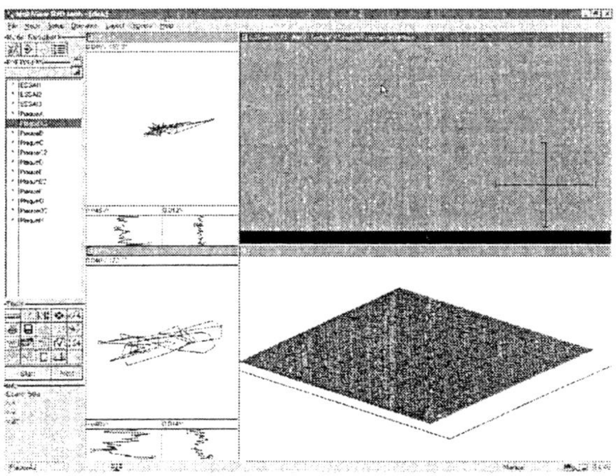

Figure 11: EC array scan of panel with no defect.

2. Containing a 0.75 mm (0.030") deep kissing bond.

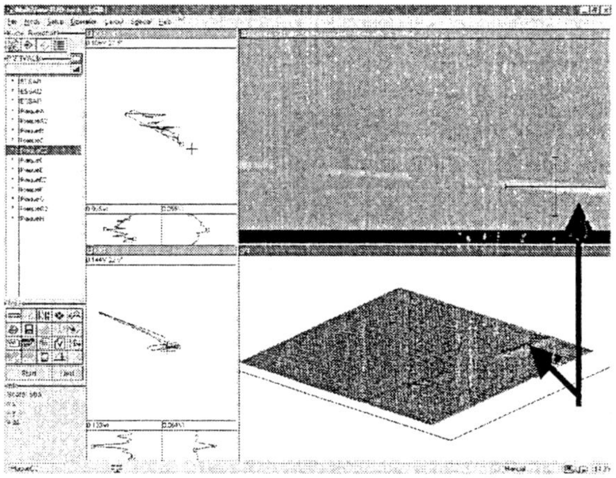

Figure 12: EC array scan of panel with 0.75 mm defect, showing discontinuous signal along weld line.

3. Containing a 1.0 mm (0.040") deep kissing bond

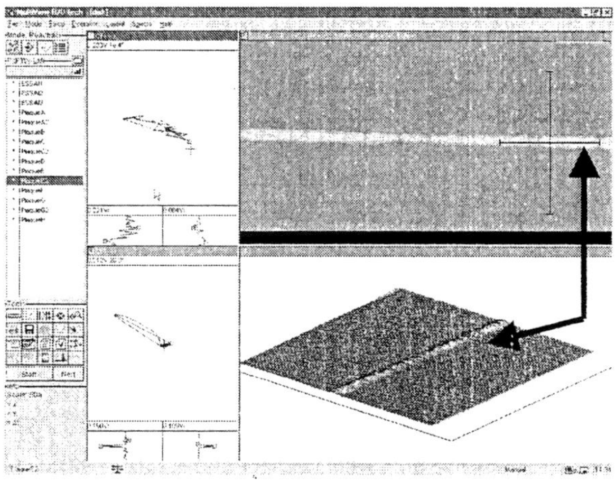

Figure 13: Eddy current array scan of panel containing 1.0 mm deep kissing bond. Signal is present but not completely along the weld.

4. Containing a 1.5 mm (0.060") deep kissing bond

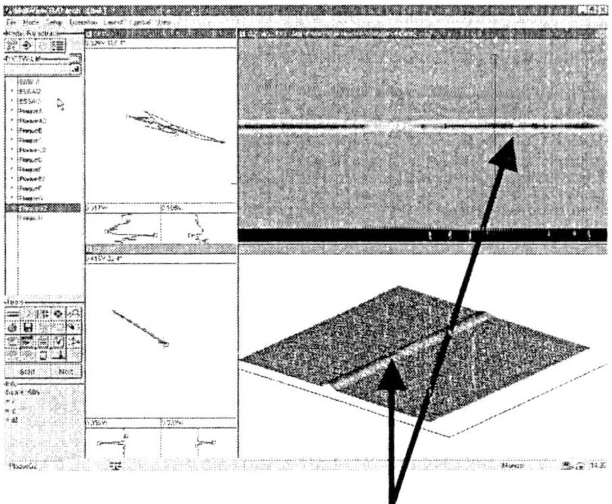

Figure 14: Eddy current array scan of panel with 1.5 mm deep kissing bond. The flaw is detected easily, but the signal amplitude is lower around the center of the weld.

Discussion

Pulse echo inspections of FSWs can detect all volumetric-type defects and kissing bonds intermittently. Phased arrays offer the advantages of optimizing inspections, e.g. by focusing and angle selection, as well as offering high speeds. For the critical kissing bonds, a combination of attenuation methods and eddy current offer considerable detection reliability. The attenuation measurements, well grounded in physics, reliably detect the conditions for forming kissing bonds, if not the actual defects, while eddy current appears capable of reliably detecting kissing bonds greater than 1 mm deep. However, both the attenuation measurements and the eddy current results are based on limited data samples, and need more evaluation. Overall, with three essentially independent NDE techniques, the probability of missing a significant kissing bond is very low.

Conclusions

1. Phased arrays provide a wide range of inspection angles, and detected all deliberate defects in these Friction Stir Welds.
2. Suitable angle selection (35° for the crown and 35 & 65° for the root) maximizes Probability of Detection for these defects for this parameter set.
3. Transverse defects can be detected using lateral and sectorial scans.
4. Electronic scanning offers the following advantages:
   - It provides the coverage and speed required.
   - It compensates for positional errors in the delivery system.
   - It eliminates the need for a second mechanical axis as scanning is performed electronically.
5. Samples containing different types of defects as well as very thin kissing bonds were manufactured, and led to the development of a new processing technique based on the attenuation differences between the weld root and the weld nugget signal amplitudes.
6. Preliminary results demonstrated that this noise attenuation ratio provides a good indication of the presence of very tight kissing bonds.
7. Based on a limited sample, the eddy current array probe detects kissing bonds above ~1 mm.
8. Between the three NDE techniques (ultrasonic pulse echo, ultrasonic attenuation ratios and eddy current array), detection of all defects including kissing bonds is highly reliable, though more evaluation needs to be performed.

References

1. See for example, "Friction Stir Welding-Equipment", The Welding Institute, UK, www.twi.co.uk/bestprac/datashts/fswequip/html and "Application of Friction Stir Welding to Automotive Lightweight Structures", The Edison Welding Institute, Columbus, Ohio, www.ewi.org/ewi/gsp/gspstir/.
2. N. Goldfine, A.P. Washabaugh and W. Arbegast, "Friction Stir Welding On-Line Process Monitoring and Post-Weld Quality Assessment with MWM-Eddy Current Sensor Arrays", AeroMat '99, 10th Annual Advanced Aerospace Materials & Processes, Dayton, Ohio, June 21-24, 1999.

3. G. Lafontaine and F. Cancre, "Potential of Ultrasonic Phased Arrays for Faster, Better and Cheaper Inspections", NDT.net, vol. 5, no. 10, October 2000, www.ndt.net/article/v05n10/lafont2/lafont2.html.
4. M. Moles, E. Ginzel and N. Dubé, "Phased arrays for pipeline girth weld inspections", Insight, vol. 44, no. 2, February 2002, p. 86.
5. M. Moles and A. Lamarre, "Phased array ultrasonic inspection of friction stir welds", 4th International Symposium on Friction Stir Welding, Park City, Utah, USA, May 14-16, 2003.
6. C. Bird, "Quality Control of friction stir welds by the application of non-destructive testing", 4th International Symposium on Friction Stir Welding, Park City, Utah, USA, May 14-16, 2003...

Proceedings from Joining of Advanced and Specialty Materials, 13–15 October 2003, Pittsburgh, Pennsylvania, ASM International, 2004

Nondestructive Inspection Techniques for Friction Stir Weld Verification on the Space Shuttle External Tank

Authors

Michael William Suits
NASA Marshall Space Flight Center
M/S ED32 Huntsville, Alabama, USA 35812
Phone: (256) 544-8336; Fax: (256) 544-9326; email: mike.suits@nasa.gov

Jeffery Leak
NASA Marshall Space Flight Center
M/S ED32 Huntsville, Alabama, USA 35812
Phone: (256) 544-9322; Fax: (256) 544-9326; email: Jeffery.L.Leak@nasa.gov

Dwight Cox
Lockheed Martin Michoud Space Systems
Department 3700 P/A 3741, Huntsville, Alabama, USA 3581270189
Phone: (256) 544-5017; Fax: (256) 544-9326; email: dwight.cox@maf.nasa.gov

Abstract

Friction Stir Welding (FSW) has gained wide acceptance as a reliable joining process for aerospace hardware as witnessed by its recent incorporation into the Delta Launch vehicle cryotanks. This paper describes the development of nondestructive evaluation methods and techniques used to verify the FSW process for NASA's Space Shuttle.

Friction Stir Welding Application

Since 1995, the Marshall Space Flight Center, located in Huntsville, Alabama, the main propulsion center for NASA, has had a keen interest in the Friction Stir Welding (FSW) process, which was developed by The Welding Institute (TWI) in Cambridge, England. The purpose of using the solid state process was to eventually replace the current fusion welds employed to join ogives, barrels, and domes which make up the hydrogen and oxygen tanks of the Space Shuttle External Tank (ET). Figure 1 shows the welds used in the ET configuration. This tank, not only provides fuel for the Space Shuttle Main engines, but serves as the structural backbone of the Shuttle stack. [1]

As has been written in many papers recently, friction stir welding is an alternative to fusion welding and provides even better strengths with fewer defects. The process joins two workpieces via a tool moving in a cyclic manner relative to the workpieces. This tool enters the joint region, locally plasticizes it and moves along the interface, thus resulting in a robust joint between the workpieces. The FSW process can produce welds with no porosity and very low distortion and shrinkage. The solid state process also has excellent mechanical properties. [2]

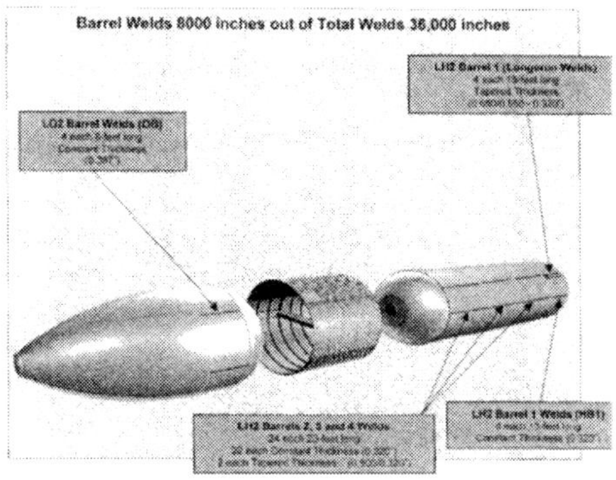

FIGURE 1

In the mid '90s the External Tank project office made the decision to study this FSW process and to determine its suitability for welding on ET. This feasibility study led to manufacturing trials in 1996 and 1997 by NASA and Lockheed Martin of New Orleans. Because of the unique types of defects which could occur when the process was not optimized, a task was funded through the NASA NDE Working Group (NNWG) by NASA Headquarters, to survey and downselect suitable NDE methods for the verification of FSWelds. This led to the development of techniques in phased array ultrasonics, creeping wave ultrasonics, meandering winding magnetometers, acoustic emission (in-process), along with modified conventional techniques in radiography, penetrant and eddy current. These results were used directly in the ET FSW allowables program that started in 1997. During this period, Boeing NDE personnel became interested in the preliminary work NASA had completed, and within 3 years they had full development of FSW for the fuel tanks on the Delta IV launch vehicle. Boeing uses automated phased array ultrasonics to verify their constant thickness welds on tanks manufactured at their Decatur, Alabama facility.

Other studies performed later by TWI in England gave further proof that these and similar techniques were optimal for these types of welds. This paper gives a background in two areas of friction stir weld processing, conventional FSW with a backside anvil and self-reacting FSW, and summarizes the critical NDE development required to verify both approaches.

Conventional Friction Stir Welding

Figure 2 shows the basic configuration of the FSW process. The base metal of the workpiece is clamped to a backing anvil. A FSW pin tool consists of a concave shoulder and a pin with a length approximately equal to the material thickness. The pin tool is rotated and plunged into the material until the shoulder penetrates below the top surface. Frictional heating from the rotating tool plasticizes the material between the anvil and the shoulder. The rotating tool is then traversed along the weld seam, generating a combination of extrusion and forging between the tool shoulder and the anvil, resulting in a ductile, high strength, solid state weld. Instead of fusion weld parameters such as current, voltage, wire feed rate, plasma gas, etc., FSW has parameters such as plunge depth/load, rotation speed, weld traverse speed, centerline position and pin length. Compared to fusion welds, FSW has improved strength and toughness properties, improved cryogenic enhancements, reduced peaking and mismatch and reduced rework and repairs.

Initial process development and allowables testing at Marshall and Lockheed Martin Michoud for Conventional FSW

During the two year period 1996-97, friction stir weld tools at the Marshall Space Flight Center were used to perform characterization and sensitivity studies at the limits of weldability. Parameter tolerances were developed along with A-basis strength allowables.

Material process maps were created by the NASA/Lockheed Martin team, for welding 0.320" and 0.650" thicknesses of aluminum weldalite alloys such as 2219 and the aluminum lithium alloy 2195. These are the most common thicknesses found in the External Tank. During early allowables testing and tool optimization, the regions near the edge of the optimum parameter "box" or process envelope, led to the formation of internal defects such as "wormholes" and shoulder voids due to inadequate tool heel plunge depth, plunge loads and welding too fast (traversing the workpieces). The volumetric flaws such as these were usually found with conventional radiography and required no new investigations. However, because of the nature of the mixing process, there are edges of the process envelope, such as inadequate pin tool penetration, which lead to the formation of tight, laminar flaws such as a lack of penetration (LOP). A similar type of laminar flaw that we called "lack of inadequate forging" or LAF for short, was found near the crown surface and was the formative defect that eventually led to the more gross shoulder voids which could be detected visually. Another surface defect, on the crown side, was that of galling which could be detected visually and with penetrant.

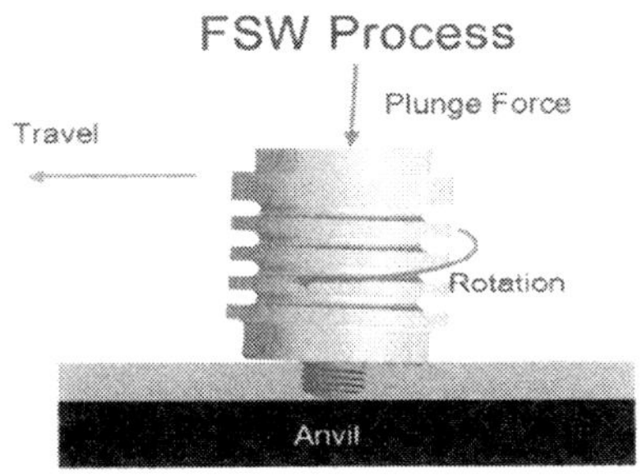

FIGURE 2 FSW Pin tool

The NDE development, therefore, concentrated on the detection of these more difficult laminar flaws which ranged from "kissing bonds" (0.007" separation) to much larger separation volumes. The smaller the length of the flaw, the less interstitial separation there was. Another problem with detecting the LOP flaw was the existence of compression on the root side which would tended to disguise the visual or penetrant detection of the LOP. It was found that by etching the root side between 0.002" to 0.004", the LOP could then be detected with penetrant. A visual technique, also, seemed to correlate well to the existence of LOP by detecting the presence of a DXZ after the similar etching process.

Though the detection of LOP utilized straight forward methods, the depth of such anomalies could not be determined. Since the depth of this flaw was critical in determining whether it became a critical flaw (unable to successfully meet the safe-life requirement for aerospace vehicles), and since the visual methods are quite intensive and depend on human factors, an automated system was very desirable for detection and sizing. Phased array ultrasound is easily automated and was found to detect this type of flaw reliably (90/95 reliability/confidence) at around 0.075". This was below the critical flaw length. It was found that as the length of the LOP flaw got smaller, that the compression on the back surface increased and the flaw was much tighter, impeding its reliable detection. Below 0.050", there was no method that could reliably find LOP.

The LAF flaw was caused by inadequate plunge loads and it was routinely missed by radiographic inspections. Later determinations of critical flaw sizes and zoning of this particular flaw showed that radiography could detect a critical flaw. However, phased array was much more effective in sizing and determining depth, and could detect LAF much smaller than the critical flaw size, making this automated method much more desirable for this type of defect also.

Self Reacting Friction Stir Welding

To improve the process of Friction Stir welding on circumferential welds such as those that join domes to barrels in the External Tank, a new process was developed. The Self reacting FSW process basically created a “crown” on both sides with no “root” and required no backing anvil, which greatly simplified tooling. A threaded nut on the backside of the workpiece, actually was threaded to the FSW pin on the opposite side, creating its own anvil action. This prevented the formation of the LOP flaw. Process envelopes, once again, were established.

Figure 3 shows a macro section of such a weld. The microstructure usually has this S shaped interface along the mixed grains and serves as an interface where laminar type flaws could form if process parameters are not in the optimized envelope. The Lockheed Martin/NASA team found that the strengths of these welds were also much better than fusion welds with fewer defects. The laminar type of flaws such as LAF could occur and was easily found by phased array ultrasonics. All other volumetric anomalies were detected by conventional radiography and surface flaws by conventional visual and penetrant methods. One anomaly that is dependent only on the surface preparation is the residual oxide defect (ROD). Even though the ROD anomaly has been reported by other researchers in conventional FSW, for our studies, the only appearance of this defect was in the self-reacting FSW. This could be due to the larger forging pressure in conventional FSW, which could crush the oxides to where they are not detrimental. Regardless, ROD is a low density, linear defect which is very tight and can only be detected by phased array ultrasonics. Work continues in this area using some of the noise estimation techniques developed by RD Tech and TWI.[3]

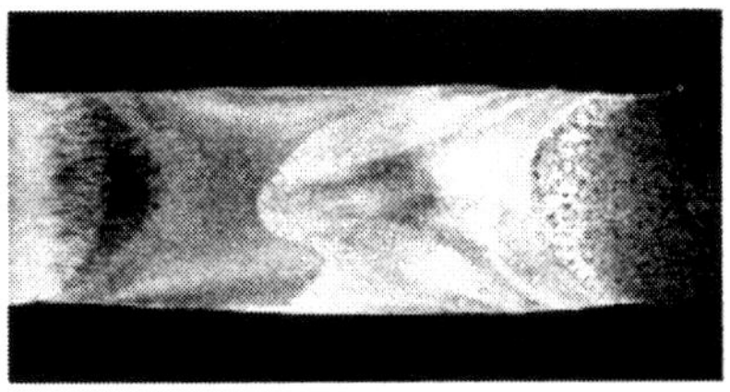

FIGURE 3. SR-FSW Macro

Implementation of Conventional FSW at the External Tank Production Facility

The 2XXX series of aluminum alloys have been used extensively in demanding aerospace applications such as the Space Shuttle External Tank. The lightweight external tank incorporated mostly 2219 and then was later upgraded to the “super lightweight” tank configuration using the new 2195 alloy. This allowed the Shuttle to carry more payload and to reach the Space Station at its higher orbit. The strength of the alloy was increased at both room and cryogenic temperatures but weldability using conventional fusion welding, especially on multiple repairs, was a major driver to look at the advantages of Friction Stir Welding.

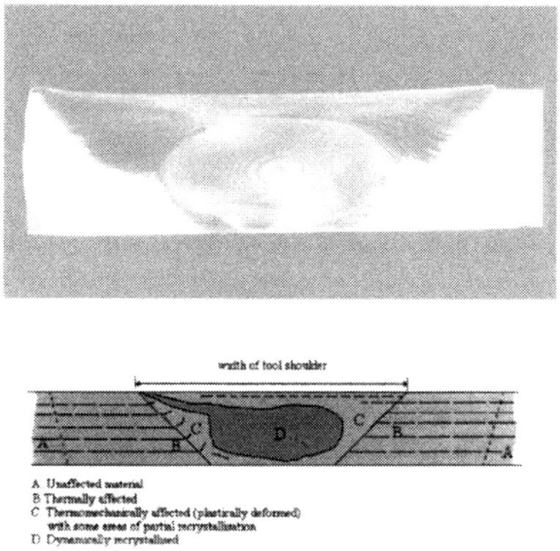

FIGURE 4 Conventional FSW

Eventually a trainer tool was procured with two production systems that performed over 800 feet of longitudinal barrel welds for the hydrogen and oxygen tanks that make up the External Tank. The radiographic, penetrant and visual methods remained in place with an addition of an automated phased array system. This automated system is being qualified and personnel are now being certified according to ASNT guidelines for ultrasonics with an additional phased array practical and specific exam.

Inspection of Friction Stir Welds

For effective verification using NDE, the engineer needs to understand the microstructure and the nomenclature of the areas of interest. Figure 4 shows an annotated micrograph of a cross section thru a full penetration conventional FSW and a cartoon of weld nomenclature.

Again, flaws observed during FSW development range from surface defects such as excess flash (material swept up above the skin line), to lack of fill under the FSW tool shoulder, to internal porosity and Lack of Penetration (LOP).

In addition, the Residual Oxide Defect (ROD) was found to occur in the self-reacting process. In every case the FSW flaw was caused by distinct FSW process condition or parameter inadequacies (outside the optimized process envelope). It was obvious from the beginning of the development programs on both FSW processes that some NDE methods were stronger at finding surface defects, some stronger at volumetric flaws, and some stronger at detecting laminar type flaws, just as is the case for fusion welds. Thus, a complimentary suite of inspection methods would be needed for reliable verification of FSW type welds.

To assess the different methods and techniques we had to grade them by many factors, including the ability to detect the critical initial flaw size (CIF), ability to detect the different types and orientations expected of the flaws, the reliability to detect certain flaws, and its maturity for implementation in a production atmosphere. There were several tests performed which compared candidate methods and measured their detectability capability. Technique variations of the chosen methods were developed as well as the accompanying procedures. The optimized methods and techniques for each flaw type was then process validated by going through capability tests which verified it as an effective fracture analysis tool, along with proof test, for validating the ability of the welds to meet fracture critical requirements.

The NDE assessment included many NDE methods such as visual, several liquid penetrant techniques, phased array ultrasound and other angle beam ultrasound techniques, radiography, and eddy current, including the new magneto winding magnetometer (MWM) technique.[4]

Visual NDE Methods [5]

The oldest of all NDE methods and the most straight forward is the visual method. In this development it was used to detect surface features such as excess flash, galling (see Figure 5), shoulder voids and features resulting from control parameters such as weld misalignment. The best way to accomplish these inspections and to help train the inspectors was to develop "workmanship" standards. Standards were made of all of these surface conditions.

The detection of these are made with the unaided eye. All of the defects that are detected visually can be traced back to weld parameters going outside the process envelope. These weld parameters include excessive travel speed, excessive rotational speed, inadequate plunge force loads and the improper seam tracking.

As mentioned earlier, the LOP or Lack of penetration flaw can lead to a critical condition and is detected by many of the NDE methods, however it was found that the resulting Dynamically Recrystallized Zone (DXZ) could also be seen clearly by the naked eye after etching. This etching process is usually performed in conjunction with penetrant inspection due to the smearing of metal in surface finishing operations. As a result of penetrant inspection surface preparation, it was discovered that the delineation between the DXZ and the Heat Affected Zone (HAZ) (see Figure 6), was very distinct and its existence is correlated with full penetration.

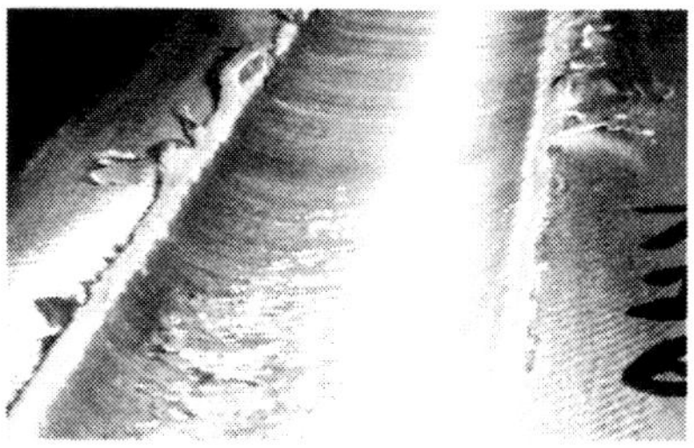

Figure 5. Galling

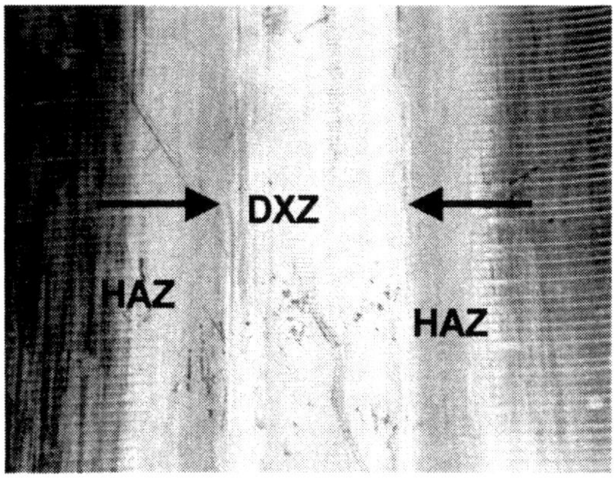

FIGURE 6. Visual Inspection for LOP

In fact, a template was made for the inspectors to measure the width of this DXZ in the post etched condition, and if it fell below a certain value, then the weld parameters would have to be adjusted, such as those that relate to heat input and material flow, and most importantly, the penetration of the pin tool in the work piece.

Penetrant NDE Method

Two types of sensitivity level III penetrants were studied for their ability to detect these types of surface flaws. The idea was to look at what is used in External Tank fusion weld program first and then if they were not sufficient, we would expand our study. The penetrants were P135E and P6F4. A test was performed to look at their capability on FSW panels in the as-welded, single etch, and double etched condition. The penetrants were applied with and without developer, while dwell times were varied. It had been established that penetrant was ineffective in the as-welded condition due to excessive background noise produced by the surface, and these test confirmed that. Also, etching was required to detect the LOP directly (unlike the correlated DXZ visual measurement), because of the compressive bonding on the root side which made the LOP very tight.

Both penetrants could detect galling, surface voids, etc, but the challenge was the LOP and there was some differences in the testing.

P135E successfully detected LOP flaws that were greater than or equal to 0.064" in depth, while P6F4 successfully detected LOP flaws that were greater than or equal to 0.050" in depth (these were measured from macros visually by measuring from the root side skin line as seen in Figure 7). Combinations of two levels of etching and various penetrant dwell times were studied along with the use of a nonaqueous developer. The single etch removed approximately 0.0002" to 0.0004" of material while a double etch removed 0.0004" to 0.0006".

The conclusion was that with standard dwell times for the penetrant (10 minutes – 30 minutes), no parameter had an effect on detection except for the etching. The P135E which is currently used in our ET production fusion weld inspections did just as well as P6F4 when it came to detecting LOP below the critical initial flaw size and the requirement of etching a minimum of 0.0004" was found to give optimum detection.

Eddy Current NDE Method

Of the typical defect types generated by the FSW process, the LOP flaw is the only tight, laminar flaw that opens to the surface when the surface is prepped sufficiently. However, to avoid all of the surface prep work, it would be nice to have another method, such as eddy current, which can detect near surface defects under smeared metal or in this case, a compressive bond. The LOP actually had adequate separation for detection under this compressive bonding on the root side

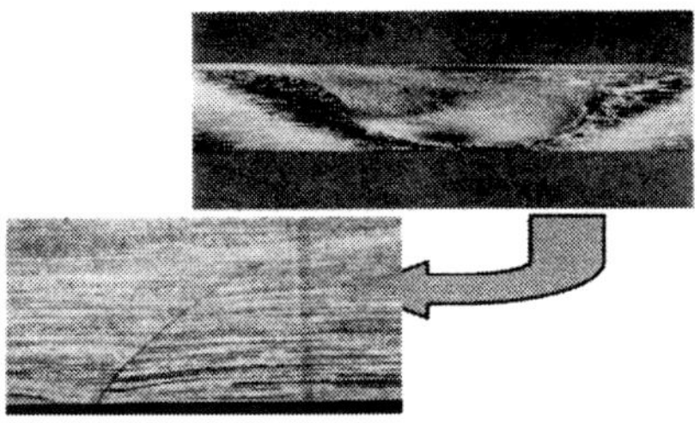

FIGURE 7. LOP Measurement from macro

Two approaches were used, one, using a 1 MHz pencil probe and 300KHz differential rotating probe, and two, a relatively new method devised by Jentek systems called Magneto Winding Magnetometer (MWM). [4]

The conventional probes had no problem detecting LOP of depths 0.065" or greater by direct measurement in FSW involving the 2195 aluminum alloy joined to 2195. However, when dissimilar alloys were joined, such as 2195 to 2219, the signal response gave many false indications and made reliable LOP detection impossible.

Much like the advantages found from other automated techniques, another method that was much easier to automate was the MWM by Jentek.

The magneto winding magnetometer is a multi-element array with 30 elements covering the FSW itself and 7 elements covering the edges of the weld land. The sensors are optimally spaced and provide absolute electrical conductivity of these aluminum alloys. This is due to their unique design over conventional coils. [6]

Conductivity is a physical attribute that has been used to relate the heat condition of an aluminum alloy. In this application, the Jentek sensors were used to map the conductivity across the weld and the surrounding weld land areas. The methodology provided superior measurements of conductivity compared to similar conductivity measurements. Figure 8 shows a C-Scan of the conductivity across an optimized FSW weld in terms of %IACS, along with a profile plot (corresponding to a transverse slice) which shows the conductivity across the weld (characteristic conductivity of the DXZ). Figure 9 shows similar results with a planar flaw such as LOP. Just like conventional eddy current, however, in dissimilar metals, the large differences in parent material conductivity make it more challenging to find planar flaws. More work needs to be done here, but slopes of these transitions, correlate well with the formation of a good weld nugget and one with LOP.

Radiography NDE Method

Conventional film radiography provided sufficient detection of wormholes and voids in general and some LOP and LAF if it was gross (usually around 0.4T). ROD was not detectable in SR-FSW. Even using a slower film such as FUJI 25, many fuzzy "dark" line indications exist on the film from these welds and could cause problems with interpretation, especially on some of the limiting cases of these laminar defects. We have seen borderline critical LAF that many feel could be mistaken by some of these fuzzy lines. What are these lines? Work continues in this area, but many feel that the lines are due to normal grain orientation of the trailing and leading edges. Cutups and macros of similar areas have shown no defects. Because of these confusing lines and the weakness of radiography to detect laminar defects, it is highly recommended to use complimentary techniques such as ultrasound, especially for providing a referee method for any dark line indications.

Ultrasonic NDE Method

Ultrasonics has historically been a complimentary technique for radiography and penetrant in fusion welds. Where radiography is better for the detection of volumetric defects such as porosity, ultrasonics is much better in detecting crack-like, planar defects. During this development, several techniques were investigated. They included the use of conventional shear waves, creeping waves, and phased array.

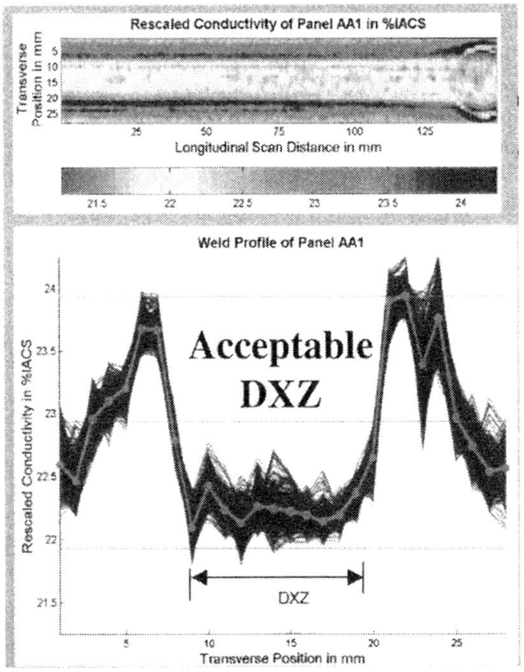

FIGURE 8. Conductivity Measurement (Good weld)

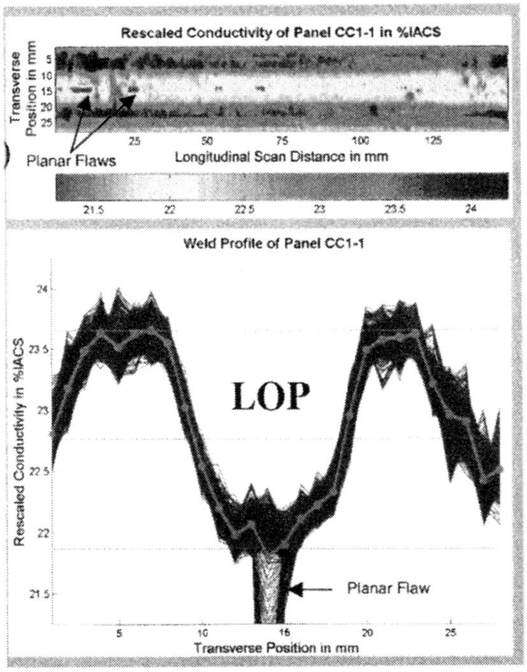

FIGURE 9. LOP conductivity correlation

Piezoelectric sensors were used in most of the tests along with some preliminary work with EMATs (electromagnetic).

In 1998, the ET program worked with The Welding Institute (TWI) in Cambridge, England to develop some of the first NDE standards that simulated some of the known defects at the time such as LOP, wormholes, and other voids. EMATs (Electromagnetic acoustic transducers) developed by Babcock & Wilcox (later MAST with TEMAT), were somewhat successful on some of the larger flaws, but the low signal to noise response on the LOP was disappointing. Since the LOP flaw was the limiting and most challenging defect in the allowables program (until the Residual Oxide Defect was found in the self-reacting welds), this became the target of our comparisons. In conventional, single pin tool/anvil configuration, the LOP of course occurs on the backside of the weld when the penetration is insufficient. There are optimum parameters which are crucial to getting a full penetration weld. The pin tool has to be some amount of distance from the back side during the stirring. This distance is called the ligament. If the ligament is not controlled, the unconsumed joint can range in physical separation from an initial gap (very easily detected) to a very tight, slightly swept to one side. LOP, that has been compressed in a pancake fashion (very difficult to detect). Because of the anvil design and the forging pressures involved, a slight compressional bond on the backside of the weld along the seam occurred which caused the lack of detection by surface methods (without etching and other techniques as was discussed earlier). Therefore it was natural to look at surface guided waves such as creeping waves. We used the 70 degree longitudinal waves and their accompanying creeping waves which are mode converted. The use of the WSY probe from Krautkramer provided excellent detection of the LOP flaw.

When the LOP (inner connected flaw) became long enough (30% thru-thickness) it became possible to use tip diffraction techniques to measure the depth.

Phased array ultrasound not only repeated the success of the WSY probes but gave us an approach that could easily be automated for production. Using the RD Tech Tomoscan system, a 32 element or 64 element array was used at 10 MHz using a linear scanning approach. This linear scanning which enabled the inspection across the entire DXZ, HAZ and weld land was accomplished by establishing focal laws (rules that determined how the individual elements work in conjunction, to direct ultrasonic energy at different transverse and skew angles). The angles studied were 45 and 60 degree. For the LOP flaw, 0.050" lengths were easily detected in friction stir welds using the original pin tool designs and tooling. However, later improvements in the tool design and the optimization of the tooling parameters caused variations in the detectability of these similar length LOP defects due to the increased "tightness" of the flaw. Further improvements and a new production training tool made the flaw even harder to detect. Regardless of tooling options, when the LOP had depth of 0.075" or greater (regardless of the thickness of the weld area, 0.320" to 0.650"), the detectability seemed to be very reliable. Thus, this length of LOP became the target for a limited reliability study with multiple operators and at least 29 LOP defects in both material thicknesses.

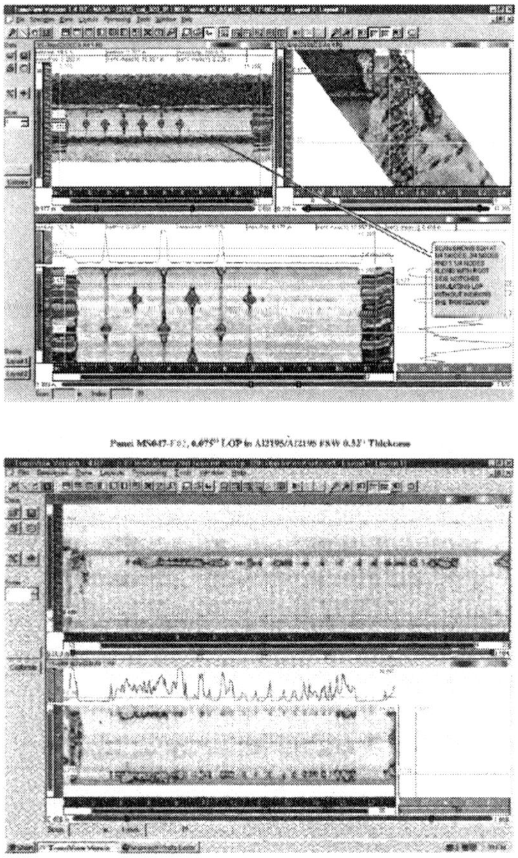

FIGURE 10. Calibration C-Scan and 0.075" LOP

Phased array ultrasound did as well as the other ultrasonic techniques mentioned, but was more easily automated. So this became the down-selected technique for further qualification studies. Figure 10 below show a typical C-Scan setup for the phased array ultrasonic method of a calibration standard and what is 0.050" LOP.

Reliability Study and Qualification of Phased Array Ultrasound

To insure the repeatability of calibration and indexing of the bubbler coupled phased array sensor, studies were performed on the applicability of using side-drilled holes, U-shaped and square shaped, surface EDM notches on the root side, and standards with friction welds and without.

A comparison of aluminum 2195-T8 transmission with a friction stir weld and without, and with transmission thru the leading edge HAZ and trailing edge HAZ was made. There was negligible change in response amplitude. A comparison was made of several types of notches and side drilled holes. Using a 45 and 60 degree angle, 10 MHz, 32 element phased array probe, it was found that although the U-shaped notches gave similar sensitivity to smaller, tighter LOP, that given our assumed reliable detection of around 0.075" depth, a side-drilled hole of 0.020" diameter was sufficient for the sensitivity needed.

Qualifying this phased array method for conventional FSW consisted of training, standardization and a probability of detection study. To facilitate the reliability study, we had five inspectors looking at more than 30 flaws in the range of 0.050" to 0.090" with at least 29 in each thickness at the assumed detection limit of 0.075" thru wall dimension. All flaws were found in the blind tests with no false calls. Though the statistics are based on a lot of assumptions (namely that the process is well characterized), this pseudo POD study did give us more confidence in the process for finding LOP. More extensive work to characterize the reliability of finding these types of flaws (LOP, LAF, ROD, etc) will be proposed as soon as the process development is finalized.

We are currently working with the Welding Institute in Cambridge, England and the Edison Welding Institute in Ohio, to develop more robust techniques.

Conclusions

Friction Stir welding has been adopted by the Space Shuttle program as an alternative to Variable Polarity Plasma Arc (VPPA) because of its superior weld properties as well as its advantages in heat repairs (thermal cracking, chemistry changes and its density effects on x-ray).

Unique NDE techniques led to the reliable detection of a Lack of Penetration (LOP) flaw and a Lack of Inadequate Forging (LAF) flaw that occurs when the process is not operating in its optimum envelope of parameters. Because of concerns of forging pressure uniformity and the resulting LOP problems found in circumferential weld tooling, a self-reacting (SR) FSW was developed by the Marshall Lockheed team. This process led to its own types of defects when operated at below optimum parameters and presented yet another challenge to NDE. With its lower forging pressure requirement, it was believed that a Residual Oxide Defect (ROD) was more prone to be found trapped in the resulting nugget of a SR-FSW. This challenging flaw was found to be detectable with phased array but detection reliability is still being determined.

As in fusion welding, one NDE method or technique is usually not sufficient and thus, multiple, often complimentary techniques, are necessary, as was determined by the Space Shuttle team in the development of two unique friction stir weld processes. Automated phased array ultrasonics, automated magneto winding magnetometers (eddy current) along with conventional film x-ray, modified penetrant techniques and unique visual inspections, provided an effective suite of verification tools for these processes.

Acknowledgments

Acknowledgment and thanks is given to Lockheed Martin personnel at the New Orleans Michoud Facility, managed by NASA's Marshall Space Flight Center, especially David Kinchen and Shan McEvoy. Thanks also go to Jentek Sensors, Inc, for their efforts.

References

1. Friction Stir Welding and AL218 Weld Wire, September 10th-13th, Program meeting VOLs S,D,VSpace, Lockheed Martin/NASA, 2001
2. Report 221061, "Friction Stir Welded Reference Specimens Containing Known Flaws for NDE Detection", TWI, January 1998.
3. Lamarre, Andre and Moles, Michael. "Ultrasound Phased Array Inspection Technology for the Evaluation of Friction Stir Welds". RD Tech, October 2003, TWI Consortium.
4. Goldfine, Neil (Jentek) and Kinchen, David (Lockheed Martin). " Friction Stir Weld Inspection Through Conductivity Imaging using Shaped Field MWM Arrays". Proceeding of the 6th International ASM Trends inWelding Conference, 15-19 April 2002.
5. Kinchen, David and Aldahir, Esma, Lockheed Martin. "NDE of Friction Stir Welds in Aerospace Applications". Inspection Trends, May 27th, 2002.
6. Zilberstein, Walrath, Grundy, Schlicker, Gooldfine (Jentek) and Yentzer (Air Force). "MWM Eddy Current Arrays for Crack Initiation and Growth Monitoring". International Journal of Fatigue, June 2003.

Proceedings from Joining of Advanced and Specialty Materials, 13–15 October 2003, Pittsburgh, Pennsylvania, ASM International, 2004

Ultrasonic Phased Arrays for Thick Section Weld Inspections

Michael Moles, Noël Dubé and Frédéric Jacques

Michael Moles
R/D Tech Toronto

Noël Dubé,
R/D Tech Québec

Frédéric Jacques
R/D Tech Québec

Abstract

Major improvements in weld inspection are obtained using Phased Array technology with capability for beam steering, electronic scanning, focusing and sweeping the ultrasonic beams. Electronic scanning is much faster than raster scanning, and can optimize angles and focusing to maximize defect detection. Inspections typically use "top, side, end" or "top, side, TOFD" views, though other imaging is possible. Special inspections can be performed, e.g. for specific defects, or increased coverage. Defects can be sized by pulse-echo as per code, by TOFD or by back diffraction.

The presentation will describe the application of phased arrays for different types of inspections in pressure vessels. The two main types of inspections are zone discrimination developed for pipeline AUT and ASME-type raster scans. The advantages of using phased arrays are explained.

Introduction:

Ultrasonic phased arrays are a novel method of generating and receiving ultrasound. They use multiple ultrasonic elements and electronic time delays to create beams by constructive and destructive interference. As such, phased arrays offer significant technical advantages for weld inspections over conventional ultrasonics. The phased array beams can be steered, scanned, swept and focused electronically. Beam steering permits the selected beam angles to be optimized ultrasonically by orienting them perpendicular to the predicted defects, for example Lack of Fusion in automated welds. Electronic (linear) scanning permits very rapid coverage of the components, typically an order of magnitude faster than a single transducer mechanical system. Beam steering (usually called sectorial or azimuthal scanning) can be used for mapping components at appropriate angles to optimize Probability of Detection of defects. Sectorial scanning is also useful for inspections where only a minimal footprint is possible. Electronic focusing permits optimizing the beam shape and size at the expected defect location, as well as optimizing Probability of Detection. Overall, the use of phased arrays permits optimizing defect detection while minimizing inspection time.

Ultrasonic phased arrays

How phased arrays work

Ultrasonic phased arrays are similar in principle to phased array radar, sonar and other wave physics applications. However, ultrasonic development is behind the other applications due to a smaller market, shorter wavelengths, mode conversions and more complex components. Several authors have reviewed applications of ultrasonic phased arrays (Clay et al, 1999, Wustenberg et al., 1999, Lafontaine and Cancre, 2000), though industrial uses have been limited until the last few years.

From a practical viewpoint, ultrasonic phased arrays are merely a method of generating and receiving ultrasound; once the ultrasound is in the material, it is independent of generation method, whether generated by piezoelectric, electromagnetic, laser or phased arrays. Consequently, many of the details of ultrasonic inspection remain unchanged; for example, if 7.5 MHz is the optimum inspection frequency with conventional ultrasonics, then phased arrays would typically use the same frequency, aperture size, focal length, and incident angle.

Phased arrays use an array of elements, all individually wired, pulsed and time-shifted. These elements are typically pulsed in groups from 4-32, usually16 elements for welds. In order to make a user-friendly system, a typical set-up calculates the time-delays from operator-input, or uses a pre-defined file: inspection angle, focal distance, scan pattern etc (see Figure 1).

The time delay values are back calculated using time-of-flight from the focal spot, and the scan assembled from individual "Focal Laws". Time delay circuits must be accurate to around 2 nanoseconds to provide the accuracy required.

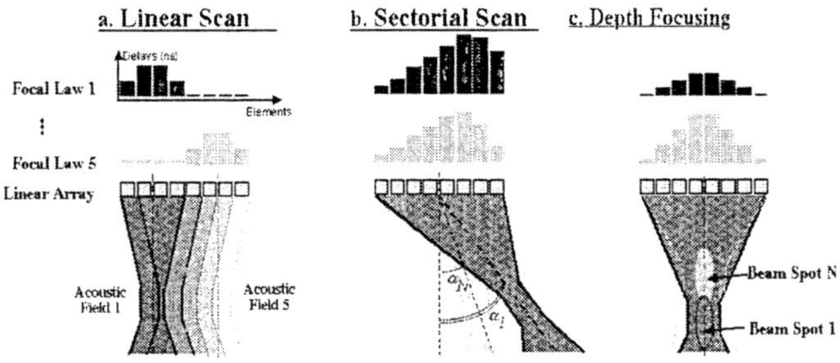

Figure 1: Schematic showing generation of linear and sectorial scans using phased arrays.

While it can be time-consuming to prepare the first set-up, the information is recorded in a file and only takes seconds to re-load. Also, modifying a prepared set-up is quick in comparison with physically adjusting conventional transducers.

Types of scans

Using electronic pulsing and receiving provides significant opportunities for a variety of scan patterns.

Electronic (Linear) Scans

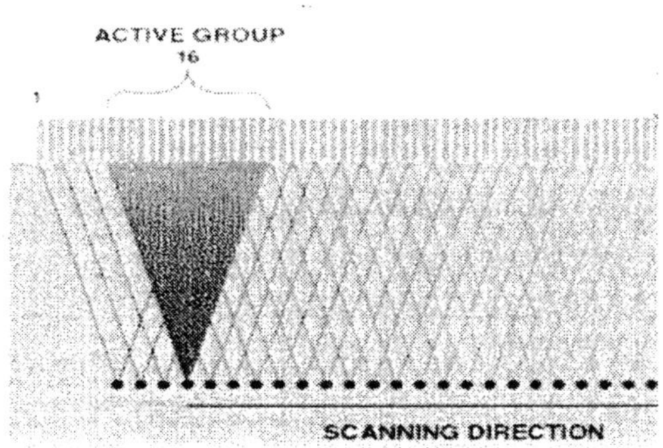

Figure 2: Schematic illustration of electronic (linear) scanning.

Linear scans are performed by multiplexing along an array (see Figure 2). Typical arrays have up to 128 elements, pulsed in groups of 8 to 16. Linear scanning permits rapid coverage with a tight focal spot. If the array is flat and linear, then the scan pattern is a simple B-scan. If the array is curved, then the scan pattern will be curved. Linear scans are straightforward to program. For example, a phased array can be readily programmed to inspect a weld using 45° and 60° shear waves, which emulate conventional manual inspections.

Sectorial (Azimuthal) Scans

Sectorial scans use the same set of elements, but alter the time delays to sweep the beam through a series of angles (see Figure 3). Again, this is a straightforward scan to program. Applications for sectorial scanning typically involve a stationary array, sweeping across a relatively inaccessible component like a turbine blade root (Ciorau et al, 2000), to map out the features (and defects). Depending primarily on the array frequency and element spacing, the sweep angles can vary from ± 20° up to ± 80°.

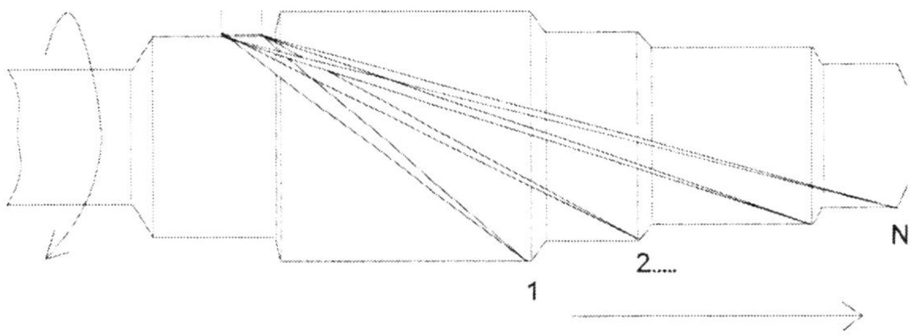

Figure 3: Schematic showing sectorial scanning used on turbine rotor.

Combined Scans

Combining linear scanning, sectorial scanning and precision focusing leads to a practical combination of displays (see Figure 4). Optimum angles can be selected for welds and other components, while scanning permits fast and functional inspections. For example, combining linear and sectorial scanning permits full ultrasonic inspection of components over a given angle range, e.g. ± 20°. This type of inspection is useful when simple normal beam inspections are inadequate, e.g. for titanium castings in aerospace where defects can have random orientations. A related approach applies to weld inspections, where specific angles are often required for given weld geometries; for these applications, specific beam angles are programmed for specific weld facets at specific locations.

Figure 4: top, ultrasonic scanning pattern using sectorial and linear scanning; bottom, ultrasonic image using all data merged together.

Linear scanning of welds

Manual ultrasonic weld inspections are performed using a single transducer, which the operator "scans" back and forth to cover the weld area (see Figure 5a). Many automated weld inspection systems use a similar approach, with a single transducer scanned back and forth over the weld area. This is very time consuming, since the system has dead zones at the start and finish of the raster.

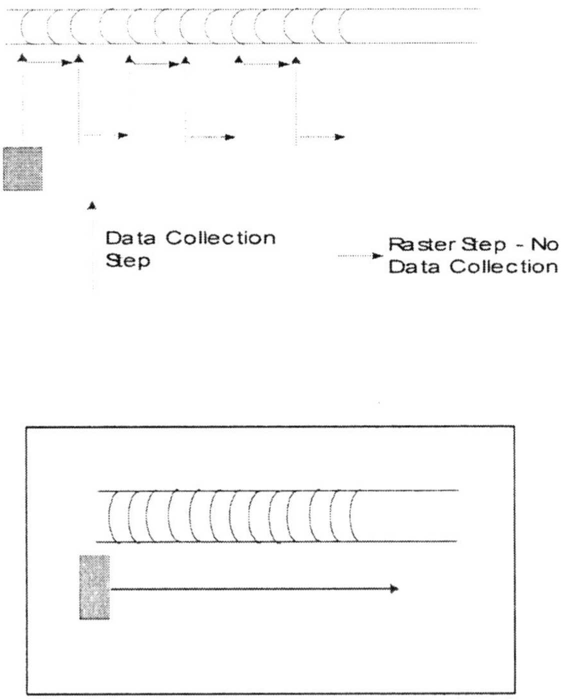

Figure 5a (top). Conventional raster scanning; 5b (bottom). Linear scanning.

In contrast, many multi-transducer systems and phased arrays use a "linear scanning" approach (see Figure 5b). Here the probe pan is scanned linearly round or along the weld, while each transducer sweeps out a specific area of the weld. The simplest approach to linear scanning is found in pipe mills, where a limited number of transducers inspect ERW pipe welds. AUT as practiced on pipelines uses up to twenty-four transducers, which makes pipeline AUT much faster than single transducer techniques.

Phased arrays for linear weld inspections operate on the same principle as the multi-transducer approach; however, phased arrays offer considerably greater flexibility than conventional AUT. Typically, it is much easier to change the set-up electronically, either by modifying the set-up or reloading another; often it is possible to use many more beams (equivalent to conventional transducers) with phased arrays; special inspections can be implemented simply by loading a set-up file.

Typical Applications

Realistically, there is no "typical application" for phased arrays. Ultrasonic phased arrays are being installed in a wide variety of industries, where the technology has inherent advantages. These industries include: aerospace, nuclear power, steel mills, pipe mills, petrochemical plants, pipeline construction, plus a selection of special applications. All these applications take advantage of one or more of the dominant features of PAs:

- **Speed**: scanning with phased arrays is an order of magnitude faster than single transducer conventional mechanical systems, with better coverage;
- **Flexibility**: set-ups can be changed in a few minutes, and typically a lot more component dimensional flexibility is available;
- **Inspection angles**: a wide variety of inspection angles can be used, depending on the requirements and the array;
- **Small footprint**: small matrix arrays can give infinitely more flexibility for inspecting restricted areas than conventional transducers.

Each feature generates its own applications. For example, "speed" is important for pipe mills and pipelines, plus some high volume applications. "Flexibility" is important in pressure vessels and pipeline welds due to geometric changes. "Inspection angles" is key for pipelines, some pressure vessel and nuclear applications. "Small footprint" is applicable to some pressure vessels, and possibly for pipelines to minimize coating cutback.

Phased arrays are now commercially competitive with multi-transducer systems. However, the technology is new and still requires some set-up efforts, especially for complex 3D applications. 2D pipeline set-ups are generally straightforward.

Applications on Welds

Zone Discrimination Approach

One technique which provides 100% coverage and rapid inspections is "zone discrimination". This applies primarily to narrow gap welds and pipelines. Gas pipelines have major inspection requirements: high speed, high quality, high detection capability (Gross et al., 1990), rapid data interpretation. Also, the use of Engineering Critical Assessment (Fracture Mechanics) makes major demands on sizing capability. Radiography is rapidly being replaced by ultrasonics worldwide because Automated UT can provide these capabilities, plus can act as process control. This typically reduces the reject rate significantly (Connelly, 1999), and saves substantial costs.

AUT for pipelines uses four special features: zone discrimination, calibration blocks, dual gate output display and defect sizing. Figure 6 shows the concept of zone discrimination with linear scanning, and Figure 7 a typical code-approved calibration block.

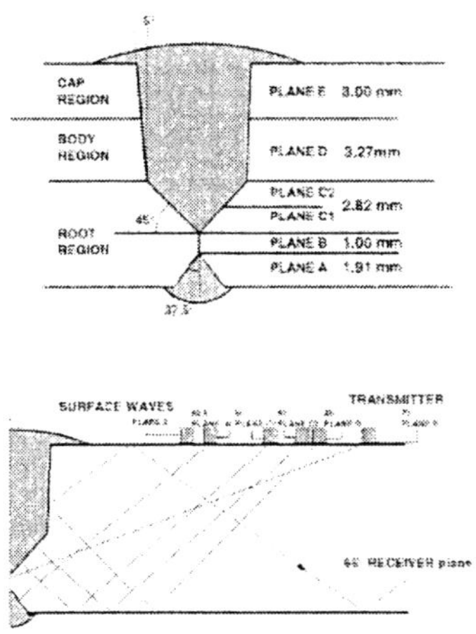

Figure 6: Schematic of zone discrimination. Top: selection of zones. Bottom: position and angles of transducers for zone discrimination inspection.

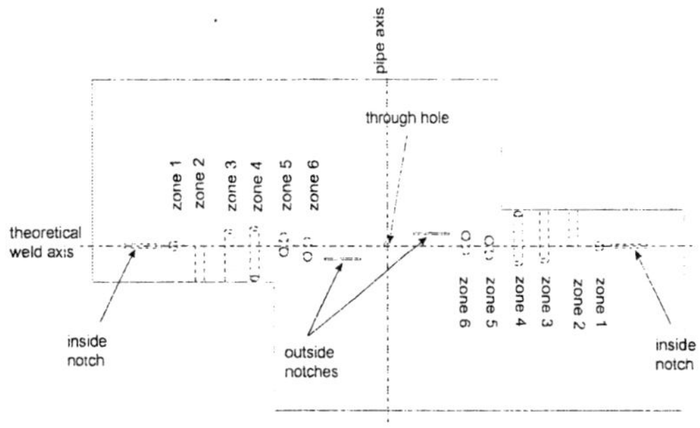

Figure 7: typical AUT calibration block, as per code ASTM E-1961.

Phased arrays over significant advantages over conventional AUT:

- Smaller and lighter probe pans;
- Faster inspections;
- Significant flexibility in pipe diameters, wall thicknesses and weld bevel geometries;
- Rapid set-ups,
- Improved sizing and imaging, plus other advantages described below.

Figure 8 below shows a typical AUT strip chart display, with dual gates for each zone (amplitude and Time-Of-Flight), plus TOFD, B-scans maps for porosity, coupling charts and position round the weld.

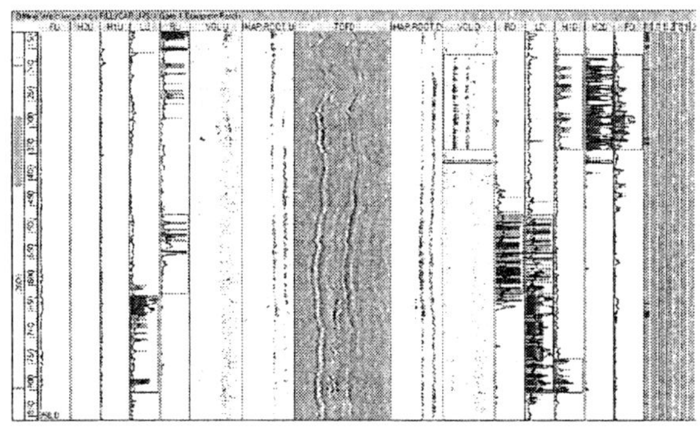

Figure 8: typical AUT display, with dual gate strip charts, mapping channels, TOFD, position and coupling. This display shows multiple boxed defects in red.

Defect sizing is performed rapidly by counting the number of zones that the defect is detected on. The TOF signals change to red once over above-threshold defects are detected (usually 40% screen height). TOFD (Time-Of-Flight Diffraction) is used to aid sizing, and also to confirm the presence (or absence) of a defect. There are now several codes for pipeline AUT weld inspections. ASTM E-1961-98 (ASTM, 1998), specifically for pipeline girth welds and based on the TCPL/NOVA specifications, was published in 1998. API 1104 19th Edition was published in 1999 (API, 1999). This code is very specific on zone discrimination, displays and calibration, whereas the API code is a lot more flexible. The optimum solution is to combine the two codes. There is also a DNV code (OS-F101) for offshore use.

ASME Raster Scans on Thick Section Welds

The petrochemical industry is an extensive user of thick section pressure vessels, almost all of which are welded. Most are inspected using radiography, which has limitations on the type of defects it can detect. With the arrival of ASME Code Case 2235 (ASME 2001), ultrasonics is now an acceptable alternative. Besides the new zone discrimination approach above, thick section welds can be inspected using traditional 45° and 70° shear waves.

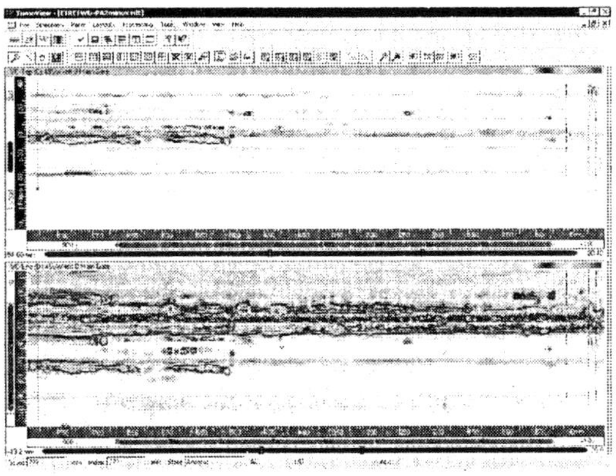

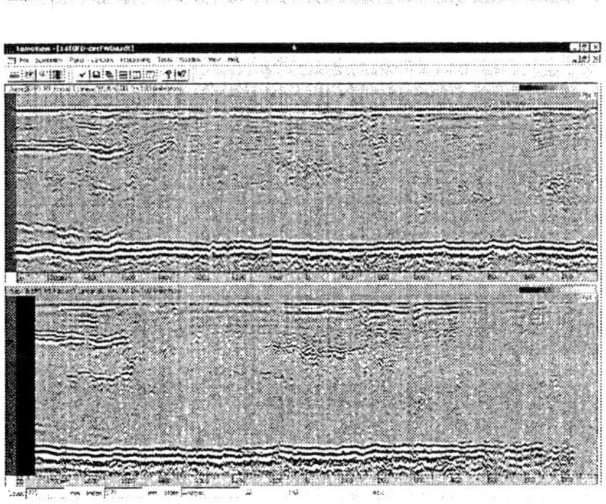

Figure 9 above shows a typical "top, side" view, combined with a twin TOFD view below. The "top, side" view is made by merging both the 45 and 60° data.

Different views are possible, e.g. "top, side, end" view, unmerged data, sectorial scans or S-scans, A-scans. Thick section welds can be inspected using a number of delivery systems, e.g. magnetic wheel scanners, vessel rotation, custom scanners. Figure 10 below shows a pressure vessel inspected using rotation. The advantages of phased arrays are programmable scans, data merging, small footprint, great flexibility, and high speed.

Figure 10. PV inspection using phased arrays.

TOFD

Time-Of-Flight Diffraction (TOFD) is a relatively recent arrival in the petrochemical business, and so far has primarily been used in pipelines. This technique uses the same arrays as pulse-echo, or can use dedicated TOFD transducers at higher frequency. TOFD essentially gives a through wall image of the weld, with OD, ID and any defects. TOFD detection capability is very good, though there are dead zones at the outside and inside surface. TOFD is an excellent sizing technique, with significantly better sizing than amplitude approaches. Figure 11 below shows a typical TOFD image of a weld.

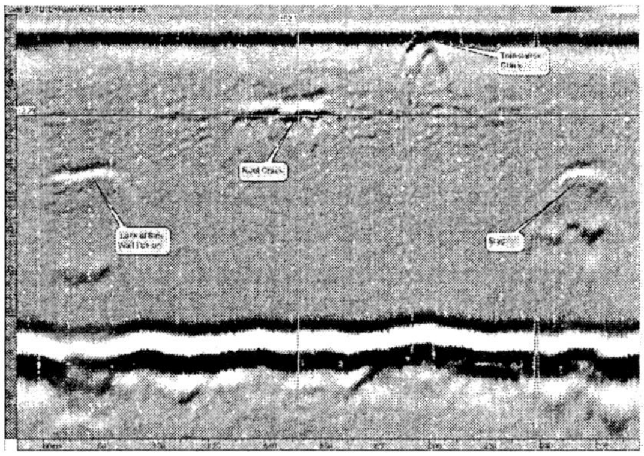

Figure 11: Typical TOFD image showing OD, ID and top and bottom of defects.

R/D Tech strongly recommends using both pulse-echo and TOFD for improved defect detection and sizing. Back diffraction offers some sizing advantages over TOFD (or forward scattering): smaller dead zones and better sizing for small defects.

Conclusions

Ultrasonic phased arrays have several advantages over conventional AUT and radiography for pressure vessel weld inspections.

1. In general, phased arrays:
    a. provide faster inspections;
    b. are much more flexible, both in set-ups and in component geometries;
    c. can optimize Probability of Detection by tailoring inspection angles and focal spot size; and
    d. have a smaller footprint.
2. For zone discrimination scans, probe pans are smaller, lighter, faster, can scan almost any diameter and wall thickness, can provide improved imaging and special set-ups, and use automated set-ups for convenience.
3. For ASME raster inspections of thick-section welds, phased arrays offer flexibility, imaging, and speed.
4. In all cases, TOFD is strongly recommended as a "no cost" addition.

References

API, American Petroleum Institute Standard 1104, "Welding of Pipelines and Related Facilities", Nineteenth Edition, September 1999.

ASME, Code Case 2235-4, "Use of Ultrasonic Examination in Lieu of Radiography, Section I and VIII, Divisions 1 and 2", November 30, 2001, American Society of Mechanical Engineers.

ASTM, American Society for Testing and Materials, ASTM E-1961-98, "Standard Practice for Mechanized Ultrasonic Examination of Girth Welds Using Zonal Discrimination with Focused Search Units", September 1998.

Ciorau P., D. MacGillivray, T. Hazelton, L.Gilham, D. Craig and J.Poguet, "In-situ examination of ABB l-0 blade roots and rotor steeple of low-pressure steam turbine, using phased array technology", 15th World Conference on NDT, Rome, Italy, October 11-15, 2000.

Clay A.C., S-C. Wooh, L. Azar and J-Y. Wang, "Experimental Study of Phased Array Beam Characteristics", Journal of NDE, vol 18, no. 2, June 1999, page 59.

Connelly T., "Update on the Alliance Pipeline", International Conference on Advances in Welding Technology, October 26-28, 1999, Galveston, Texas, sponsored by EWI and AWS.

Dubé, N., "Electric Resistance Welding Inspection", 15th WCNDT, Rome, Italy, October 2000.

Gross B., J. O'Beirne and B. Delanty, "Comparison of Radiographic and Ultrasonic Inspection Methods on Mechanized Girth Welds", Pipeline Technology Conference, 15-17 October, 1990, Ostend, Belgium.

Lafontaine G. and F. Cancre, "Potential of Ultrasonic Phased Arrays for Faster, Better and Cheaper Inspections", NDT.net, vol 5, no. 10, October 2000 http://www.ndt.net/article/v05n10/lafont2/lafont2.htm

Moles M.D.C, E.A. Ginzel and N. Dubé, "PipeWIZARD-PA – Mechanized Inspection of

Girth Welds Using Ultrasonic Phased Arrays", International Conference on Advances in Welding Technology, October 26-28, 1999, Galveston, Texas, sponsored by EWI and AWS.

Wüstenberg H, A. Erhard and G. Shenk, "Some Charateristic Parameters of Ultrasonic Phased Array Probes and Equipments", NDT.net, vol 4, no. 4, http://www.ndt.net/article/v04n04/wuesten/wsuesten.htm

Proceedings from Joining of Advanced and Specialty Materials, 13–15 October 2003, Pittsburgh, Pennsylvania, ASM International, 2004

Effect of Purging Gas Oxygen Levels on Structure and Mechanical Properties of GTA Welded Type 304 Stainless Sanitary Tube

Leijun Li
Mechanical & Aerospace Engineering, Utah State University, Logan, Utah

Timothy Davis
AWD Sales & Service, Inc., Payson, Utah

Abstract

Oxygen contamination in argon-purge gas used in welding stainless steels causes what is referred to in sanitary industry as sugar or dross, an oxide layer on the root surface of the weld bead that is rough, pitted, and porous. This oxide layer can trap product that leads to contamination and weaken the mechanical properties and corrosion resistance of the joints.

This study consisted of GTA welding square groove butt joint of Type304 stainless steel tube. Argon with a varying percentage of oxygen content was used to purge the inside surface of the tube. Weld discoloration comparisons, surface analysis, and tensile tests of the weld joints were performed.

The result shows that to prevent excessive oxidation and maintain mechanical properties of the welded joints, the argon purge should contain less than 0.1% by volume of oxygen.

Introduction

Sanitary processing systems are those that handle products for human and animal consumption. Examples of these systems include pharmaceutical, dairy, meat, beverage, fruit, vegetable and various other products used by humans and animals. In the industry, austenitic stainless steel tubes are used to transport product between the processors and from the manufacturer to product containers. This tubing has been electro-polished both on the inside diameter and the outside diameter to create a smooth surface which will prevent product build up and contamination. Contamination can lead to the formation of bacteria in the product.

In welding austenitic stainless steels, the oxidation due to poor purge may lead to a defect of very uneven surface finish for the weld fusion and heat-affected zones generally known as sugar or dross. This oxide layer is rough and porous and has the potential to trap product. Stringent requirement of argon purge is usually required during welding to guarantee quality surface finish of the weldment.

In the AWS D18.1 code for welding sanitary stainless steel tube, visual inspection and welder qualification have been addressed. Literature is available on the effect of heat tint on corrosion resistance of the joints [1]. More studies have been done on the effects of chemical compositions of the base metal on oxidation behavior of the alloys undergoing different thermal histories [2-5]. The effect of oxidation on impact toughness of the weld metal has been discussed by a number of studies [6-8].

However, detailed information about the heat tint or dross as influenced by quality of purge is not generally available, and there is little information on the mechanical properties of welded joints as influenced by formation of such dross. By simulating the processes and procedures used in industry on job sites, the aim of this study was to determine the critical point where the oxygen content in the purge gas may lead to dross formation, and the effect of dross on strength and ductility of the welded tube joints.

Experimental Procedures

Samples of Type 304 sanitary stainless tube with an outer diameter of 101mm (4 in.) and a wall thickness of 2.1mm (0.083 in.) were used for the test pieces. The chemical compositions for the test material are the following: 0.019% C, 1.790% Mn, 0.026% P, 0.015% S, 0.40% Si, 18.39% Cr, 8.17% Ni, 0.18% Mo, 0.17% Cu, 0.090% Co, 0.056% N, and balance Fe. The tube was saw-cut to 228 mm (9 in.) long sections, with all burs and sharp edges removed. Each section was cleaned in detergent then dipped in acetone before welding. All brushes, files, and miscellaneous equipment were used only on the test stainless steel to avoid any contamination. This procedure was done in conformance with ANSI/AWS D18.1, and the heat tint was compared with the "weld discoloration level chart" of ANSI/AWS D18.2.

The test pieces were assembled to form a square-groove butt joint, and were autogenously welded using a Miller Maxstar 140 GTA welder. Pure argon and oxygen were mixed in a gas diffuser to create the purge gas that was fed to one end of the tube workpiece. A masking tape was used to seal the ends of the tube. An oxygen meter was connected to the other end of the tube through which the purge gas exhausts to the atmosphere. The oxygen content in the argon purge was varied at the following levels as measured by the oxygen meter: 20, 10, 5, 2.5, 1, 0.5, 0.1, and 0.005% by volume. The overall experimental setup is schematically shown in Figure 1. The tube was welded in the 5G position with a vertical up progression. The welding current was 60 A, with a travel speed controlled to about 90 mm/min (3.5 inch per minute), and a tungsten stick-out of 6 mm (1/4 in.). "Walking the cup", a technique that utilizes resting the nozzle of the GTA torch on the solidified weld bead and rocking it slightly, was used for forward progression around the tube while welding. This procedure was chosen because it is the standard for welding tube in industry. The 5G weld position was chosen because it also is the most frequently welded position.

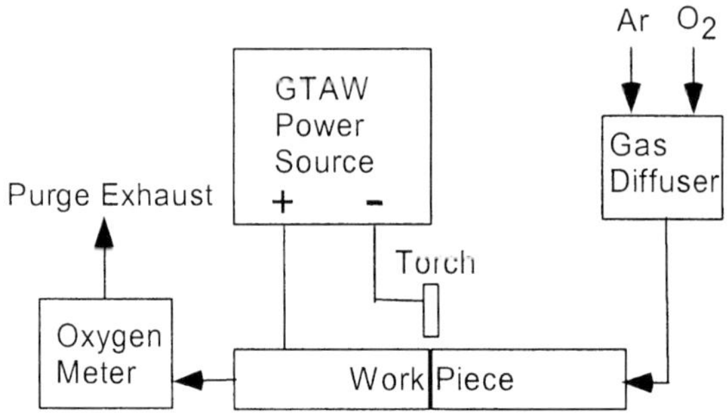

Figure 1. Experimental setup.

After the welding was completed, four 25x304 mm (1x12 in.) tensile test coupons were machined from each of the welded tube samples according to the schematic shown in Figure 2E. The tensile test coupons were prepared and tested according to ANSI/AWS B4.0-98 "Standard Methods for Mechanical Testing of Welds". The results from the four test coupons were averaged for each welded tube sample. Metallography samples were prepared for microstructural analysis. A chemical etchant consisting of 15 ml hydrochloric acid, 10 ml nitric acid, and 10 ml acetic acid was used to prepare the surface for metallographic observation. An S-4000 Hitachi Scanning Electron Microscope was used to analyze the surface morphology and to qualitatively determine the surface chemistry by EDS.

Results and Discussion

The effect of oxygen content in the purge argon on the inside surface of welded tube is clearly seen in Figure 2. As the oxygen content increased from 0.005% to 20%, the degree of oxidation in the weld zone showed a drastic increase. If the argon purge gas had less than 0.1% oxygen by volume, the weld appeared bright and silver with minimal heat tint on the weld bead and the heat effected zone (Figures 2A and 2B). When compared to the weld discoloration level chart of AWS D18.2, the heat tint shown in Figures 2A and 2B correlated nicely to the chart.

If the oxygen content in the purge gas is greater than 0.5% by volume, a dark oxide layer (dross) on the weld fusion zone and heat tint on the weld heat-affected zone can be seen on the inside diameter surface (Figure 3C). The formation of dross on the surface of the welds continued to aggravate with increased oxygen contents until the formation of rough and porous dross on the weld with 20% oxygen purge (Figures 2D to 2F).

When observed under the SEM, the weld root surface morphology showed more detailed features. For the weld made with 0.005% oxygen in the purge, the weld root surface is smooth with weld ripples and dendritic relief clearly seen, and without any dross formation (Figure 3A). For the weld made with 0.1% oxygen in the purge, the weld root surface showed similar features, with weld ripples and dendritic relief clearly seen, and without any dross formation (Figure 3B). However, there were signs of oxide scale formation in the HAZ, which also showed as a slight heat tint in Figure 2B.

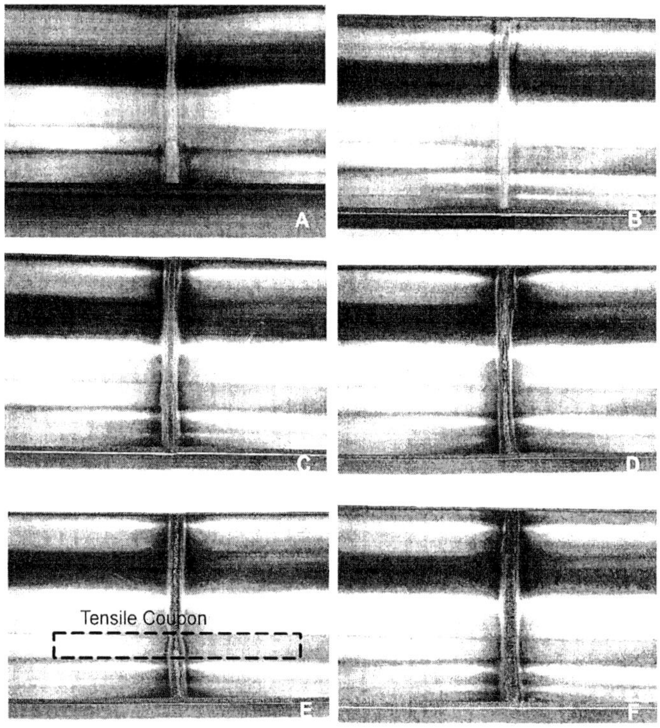

Figure 2. Heat tint formed on the inside surfaces of the tube joints welded with various oxygen contents in the purge argon gas: (A) 0.005% O_2, (B) 0.1% O_2, (C) 0.5% O_2, (D) 1.0% O_2, (E) 2.5% O_2, and (F) 20.0% O_2.

As the oxygen content in the purge increased, there has been a clear morphology change. Figure 3C shows the weld root surface welded with 0.5% oxygen in the purge. The weld ripples and dendritic relief in the fusion zone had disappeared; the fusion zone has a rough surface, with clear indication of dross formation. The heat-affected zone was also seen to be covered with an oxide scale.

The dross forms excessively in the presence of 20% oxygen in the purge (Figure 3D). The "dross" is consisted of out-of-plane dendrites covered with an oxide scale. The severe oxidation of the root surface must have drastically changed the surface chemistry and decreased the surface tension of the solidifying weld pool to promote such dendritic growth. More evidence on the weld fusion zone shape changes will be provided later. Obviously such rough dendrites on the weld root surface could become ideal traps for product being processed in the sanitary equipment.

(A)

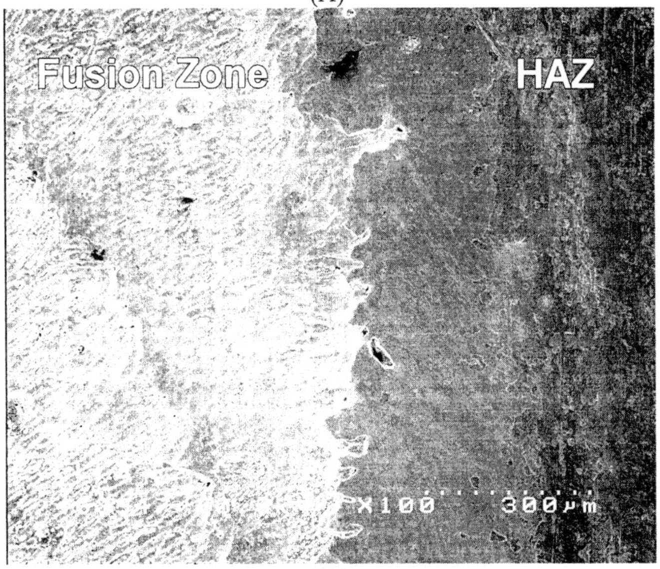

(B)

(C)

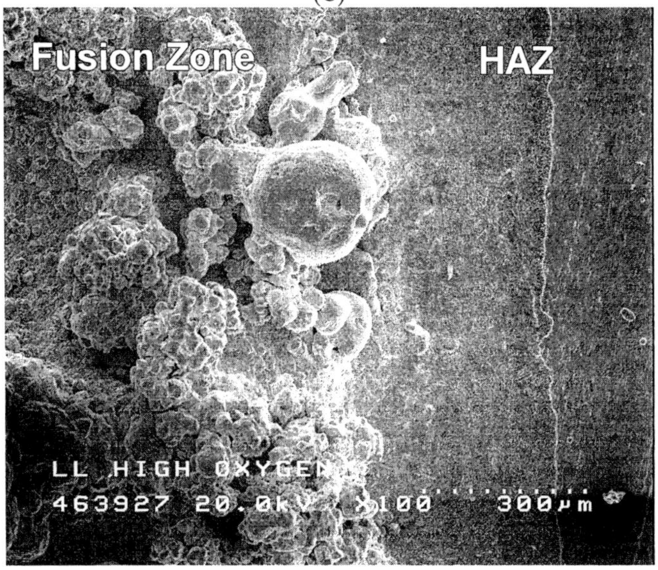

(D)

Figure 3. SEM secondary-electron images of as welded tube inside surface. (A) was welded with a purge containing 0.005% O_2, (B) with a purge containing 0.1% O_2, (C) with a purge containing 0.5% O_2, and (D) with a purge containing 20% O_2.

The heat tint area in the HAZ of the weld root surface was analyzed in more detail under the SEM. Figure 4 shows the comparison of the sample welded with a purge containing 0.1% O_2 and the sample welded with a purge containing 0.5% O_2. The original oxide scale of the base metal surface was partly stripped by the welding arc, exposing the matrix, which shows some degree of oxidation. The sample with more oxygen (Figure 4B) in the purge shows much more severe degree of oxidation not only on the exposed matrix, but also on the original oxide scale. The central areas in Figure 4 are shown in a higher magnification (Figure 5) to reveal the sizes of oxidation product on the sample surface. In Figure 5A, oxide particles approximately 0.3 micro-meter in diameter have formed on the surface. The oxidation seems to be in its early stage when there are areas in the matrix that are not covered by the growing particles.

For the sample welded with more oxygen in purge (Figure 5B), the entire surface has been covered with particular oxidation products. These particles are about 1 micro-meter in diameter and have piled up to more than 3 layers. Oxidation and scale development on austenitic stainless steels was believed to be occurring at the scale-metal interface, through the formation of Cr, Fe and O rich hematite, magnetite, and spinel [9].

A qualitative chemical analysis of the heat tint by EDS shows the oxide particles on the surfaces are enriched with Cr. Figure 6A is for the sample welded with a purge containing 0.1% O_2, and Figure 6B is from the sample welded with a purge containing 0.5% O_2. For the lower oxygen sample (Figure 6A), the Fe peaks from the matrix are the strongest, with strong Cr peaks and an indication of an oxygen peak. This result supports the direct observations made from Figure 5 that the oxidation is still in its early stage. The higher oxygen sample, in contrast, shows dominant Cr peaks and a strong O peak (Figure 6B). Since both samples were welded using the same procedures and welding parameters (therefore, the same thermal history), the difference in degree of oxidation should be attributed to the differences in oxygen contents in the purge argon.

(A)

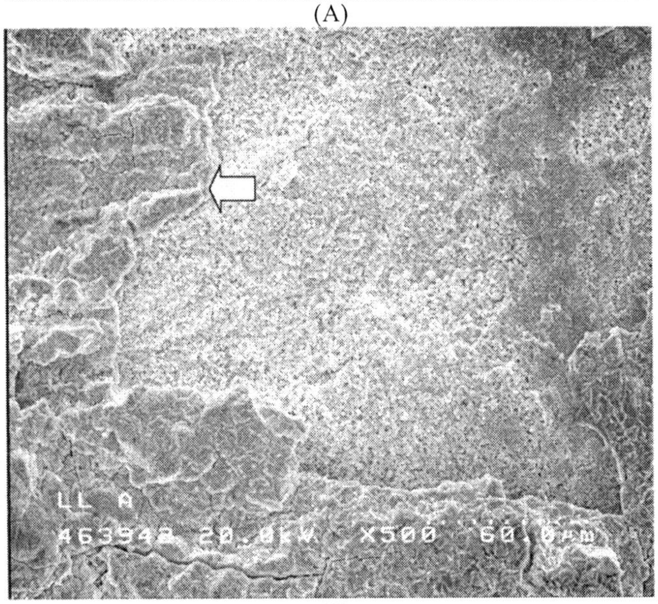

(B)

Figure 4. The heat tint area in the heat-affected zone of the weld root surface of (A) sample welded with a purge containing 0.1% O_2, and (B) 0.5% O_2. The arrows point to the remnants of the original oxide scale on the base metal.

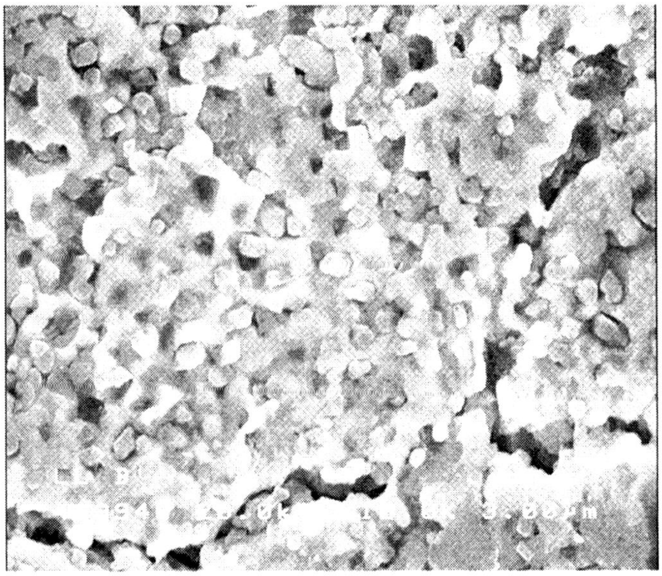

(A)

(B)

Figure 5 Higher magnification of Figure 4. (A) from the sample welded with a purge containing 0.1% O_2, and (B) with 0.5% O_2.

The macroscopic views of the cross-sections of welded joints are shown in Figure 7. Both the OD and ID surfaces of the weld fusion zones are smooth for the lower oxygen samples. However, starting with the samples welded with 1.0% oxygen in the purge, the weld bead becomes narrower, with more weld reinforcement on either the ID or OD. There has been documented evidence to show the effect of oxygen on decreasing the surface tension and viscosity of austenitic steels [10]. It seems plausible to contribute the narrower weld bead to the effect of decreased surface tension, which may have led to the increased reinforcement, all under the condition of the same weld pool volume from the same heat input.

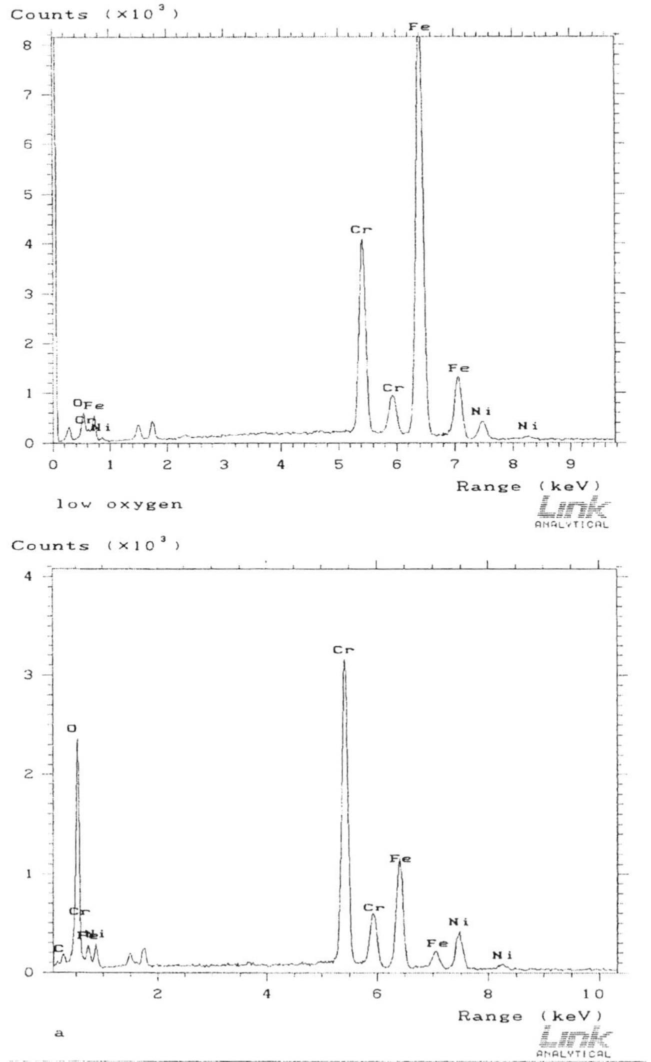

Figure 6. EDS analysis of oxide layers shown in Figure 5. (A) is from the sample welded with a purge containing 0.1% O_2, and (B) with a purge containing 0.5% O_2.

Strength of the joints as a function of the oxygen contents in the argon purge is shown in Figure 8. No significant variations in strength can be seen among the samples welded with different oxygen-containing purge, although the strength of all the joints is lower than that of the base metal by approximately 15%. However, the ductility of the joints clearly showed a significant decreasing trend as the oxygen content in the argon purge increases (Figure 9). The more ductile joints were made with less than 0.1% oxygen in the purge. More than 25% of reduction in ductility can be seen from the joint made with 0.1% oxygen to the joint made with 5% oxygen.

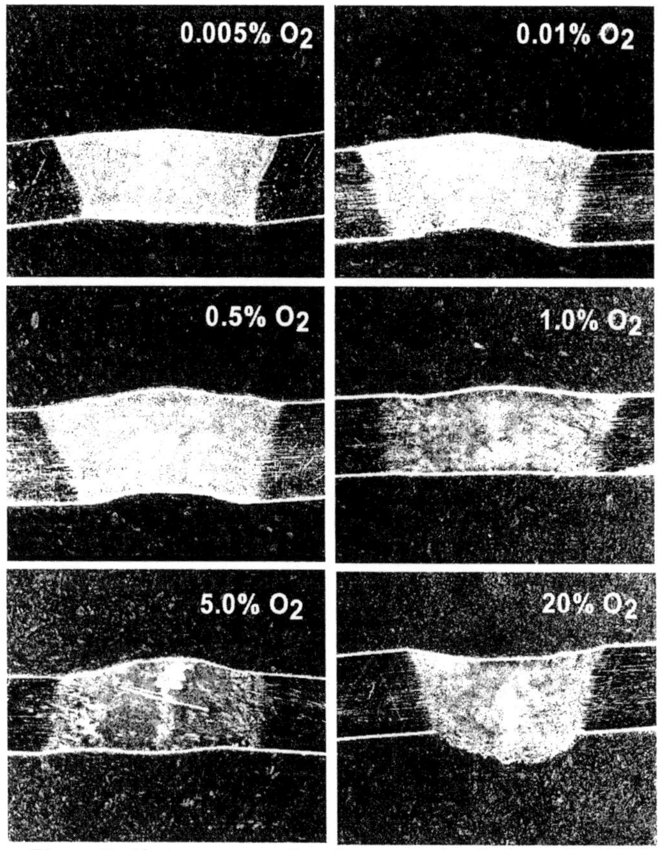

Figure 7. Photo macrographs of cross-sectional view of welded joints with different oxygen contents in the purge. All photos are oriented so that the OD surface is on top. 8X.

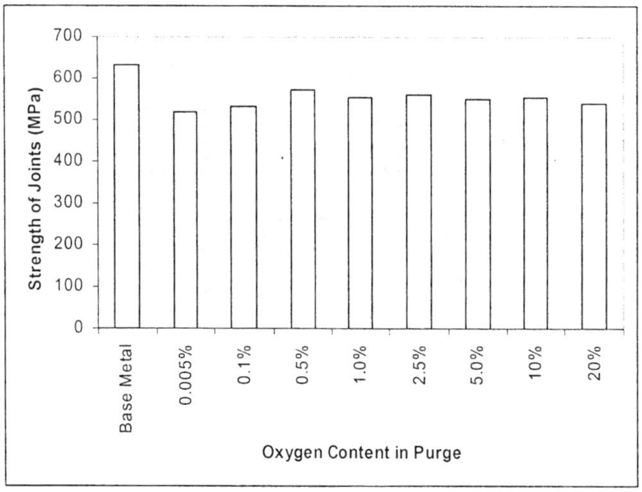

Figure 8. Strength of the joints as a function of oxygen content in the Ar purge.

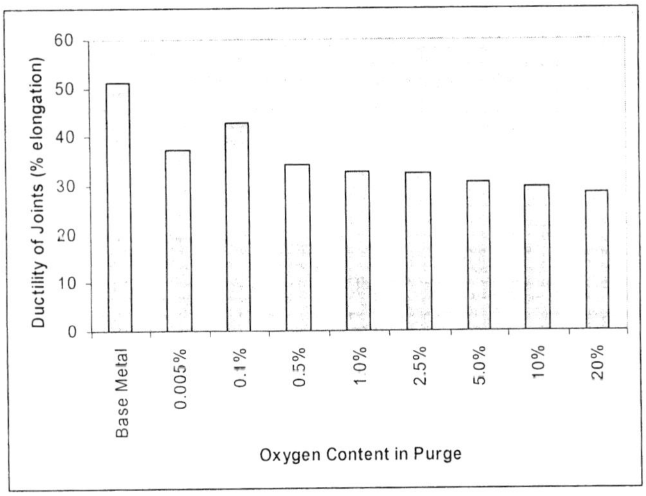

Figure 9. Ductility of the joints as a function of oxygen content in the Ar purge.

The decrease in weld joint ductility can be explained by examining the weld microstructure. From the microstructure of 0.005% oxygen by volume (Figure 10A), it can be seen that the fusion zone is consisted of a uniform structure of darker ferrite on austenitic matrix. The weld made with higher oxygen content in the purge has many inclusions that are distributed dispersedly (Figure 10B). The negative effect of inclusions on the ductility of weld metals is well known. The notch toughness of the welds was also found to be decreased by increasing amount of oxide inclusions [8].

Conclusions

The amount of oxygen present in the purge gas affects the heat tint and can lead to formation of dross, the worst case of which is identified to be the surface dendritic structure covered with oxides. The tensile strength of the weld joints appears to be not affected by the oxygen content in the purge. However, the ductility of the welded joints was negatively affected if the oxygen in the purge is more than 0.1%. Under the conditions investigated in this study, unacceptable oxidation at the weld root surface can be avoided by using an argon purge containing less than 0.1 vol. % oxygen. For control of both mechanical properties and formation of dross, it is recommended to limit the oxygen content in the purge to no more than 0.1% by volume.

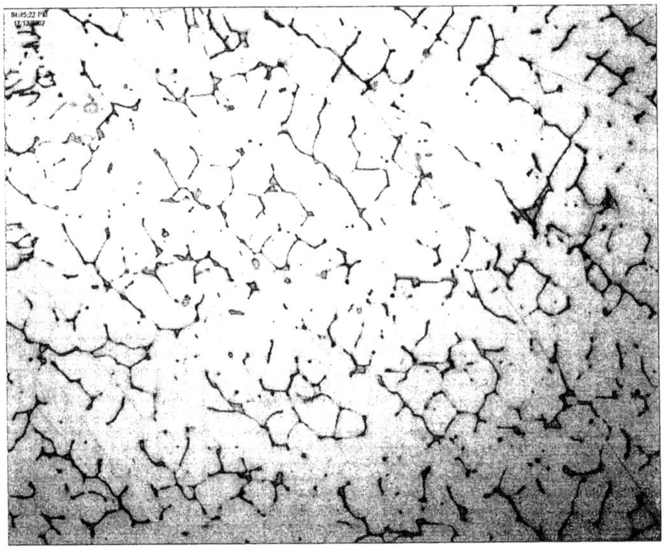

(A)

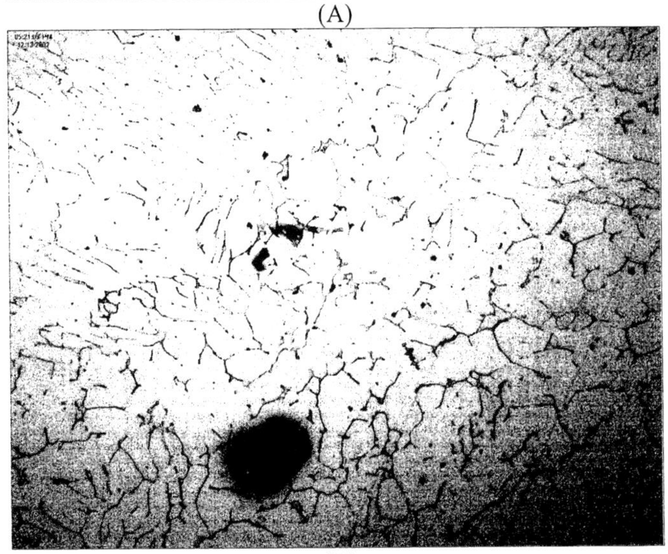

(B)

Figure 10. Microstructure of weld fusion zone at 500X magnification. (A) is from the sample welded with 0.005% oxygen in the puge, (B) is from the sample welded with 20% oxygen in the purge.

References

1. W. Huang, M. Mckeown, R. Paciej, and L. Link, *Proc 6 Int Symp Environ Degrad Mat Nucl Power Syst Water React*, p 387-390 (1993)
2. H. Fujikawa,. T. Morimoto; Y. Nishiyama; S. B. Newcomb, *Oxidation of Metals*, 59(1-2), 23-40 (2003)
3. A. Vesel; M. Mozetic; A. Zalar, *Appl. Surface Sci.*, 200(1-4), 94-103 (2002)
4. S. Hertzman, R. J. Pettersson, R. Blom, E. Roland; J. Kivineva, J. Eriksson, *ISIJ International*, 36(7), 968-976 (1996)
5. G. Lorang; B. M. Da Cunha; A.M.P. Simoes, M.G.S Ferreira, *J. of the Electrochemical Soc.*, 141(12), 3347-3356 (1994)
6. J.H. Kim, B.W. Oh; J.G. Youn; G-W. Bahng; H-M. Lee, *Advances in Cryogenic Engineering*, 36, 1339-1346 (1990)
7. J. Kuna, *International J. for the Joining of Materials*, 2(3), 94-99 (1990)
8. M.T. Liao; W.J Chen, *Materials Chem. and Phys.*, 55(2), 145-151 (1998)
9. F.B. Waanders; S.W. Vorster; A. Engelbrecht, *Scripta Materialia*, 42(10), 997-1000 (2000)
10. C.R. Heiple, J.R. Roper, R.T. Stagner, R.J. Aden, *Welding J.*, 62(3) 72s-77s.

Plasma Welding of NiTi to NiTi, Stainless Steel and Hastelloy C276

Casper van der Eijk, Hans Fostervoll, Zuhair K. Sallom and Odd M. Akselsen
SINTEF Materials Technology, Richard Birkelands vei 3A, N-7465, Trondheim, Norway

Abstract

NiTi alloys are known for their shape memory effect and their superelastic behavior. The past two decades, a lot of research has been done on these alloys. The number of applications where the material is used is however still limited to some niche markets, like the medical one. Widespread application of NiTi is partly restricted by the lack of knowledge on how to weld NiTi to NiTi and to other materials.

In this paper, the results form welding experiments of NiTi to NiTi, stainless steel and Hastelloy C276 are presented. The welds are characterized by DSC, optical microscopy and Field Emission SEM. The investigation of the NiTi-NiTi welds show that there is no compositional variation of the material through the weld. The mechanical properties are however significantly deteriorated after welding. The investigation of the dissimilar welds shows that the mixed zone of these welds contains a number of brittle phases, deteriorating the quality of the weld.

Introduction

Successful commercialization of products using shape memory alloys will require that these alloys can be joined using processes such as welding, brazing and soldering. The development of joining processes for shape memory alloys will open up for new low cost applications. For example, by joining two shape memory alloys with different transformation temperatures, a complex actuator which works at various temperatures can be produced in a simple tiny component[1]. Joining of shape memory alloys is however challenging.

In the case of welding of NiTi to NiTi, the properties to be matched between the base metal and joint region are the chemical composition and the microstructure, which dictate the shape memory response. This is where, fundamentally, the joining of shape memory alloys is different from that of conventional structural alloys such as steel, aluminum and titanium alloys. In conventional alloys, successful joints are obtained despite differences between the chemical compositions and the microstructure of weld metal and heat affected zone on one hand and the base metal on the other. In the case of a NiTi-NiTi weld, this is not the case, matching of chemical composition, microstructure and transformation temperatures is very important in addition to matching the mechanical properties and corrosion resistance[2].

During fusion welding of NiTi alloys, embrittlement may occur due to the reaction with oxygen, nitrogen and hydrogen at high temperatures. In addition, precipitation of brittle intermetallic compounds such as Ti_2Ni and $TiNi_3$ during solidification can have adverse effects on both strength and shape memory characteristics of the material[3].

Ikai *et al.* describe the TIG welding of NiTi wire and sheet. Heat treatment and training of the wire close the weld are necessary to recover the shape memory effect after welding[1]. Good results have been reported using laser welding. Only minor changes in the transformation temperatures, no deterioration in shape recovery and a higher strength, but lower ductility are observed after laser welding of 2 mm thick NiTi sheet with a 5 kW continuous-wave CO_2 laser[4]. When thin (0.5 mm) sheets of NiTi are welded using a Nd-YAG laser, the shape memory effect is preserved. However, the strength is decreased and the temperature window for use of the pseudoelastic effect is narrowed[5]. Schlossmacher *et al.* describe the Nd-YAG laser welding of thin (0.5mm) pseudoelastic Ti-51.5at%Ni sheet. They show that the pseudoelastic plateau is preserved after welding. The ultimate tensile strength of the welded material is still 80% of the base material with an elongation to fracture of 8.4%[6].

Solid state welding processes, like friction welding could be considered as a suitable method since prevention of grain growth and good mechanical properties can be expected. In addition, friction welding of NiTi can be carried out in air without inert gas shielding whereas fusion welding is impossible without proper gas shielding. Friction welding of 6 mm rods has shown that satisfactory welds can be obtained provided that heat treatment is performed after welding to minimize the variation in the transformation temperatures of the welds with respect to those of the base metal[3].

Materials and Experimental Procedure

Materials

Nickel and titanium were weighed according to the target compositions of 50/50 at% (equi-atomic) ratio which corresponds with 55wt% Ni and 45wt% Ti. The Ni had a 99% purity with 0.0075%C, 0.0062%O and 0.0004%N. The titanium pieces (ASTM B 265-90 grade 2) contained 0.02%C, 0.14%O, 0.010%N and 0.07%Fe. The alloys were made by melting in a graphite crucible 50 mm in diameter and 180 mm height, using a Balzer vacuum induction furnace. The melt was cast in a copper mould containing 6 cylindrical cavities of 15 mm in diameter, 98 mm height. After casting, the material was heat treated in vacuum at 950°C for 72 hours.

The chemical analysis of the NiTi alloy is shown in Table 1. Some Ni has been lost during the melting process. NiTi strips (1mm thick, 15mm wide) are machined from rods using electrical discharge machining. The strips are sand blasted and then ground using 500 mesh emery papers to remove any surface irregularities. Standard Hastelloy C276, which is a Ni-based alloy containing Mo, Cr, Fe and W, and an ASTM A240 stainless steel were used for the dissimilar welds. A filler wire was used in one of the NiTi-stainless steel welds. The composition is shown in Table 2. Welded specimens were examined by optical microscopy and SEM (Hitachi 4300 FE-SEM).

Table 1. Analyzed composition of the NiTi casting (wt%)

Ni	Ti	C	O	N
53.7	46.2	0.050	0.048	0.0003

Table 2. Composition of the filler wire used for the NiTi-stainless steel joint (wt%)

Ni	Cr	Mo	Fe
63	21	9	7

Welding Procedure

The welding tests were performed in a robotic welding station using the plasma arc welding process. The variant of plasma welding torch used was PPAW (Powder Plasma Arc Welding), where the filler metal can be added as powder. In the current welding tests, filler metal was used in only one of the experiments. An extra outer shielding gas cup was mounted on the torch to improve the gas shielding. The welding was performed in the flat position in a square groove with no gap. The welding parameters are listed in Table 3.

Table 3. Welding parameters

Welding current	42/15 A, pulsed 0.2s/0.2s
Welding speed	2.8 mm/s
Plasma nozzle dim.	ø 1.5
Electrode tip angle	30°
Plasma gas type, flow	Argon 99.996 %, 0.9 + 0.9 l/min
Shielding gas type, flow	Argon 99.996, 5 l/min
Backing gas + outer shielding	Argon
Work distance, torch angle	6 mm, 10° forehand

Differential Scanning Calorimetry (DSC)

In order to compare the SMA phase transformation temperatures of the base metal with those of the molten metal in the NiTi-NiTi weld, DSC measurements were carried out on both the base metal and the metal in the weld. Round specimens with a mass of about 70 mg were cut, polished and run in a DSC apparatus in the temperature range (-10 to 690)°C, heating and cooling rates were 5°C/min.

Tensile testing

Tensile tests were performed on the NiTi base material, on the NiTi-NiTi weld and on the NiTi-Hastelloy C276 weld. Tensile specimens had a surface of 1mm x 8mm in the transverse direction. The welded specimens had the welded section in the middle transverse to the tensile direction.

Results and Discussion

Characterization of the NiTi material

Characterization of the NiTi material reveals a number of flaws. Besides some cracking, a large number of second phase particles, as shown in Fig. 1, are observed in the material. EDS analyses on these particles show that these have a composition of 67at% Ti and 33at% Ni, indicating that it is $NiTi_2$ phase which has a lower melting point than NiTi. $NiTi_2$ is present as large (ca. 5 μm) cubic particles and as smaller globular precipitates (ca. 1 μm). The cracks in these cubes, as visible in Fig. 1, are an indication of their brittleness. The loss of Ni during the melting process might be the cause of their presence. The material has a high carbon content which leads to the precipitation of titanium-carbides. Such a titanium carbide is shown in Fig. 2. The shape and size of the titanium carbides suggests that these precipitate before or during solidification.

NiTi-NiTi plasma weld

The microstructure of the weld is shown in Fig. 3. The transformation temperatures of the base metal and the weld metal are shown in Table 4. The A_s and A_f temperatures do not show very large differences between the base metal and the molten metal. A large difference is observed during cooling in which the base metal starts transforming at a lower temperature and the transformation is finished at a much higher temperature than in the molten metal.

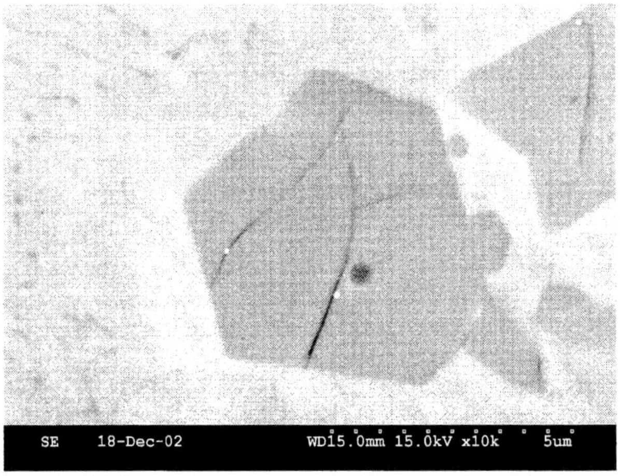

Fig. 1. SEM secondary image of $NiTi_2$ phase in NiTi matrix

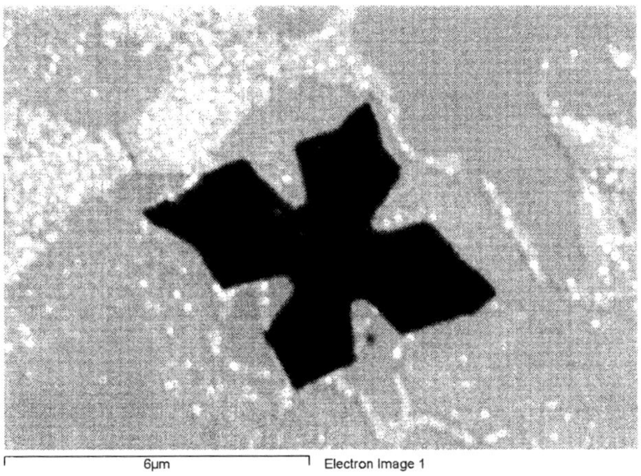

Fig. 2. SEM backscattered image of a Titanium-carbide in the NiTi material

The hardness profile through the weld is shown in Fig. 4. It appears that there is a slight increase in the hardness in and around the weld. The ratio between Ti and Ni, measured by EDS, is also shown in Fig. 4. It does not seem to be influenced by the welding process indicating that there is no loss of Ti or Ni during welding. The stress-strain curves of the NiTi material and the average of two different NiTi-NiTi welds are shown in Fig. 5. It is surprising how strong and ductile the base metal is, taking into account, the cracking, $NiTi_2$ phase particles and Ti-carbides, present in the material. It is clear that the strength of the welded material is much less than that of the base metal.

Table 4. Results of the DSC measurements on the base metal and the molten metal in the NiTi-NiTi-weld

	A_s (°C)	A_f (°C)	M_s (°C)	M_f (°C)
Base metal	63	98	53	35
Weld metal	59	114	65	5

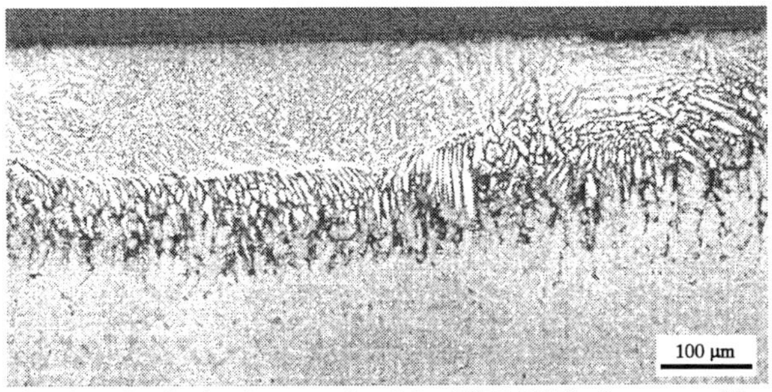

Fig. 3. Optical micrograph of the NiTi-NiTi plasma weld

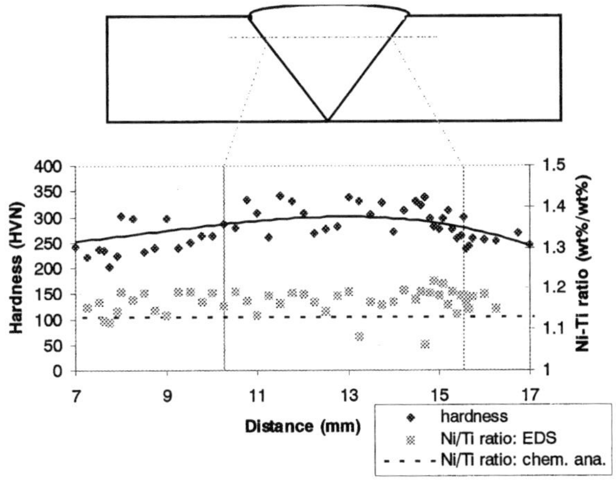

Fig. 4. Hardness profile and chemical composition through the NiTi-NiTi weld

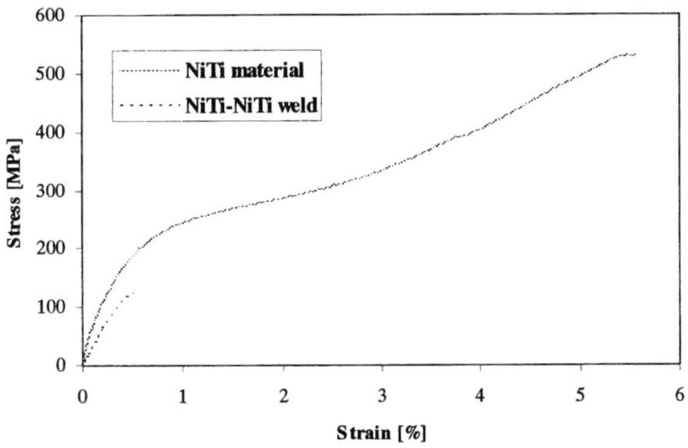

Fig. 5. Stress-strain curves of the NiTi and the NiTi-NiTi welds

NiTi-Hastelloy C276 plasma weld

Fig. 6 shows an optical macrograph of a weld between NiTi and Hastelloy C276, with the corresponding SEM backscattered image of the specimen with Hastelloy on the left side and NiTi on the right hand side shown in Fig. 7. A number of different phases are distinguishable in the NiTi. EDS analyses were done into the NiTi side of the weld to investigate how large the mixed zone extends. At a distance of 500µm from the fusion line, the NiTi still contains elements, e.g. Cr, originating from the Hastelloy, indicting that the size of the mixed zone is about 0.5mm. NiTi has a lower melting point (ca. 1310°C) than Hastelloy C276 (ca. 1370°C) which can explain the clear border between the Hastelloy and the weld. This also is the cause for the absorption of elements from the Hastelloy into the NiTi. Fig. 8 shows a backscattered image close to the fusion line. Six different phases are observed. Their composition, as measured by EDS, is listed in Table 5. Tensile testing of this weld shows that the strength has decreased to about 70 MPa with a negligible strain. The presence of the different phases close to the fusion line, of which many are expected to have a brittle nature, is the most probable cause for the deteriorated mechanical properties.

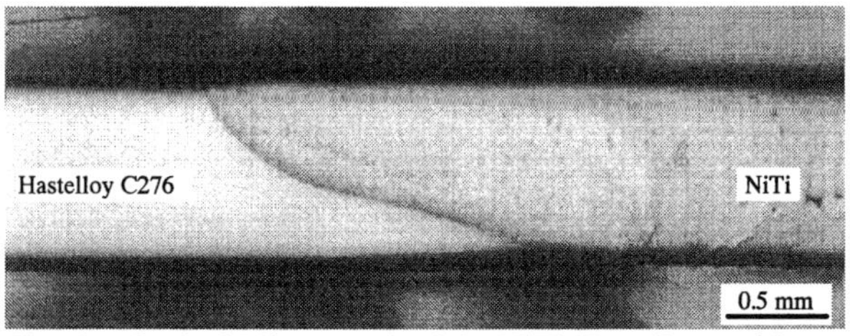

Fig. 6. Optical Micrograph of the NiTi-Hastelloy weld

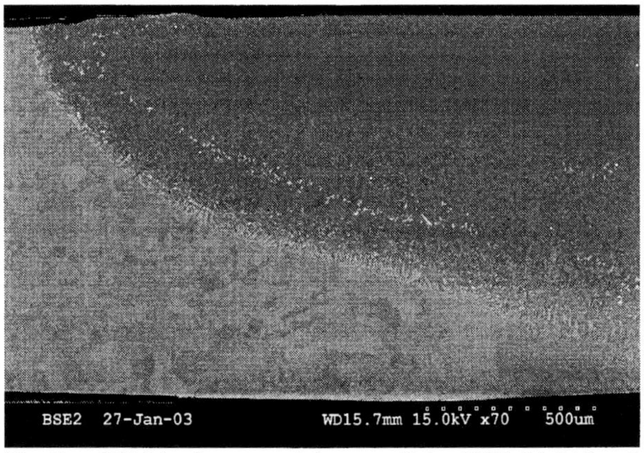

Fig. 7. SEM backscattered image of the NiTi (right hand side)-Hastelloy (left hand side) weld

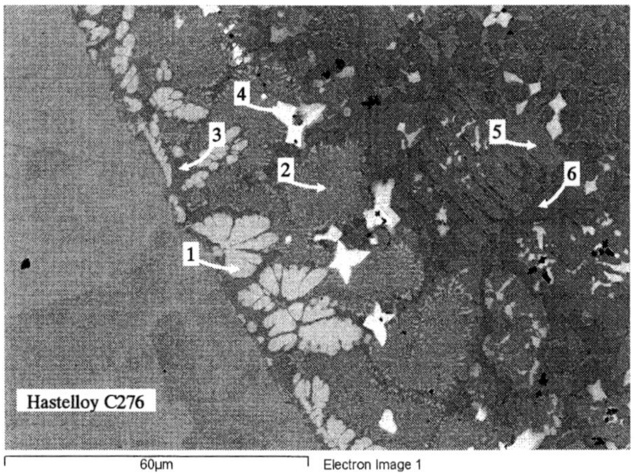

Fig. 8. SEM backscattered image of the fusion line of the NiTi-Hastelloy weld

Table 5. Composition of different phases marked in Fig. 8 (wt%)

No.	Ti	Cr	Fe	Ni	Mo	W	Zr	Mn
1	4.4	20.7	4.2	30.3	30.7	9.6		
2	15.1	11.1	3.5	58.9	9.6	1.8		
3	16.3	7.3	3.1	68.4	3.8	1.6		
4	6.7	9.5		3.5	52.4	26.5	1.5	
5	21.8	1.4	1.3	75.5				
6	33.1	5.7	3.3	55.1	2.3			0.6

NiTi-stainless steel plasma weld

Without Filler Wire

A macrograph of the NiTi-stainless steel weld is shown in Fig. 9. Cracking is observed on the NiTi side of the weld. During the welding process, a much larger amount of NiTi than steel seems to melt resulting in the absorption of elements from the steel into the NiTi. The mixed zone extends over several hundreds of microns. The cracking is observed in this mixed zone. Fig. 10 shows an example of such a crack in the NiTi. Table 6 lists the EDS analyses of positions along the crack shown in Fig. 10. It becomes clear form the analyses that the crack is arrested in the region where the composition of the NiTi is nearly free of foreign elements. The black spots in Fig. 10 are Ti-carbides. NiTi intermetallic compound and the used stainless steel have a thermal expansion coefficient of 12 and $17x10^{-6}/K$ respectively[7]. The stresses caused by this difference can easily cause cracking in the brittle phases present in the mixed zone. Hence, it is vital to keep the mixed zone as small as possible. An alternative approach is to use a filler wire which can function as a buffer between the NiTi and the stainless steel.

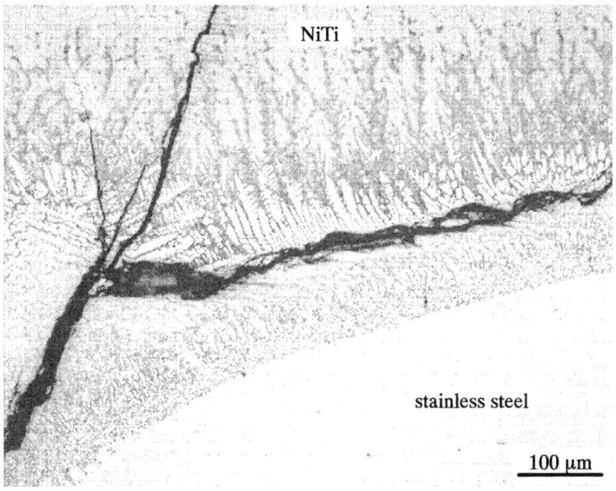

Fig. 9. Optical micrograph of cracks in the NiTi-stainless steel weld

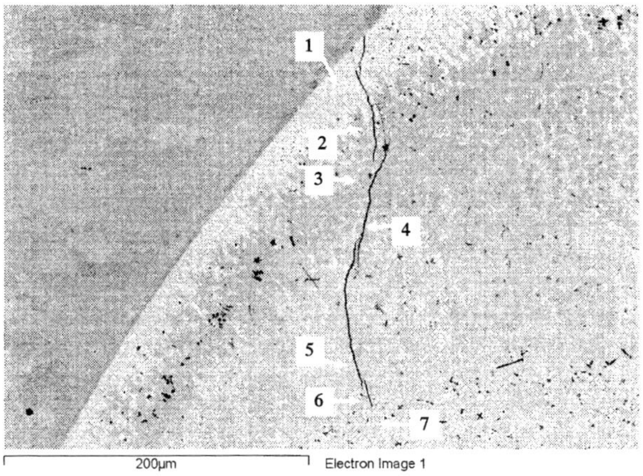

Fig. 10. SEM Backscattered image of a crack in the NiTi-stainless steel weld. The numbers indicate the positions of EDS analyses

Table 6. EDS analyses of the positions shown in Fig. 10 (wt%)

No.	Si	Ti	Cr	Mn	Fe	Ni
1	0.4	14.1	17.3	1.2	53.1	13.9
2		31.8	8.8	1.0	34.1	24.3
3	0.6	26.9	11.0		43.2	18.3
4	0.6	28.3	9.5		38.2	23.5
5		36.5	1.6		10.2	51.7
6		40.8	0.6		5.3	53.4
7		41.3	0.4		3.2	55.1

With Filler Wire

Under the prevailing conditions, the welding of NiTi to stainless steel with a Ni-rich filler wire does not eliminate the formation of brittle phases in the mixed zone. There is still a mixing of the elements especially into the NiTi. A backscattered micrograph of the mixed zone is shown in Figure 11. The matrix phase (spectrum 1) is a mixture of Ti, Ni and Fe while the eutectic phase (spectrum 1) is a Ni-rich phase. The darkest phase (spectrum 3) are Ti-carbonitrides.

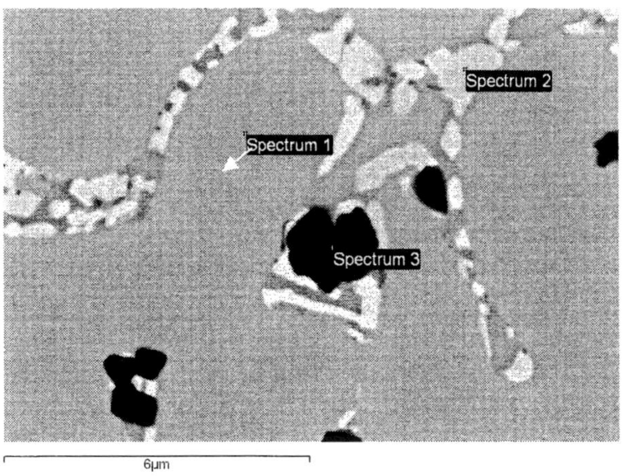

Fig. 11. Mixed zone of the stainless steel-NiTi weld with Ni filler wire

Summary and Conclusions

Plasma welding of NiTi to NiTi does not influence the Ni/Ti ratio in the weld. The phase transformation temperatures are however altered while the mechanical properties deteriorate.

Fusion welding of NiTi to Hastelloy C276 and stainless steel is extremely difficult. The NiTi has a lower melting point than the other materials. Brittle phases are formed close to the fusion line and there is a tendency of the NiTi to absorb elements from the Hastelloy C276 and steel.

References

1. A. Ikai, K. Kimura and H. Tobushi, *Journal of Intelligent Material Systems and Structures*, 7, 646-655 (1999)
2. N.B. Potluri, *Welding Journal*, 78, No. 3, 39-42 (1999)
3. T. Shinoda, T. Tsuchiya and H. Takahashi, *Trans. Jap. Weld. Soc.*, 22, 30-36 (1991)
4. Y.T. Hsu, Y.R. Wang, S.K. Wu and C. Chen, *Metall. and Mater. Trans. A*, 32A, 569-576 (2001)
5. A. Tuissi, S. Besseghini, T. Ranucci, F. Squatrito and M. Pozzi, *Mater. Sci. Eng.*, A273-275, 813-817 (1999)
6. P. Schlossmacher, T. Haas and A. Schlüssler, *J. Phys. IV France 7*, C5, 251-256 (1997)
7. T. Owa, T.Shinoda and Y. Kato, *Welding International*, 16, No. 4, 276-283 (2002)

Proceedings from Joining of Advanced and Specialty Materials, 13–15 October 2003, Pittsburgh, Pennsylvania, ASM International, 2004

Theoretical Models to Describe Properties of Magnesium Alloy Fusion Welds

M. Marya, G.R. Edwards, D.L. Olson

Center for Welding, Joining and Coatings Research
The G.S. Ansell Department of Metallurgical and Materials Engineering,
Colorado School of Mines (CSM)
Golden CO 80401 (USA)

ABSTRACT - The characteristics of magnesium alloy fusion welds produced with different processes are analyzed. Theoretical and phenomenological models as well as semi-empirical correlations are introduced to demonstrate that the performances of magnesium alloy fusion welds are largely predictable. Interactions between magnesium and the periodic table, fusion zone microstructure, porosity, solidification cracking susceptibility, alloying element evaporation, corrosion, and mechanical properties of magnesium alloy fusion welds are all documented and explained. Effects of welding variables on properties of weldments (particularly in the Mg-Al based alloys) are also discussed for several processes.

1. INTRODUCTION

The transportation industry has relentlessly searched for means to reduce mass and improve energy efficiency without decreasing structural integrity. Magnesium's density is only 1.74 and approaches 1.83 for its alloys, a value about 1.5 times less than that for aluminum alloys and 2.5 times less than the values for titanium alloys [1,2]. Magnesium alloys also yield readily, but their low density values give them greater strength-to-weight ratios than conventional engineering materials, including the aged-strengthened aluminum alloys. Although magnesium's hexagonal crystal structure restricts its ductility at ambient temperature, many magnesium alloys are castable, weldable, and sometimes thanks to fine grain structures, also superplastic at elevated temperatures. Such characteristics, added to high strength-to-weight ratios, have made magnesium alloys prime candidates for 21st century applications.

Of all joining processes, fusion welding is by far the most economic method to fabricate joints with suitable load-carrying capabilities. In fusion welding, the material's chemical composition and the thermal history left by the transient heat source are the main parameters controlling performance of weld joints. Several investigators [3-4] have reported that the properties of magnesium alloy fusion welds produced by electric arcs and laser beams differ considerably. Due to thermal cycles, the rapidly-traveling high power-density laser beams typically produce welds with higher quality [3-5]. Weld quality is also usually less in rich alloys as differences between weld and initial materials increase.

In this paper, the physical properties of pure magnesium that best describe the formation of fusion welds are first discussed. Second, the alloying of magnesium, which improves properties to engineering standards, is analyzed since knowledge of chemical interactions is of great value to understand microstructural developments, starting with alloy solidification. Both solution thermodynamics and semi-empirical models are introduced to describe chemical interactions, in particular the partitioning of the alloying elements between the liquid and the solid phases of magnesium. In multi-component metallurgical systems, this partitioning gives rise to a diversity of microstructures, microstructural gradients, and properties depending upon welding parameters. Relationships between physical properties, chemical composition, microstructure, defects, mechanical properties, corrosion, and welding parameters are discussed. Throughout this paper, well-established theoretical models are utilized to illustrate that the behavior of as-solidified magnesium alloy microstructures can be understood, and thus foreseen from fundamental knowledge.

2. IMPORTANT PROPERTIES OF PURE MAGNESIUM FOR WELDING

2.1. Thermal and Physical Properties [1,6-7]

At one atmosphere pressure, magnesium and aluminum melt at similar temperatures (649°C vs. 660°C for Al). Magnesium also requires about two-thirds of the heat of aluminum for melting and its thermal conductivity is considerably smaller (143 $W \cdot cm^{-1} \cdot K^{-1}$ vs. 238 $W \cdot cm^{-1} \cdot K^{-1}$ for Al at ambient temperatures). These properties not only make magnesium welding possible with any process used for aluminum, but they also predict that identical heat inputs will produce greater peak temperatures, wider and deeper welds, and slower cooling cycles in magnesium.

In contrast with aluminum, magnesium also possesses a considerably higher partial pressure (360 Pa vs. 10^{-6} Pa for Al). As a result, magnesium's boiling point is only 1090°C compared to 2450°C for aluminum and this has several implications. First, the use of high heat inputs must be restricted to prevent fumes and spatters. Second, magnesium's low boiling temperature is an excellent indicator of weak atomic bonding. Young's modulus, which is less in magnesium (45 GPa vs. 70 GPa for Al), confirms magnesium's relatively weak interatomic strength. Also, thermal expansion, which may be regarded as a rough estimate of how bond strength decreases with temperature, is slightly greater in magnesium (25.10^{-6} $m \cdot °C^{-1}$ vs. 23.8 $m \cdot °C^{-1}$ for Al). Young's modulus, thermal expansion coefficient and magnesium's limited strength all suggest potential problems with distortions, certainly greater then they may be in aluminum. However, volumetric change at melting is smaller in magnesium (4.2 vol. % for Mg vs. about 7 vol. % for Al) [6]. Shrinkage voids, like in resistance spot welds, will thus likely be smaller in magnesium.

Because interatomic strength is relatively small, magnesium's interfacial tension (surface free energy) is less than that of aluminum (550 $mN \cdot m^{-1}$ vs. 910 $mN.m^{-1}$ for Al at 660°C). As a result, an event like keyhole instability in high-power-density welding, which is largely controlled by interfacial tension, is expected to occur at relatively shallow weld penetrations in magnesium [8]. Like any other pure substance, the surface tension of magnesium decreases with temperature, but at a greater rate than in aluminum (0.26 $mN \cdot m^{-1}$ vs. 0.155 $mN \cdot m^{-1}$ for Al), as could be expected from magnesium's greater thermal expansion coefficient. At comparable heat inputs, surface-tension driven flow (also referred to as Marangoni flow) [9-10], coupled with magnesium's smaller heat loss by bulk thermal conduction, will thus make magnesium welds larger than aluminum welds.

Like surface tension, viscosity decreases with temperature. In the absence of surface-active elements, these two properties are related [11] and both will contribute to spatter, as in arc welding [10], and to geometric defects like underbeads in slowly solidified melt-through welds [5].

Of all physical properties, thermal conductivity is most probably the foremost property to affect welding. In the case of pure magnesium, thermal conductivity, k, and electrical resistivity, ρ_e, are related by the free electron model [12]. They can be related to each other by a sort of Wiedemann-Franz law [13], here corrected to include the contributions of phonons when temperature, T, increases:

$$k = \frac{2.26 \cdot 10^{-8} T}{\rho_e} + 0.0167T \quad (1)$$

Equation 1 indicates that the high thermal conductivity of magnesium can explain its low electrical resistivity. Practically, this feature also substantiates observations made during the resistance welding of magnesium sheets where elevated currents, but extremely short weld times, were required to generate the necessary heat to produce strong joints [14].

Although not obvious, electrical resistivity can also be useful to laser welding because absorptivity (i.e. emmisivity, ε_λ) of a substrate to a monochromatic electromagnetic radiation of wavelength λ and electrical resistivity are related, as expressed by Equation 2 [15].

$$\varepsilon_\lambda = 60.365\left[\frac{\rho_r(T)}{\lambda}\right]^{1/2} - 0.667\left[\frac{\rho_r(T)}{\lambda}\right] + 0.006\left[\frac{\rho_r(T)}{\lambda}\right]^{3/2} \quad (2)$$

Due to a greater electrical resistivity, absorptivity of a laser beam is expected to be greater in magnesium. Equation 2 also shows that a short electromagnetic wavelength, as promoted by solid-state Nd:YAG lasers, is always preferable to the longer wavelength of a gaseous CO_2 laser. Largely due to lattice vibrations [12], electrical resistivity increases linearly with temperature [13], therefore enhancing electromagnetic absorption by the metal substrate, or as in most situations, by its thin oxide layer.

2.2. Reactivity

With its atomic number of 12, magnesium can be categorized as an alkaline-earth metal and a Hume-Rothery Class I metal. Like aluminum, magnesium is silvery white and it is covered with a protective oxide. As pointed out by its smaller value of electronegativity, magnesium is more reactive than aluminum and its oxide is more refractory (2800°C melting point for MgO vs. 2015°C for Al_2O_3). Magnesium's reactivity is such that it may be ignited in air when brought to its molten state. The burning of magnesium in air generates a bright white flame leaving a mixture of magnesium oxide, MgO, and magnesium nitride, Mg_3N_2 [6-7], as described by the following reactions of oxidation (Equation 3) and nitriding (Equation 4):

$$2Mg(s) + O_2(g) \rightarrow 2MgO(s) \quad (3)$$

$$3Mg(s) + N_2(g) \rightarrow Mg_3N_2(s) \quad (4)$$

Practically, the use of shielding gases and fast-welding conditions makes magnesium's reactivity irrelevant. Only the inert gases, with argon being the most popular, can protect magnesium appropriately. In gas tungsten arc (GTA) welding as well as laser welding, pure helium is beneficial because its first ionization potential is the greatest of all elements. Helium's resistance against ionization raises the plasma temperature in current-controlled GTA welding, while it prevents energy dissipation into surface plasmas in laser welding. However, in gas metal arc (GMA) welding, where voltage is normally self-regulated, helium is undesirable because it increases the current for the spray arc transfer, and consequently it increases weld spatter; i.e. arc instability [10].

Other than air, magnesium also reacts with liquid water, but not to the extent seen with steam, where it forms magnesium oxide (MgO), or magnesium hydroxide, $Mg(OH)_2$, and hydrogen gas (H_2):

$$Mg(s) + 2H_2O(g) \rightarrow Mg(OH)_2(aq) + H_2(g) \quad (5)$$

Equation 5, which is depicted by a vertical phase boundary on the Pourbaix diagram [16], shows that magnesium hydroxide is stable in hydronium-rich solutions (Ph>11) where it reduces corrosion. In real situations, however, aqueous solutions are chemically unpredictable, and corrosion of magnesium typically proceeds rapidly. Weld surface ripples, residual and in-service applied stresses will worsen resistance against corrosion. Post-weld heat treatments may be recommended to reduce stress-corrosion susceptibility [10].

Magnesium is also reactive towards the halogens such as chlorine, Cl_2, or bromine, Br_2, with which it burns to form magnesium(II) chloride, $MgCl_2$ and magnesium(II) bromide, $MgBr_2$, respectively. Halides can be used to assist the GTA welding of magnesium [17-19], as in the A-TIG welding process, where they can disrupt magnesia before welding.

2.3. Gaseous Solubility

Porosity is a major defect shared by all fusion welding processes, which results from entrapment of dissolved gases or gas produced by chemical reactions. The pore fraction in weld fusion zones involves at least three competing time-dependent processes: (1) the nucleation, (2) the growth, and (3) the transport of pores [9-10,20,22].

Inert monatomic gases and diatomic halogen gases are in effect insoluble in magnesium. Solubility of halogen gases is only a few ppm in liquid magnesium at 650°C and it increases with temperature to reach as much as 100 ppm at 950°C [21]. Although halide inclusions may be stabilized, pore formation is highly improbable since partial pressure of soluble halogen is too low, as described by the following models. In metals, gaseous elements partition exactly like alloying elements. During solidification, a concentration of gas, C(x), builds up ahead of the moving solidification front, or solid-liquid interface. Gas concentration in the liquid phase can be approximated using traditional solidification theory [9-10,20,22] as:

$$C(x) = C_0\left[1 + \frac{1-k_0}{k_0}\exp\left(-\frac{Gx}{D}\right)\right] \tag{6}$$

where x is the distance from the solidification front into the liquid, k_0 is the partitioning coefficient, G is the solidification rate and D is the diffusion coefficient across the interface. Once a critical gaseous concentration, or partial pressure, is reached, a pore can form. The partial pressure of the gas in solution, P_g, must satisfy the criterion:

$$P_g > P_{atm} + P_{hyd} + P_{capillarity} \tag{7}$$

where P_{atm} is the atmospheric pressure, P_{hyd} is the hydrostatic pressure, and $P_{capillarity}$ is the pressure due to curvature, which for a spherical pore can be approximated as $2\gamma/r$ where γ is the interfacial tension of the liquid phase with the gas and r is the pore radius. In all probability, a pore will continue to grow when its radius exceeds a critical value, approximated as:

$$r_c > \frac{2\gamma V\phi}{kT\ln\left(P_g/P_0\right)} \tag{8}$$

where V is the molar volume of the gas phase, k is Boltzmann's constant, T is the absolute temperature, and ϕ is a correction factor for heterogeneous media. According to Equation 8, magnesium's low interfacial tension will facilitate the nucleation of pores, and therefore increase the probability of producing porosity. In welding, risk of porosity is typically high when the ratio of the solubilities at the average weld pool temperature and the melting point is high [9,23]. The amount of gas dissolving into the weld pool from surface absorption depends upon the partial pressure of the gas in the atmosphere of the heat source and the temperatures over the pool surface. Knowing that the temperatures for arc weld pool surfaces must approach the boiling temperature of magnesium (1090°C) and that pure magnesium solidifies at 649°C at equilibrium, Sievert's law can be applied to estimate the potency of soluble gas to generate porosity. For hydrogen dissolved in liquid magnesium [24], Sievert's law is:

$$[H] = 608 \cdot p^{1/2} e\left(-\frac{24{,}400}{RT}\right) \tag{9}$$

where [H] is the total volume of hydrogen in the liquid metal in ml/100g, p is the partial pressure of hydrogen in the arc expressed in atmospheres, and R is the gas constant (8.31 $J.mol^{-1}.K^{-1}$). Equation 9 predicts that the equilibrium solubility at one-atmosphere hydrogen pressure and 1090°C is 70.6 ml/100g compared to 25.3 ml/100g at 650°C. Since welding is a dynamic process, the solubility must be changed by an unknown factor and the hydrogen partial pressure may be lower than that calculated with Equation 9. However, under the assumption that equilibrium applies, if the hydrogen partial pressure at 1090°C is p atmospheres and the pool cools to 650°C, the pressure required to hold the gas in solution is $(70.6/25.3)^2$=7.8 atmospheres. Thus, if bubbles form when the gas concentration in solution is equal to the equilibrium solubility at one atmosphere hydrogen pressure, then they will form provided that the hydrogen partial pressure at 1090°C is higher than (1/7.8) = 0.13 atmosphere, that is when hydrogen constitutes at least 13 percent of the ambient gas. An analogous calculation applied to aluminum [5,9,23] would show that hydrogen partial pressure would be two orders of magnitude lower in aluminum. Therefore, hydrogen potency to form porosity is significantly less in magnesium than in aluminum, as confirmed by several investigations [5,9,23].

3. THE ALLOYING OF MAGNESIUM AND ITS IMPACTS ON WELDING

Magnesium is alloyed to improve mechanical properties, corrosion resistance, and oxidation resistance. Room-temperature strengthening is achieved by interstitial and/or substitutional solid solution, phase precipitation, and grain refinement as well described by the Hall-Petch equation. Because single-phase solid solutions are prerequisite to several of these strengthening mechanisms, the conditions for stable solid solutions through atomic substitution are first analyzed. Also, since weldment microstructures are mainly composed of the equilibrium microconstituents [4-5,26], phase equilibria in dilute magnesium systems are also important to understand both magnesium alloy weld microstructures and properties.

3.1. Rules for Alloying

Hume-Rothery [27] was first to observe that extensive solubility occurs when atomic radii of solute and solvent are within about 15 percent of each other. When atomic radii are favorable, large solid solubility is mainly affected by difference in electronegativity between solute and solvent atoms; a value less than 0.4 on the Pauling scale is usually required. For substantial solid solution, solute and solvent atoms must also have similar valence electron energies and they must share a common crystal structure. Due to these restrictions, the Hume-Rothery class I metals are most appropriate. In fact, it is known that the maximum solubility in magnesium is observed for elements of group IIB of the periodic table, namely zinc and cadmium, two elements also characterized by a hexagonal crystal structure. However, only cadmium forms a continuous solid solution with magnesium at temperatures beyond 253°C and 1-atm pressure [2,28].

For efficient solid-solution strengthening, large differences between atomic radii of solute and solvent are favorable. Precipitation hardening requires a reduction in solid solubility as temperature drops, causing the precipitation of second phases that block the slip of dislocations. From a thermodynamic standpoint, solute elements with restricted solubility possess positive enthalpies of mixing with magnesium, a characteristic that also stabilizes the high-temperature liquid phase to lower temperatures. The quasi-chemical regular solution model, represented by Equation 10, helps to understand interactions between alloying elements in terms of bond energy, ε ($\varepsilon<0$) [29]:

$$\Delta H = \Omega_{ij} x_i x_j \text{ with } \Omega_{ij} = N_{avo}\left[\varepsilon_{ij} - \frac{1}{2}\left(\varepsilon_{ii} + \varepsilon_{jj}\right)\right]z \quad (10)$$

In Equation 10, ΔH is the molar enthalpy of mixing, Ω is the interaction parameter between species "i" and "j", x_i and x_j are their molar fractions, ε is the bond energy, N_{avo} is Avogadro's number, and z is the number of bonds per atom. Equation 10 shows that the sign of the enthalpy of mixing is determined by the respective bond energy. Positive enthalpy of mixing can be associated with weak interatomic bonding between the dissimilar atoms relative to the pure phases.

The electronic models proposed by Miedema [30], justified on a quantum-mechanical basis by Chelikowski et al. [31], have been proven satisfactory to estimate the enthalpy of mixing of hundreds of binary compounds, as well as the solubility limits of binary alloy systems. In Miedema's latest model [See 29], the enthalpy of formation of a binary mixture of i and j is approximated using f and g functions to account for composition, not initially included in Miedema's early work [30]:

$$\Delta H(X_i, X_j) = \begin{array}{c} f\left(x_i^S, x_j^S\right) \cdot g\left(x_i, x_j, V_{mi}, V_{mj}\right) \cdot \\ \left[-P\left(\Delta\phi^*\right)^2 + Q\left(\Delta n_{ws}^{1/3}\right)^2 - R\right] \end{array} \quad (11)$$

In Equation 11, $n_{ws}^{1/3}$ is the electron density based upon the volume of Wigner-Seitz atomic cells, V_m, and the superscript "S" refers to the cell surface. Miedema's model [30] assumes that differences in electron density, $\Delta n_{ws}^{1/3}$, between different species of atoms always lead to local perturbations that give rise to a positive energy contribution. In Equation 11, ϕ^* is a chemical potential, represented by the macroscopic work function assuming that the differences, $\Delta\phi^*$, at the cell surfaces between different atomic species lead to an attractive term, similar to the energy gain from the formation of an electric dipole layer. P, Q and R are constants, which turned out to be universal constants [30].

Figure 1, which utilizes coordinates suggested by Miedema's models, shows that the Hume-Rothery Class I metals usually exhibit positive enthalpies of mixing when they are added to magnesium. It turns out that these elements are reasonably soluble in solid magnesium and they also separate from magnesium during solidification by eventually creating eutectic decompositions (cadmium being the exception). Figure 1 shows that the transition and rare-earth metals create exothermic reactions ($\Delta H<0$). As expected from the Hume-Rothery's rules, these elements have generally less solubility than the Class I metals and they stabilize a diversity of intermetallic phases (Part 3). Preferable alloying elements for magnesium are located on the gray band of Figure 1. This band's average slope is 3.0 and it tracks the optimal conditions for large mutual solid solubility [30]. Although aluminum, zinc, calcium and silicon, all form equilibrium intermetallic phases with magnesium, they are also the primary alloying elements with magnesium. In addition to suggesting eutectic partitionings with the Hume Rothery Class I metals, Figure 1 was therefore able to successfully identify major alloying elements in magnesium.

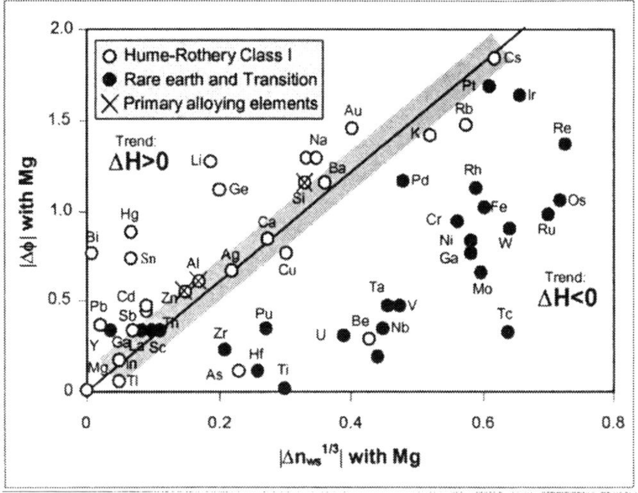

Figure 1 – Graphical representation to identify the elements with a large solubility in solid magnesium. Coordinates are based upon tabulated data from Miedema [30].

3.2. Thermodynamic Description and Alloying Element Interactions

Unless temperature is absolute zero, enthalpy, H, is insufficient to describe phase stability. The randomness within the phases, expressed by the entropy, S, must also be considered, particularly in non-dilute solutions. Gibbs free energy (G=H-TS) is therefore a more meaningful thermodynamic quantity. The CALPHAD (Calculation of phase diagrams) methods are useful to think of chemical interactions within alloys, particularly when more than two chemical species are involved. These methods try to provide a true equilibrium calculation by considering Gibbs free energy of all phases and by minimizing the total Gibbs free energy, G, which is calculated from knowledge of the fractions of each component i, X_i, their individual free energy, G_i^0, and free energy of mixing. Following developments on the regular solution models by Kaufman and Tomika [See 29], the total free energy of a system made of "n" chemical species is usually described as follows:

$$G_{Total} = \left| \begin{array}{l} \left\{ \sum_{i=1}^{n} x_i G_i^0 + RT \sum_{i=1}^{n} x_i \log_e x_i \right\}_{Ideal} + \\ \left\{ \sum_{i=1}^{n} \sum_{j=1}^{n(i \neq j)} \sum_{\nu} \left[\Omega_{ij}^{\nu} \left(x_i - x_j\right)^{\nu} \right] \right\}_{Excess} \end{array} \right| \quad (12)$$

where Ω_{ij} are temperature-independent interaction parameters that depend on the value of the binary enthalpy of mixing between the components i and j, as was shown in Equation 10.

Note that the solution becomes regular when v equals zero, and sub-regular when v equals 1. Clearly, a primary parameter to model equilibrium realistically is the interaction coefficient. Theoretical estimates of such parameters have been made using a variety of methods, including electron energy calculations and the Miedema's models [30]. In this paper, the simple model used by Molt and Kaufman [See 29] has been selected for its simplicity. Interaction coefficient is defined as the sum of an attractive term and a repulsive term:

$$L = \left| \begin{array}{l} -100\left(\varphi_i - \varphi_j\right)^2 + 0.3\left(V_{m(i)} + V_{m(j)}\right)\cdot \\ \left[\left\{\left(\dfrac{-H_{v(i)}}{V_{m(i)}}\right)^{1/2} - \left(\dfrac{-H_{v(j)}}{V_{m(j)}}\right)^{1/2}\right\} + F\left(V_{m(i,j)}\right)_{Strain}\right] \end{array} \right| \quad (13)$$

Equation 13 indicates that two elements with a large difference in electronegativity, ϕ_i, attract each other, whereas strongly bonded elements (measured by the enthalpy of vaporization, H_v, and the molar volume, V_m) resist attraction (Part 2.1 pointed out that these elements also have a high boiling point). Applying Equation 13 to the Mg-Al-Zn ternary system reveals that primary interactions are between magnesium and aluminum, which explains the fact that an equilibrium phase in AZ alloys (A: Aluminum and Z: Zinc) is the β-$Mg_{17}Al_{12}$ phase, as seen in Figure 2 [28]. Note that the β-$Mg_{17}Al_{12}$ phase is non-stochiometric near the eutectic temperatures, allowing 42 to 58-wt. aluminum percent solubility. This solubility range suggests that zinc, the second main alloying element in AZ alloys, must be distributed between the magnesium and the β-$Mg_{17}Al_{12}$ intermetallic phase, perhaps better designated as β-$Mg_{17}(Al,Zn)_{12}$.

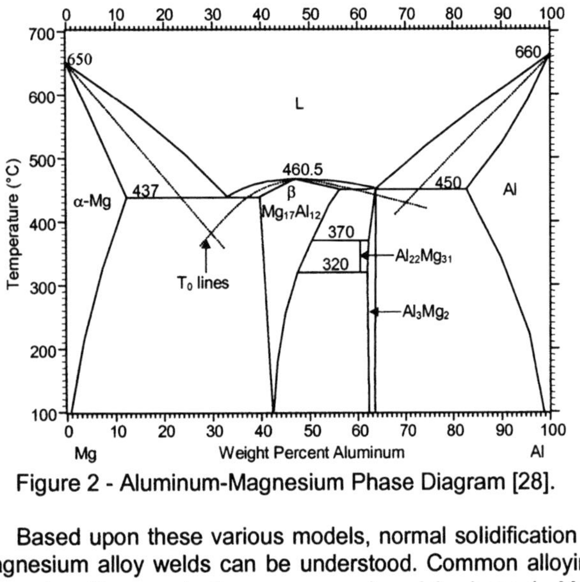

Figure 2 - Aluminum-Magnesium Phase Diagram [28].

Based upon these various models, normal solidification in magnesium alloy welds can be understood. Common alloying elements will separate from a magnesium-rich phase (α-Mg), lower the melting temperatures by generally promoting eutectic decompositions, a transformation which will also stabilize a strengthening intermetallic phase (like β in Mg-Al alloys).

3.3. Descriptive Relationships for the Solidification of Binary Magnesium Alloys

Now that we know that phase equilibria can be predicted from free energy minimization, we can expect that certain characteristics of phase equilibria must be interdependent (since they can be considered as the results of the same calculation). Figures 3 to 5 [4-5], which recap common features from binary phase diagrams with magnesium, confirm that such correlations exist. Figure 3 [4-5] shows that elements having a small solid solubility in magnesium segregate the most. Also, partitioning coefficient between magnesium and the transition metals and rare-earth metals is larger than that with other Hume-Rothery Class I metals.

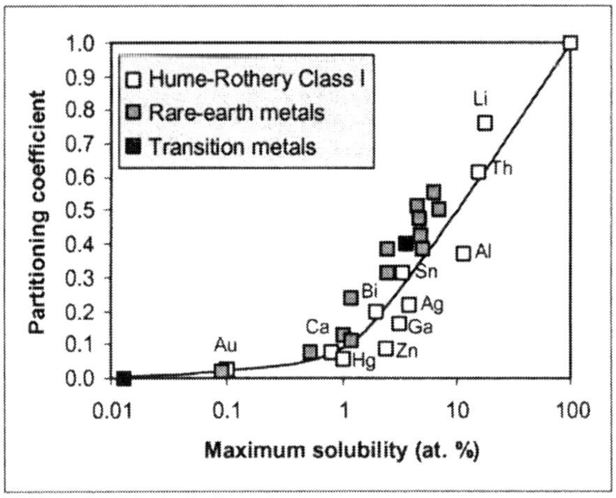

Figure 3 –Partitioning coefficient at solidification vs. maximum solubility in solid magnesium for various elements [4-5].

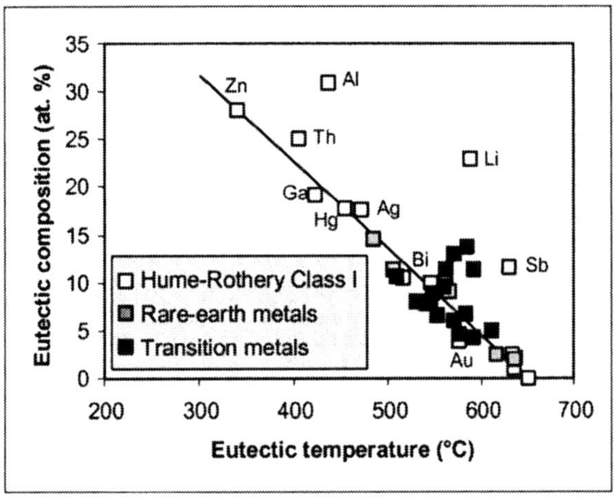

Figure 4 –Eutectic composition vs. eutectic temperature for various elements alloyed to magnesium [4-5].

For the eutectic transformation of the liquid magnesium phase into the solid α-Mg phase and an intermetallic phase, Figures 4 and 5 [4-5] reveal that the eutectic temperature correlates well with both the eutectic composition and the melting temperature of equilibrium intermetallic phases. As depicted in Figure 5 [4-5], the stability of the intermetallic phases tends to augment as the eutectic temperature increases; i.e. with elements generating small eutectic compositions (Figure 4). In Figure 5 [4-5], elements near the border with the shaded region have intermetallic phases with melting temperatures close to the eutectic temperature. When alloys are made of those elements, including zinc and aluminum, re-melting (liquation) of the intermetallic phases is likely, and therefore risks of cracking are increased in weld heat-affected zones if thermal, residual, and applied stresses are also large.

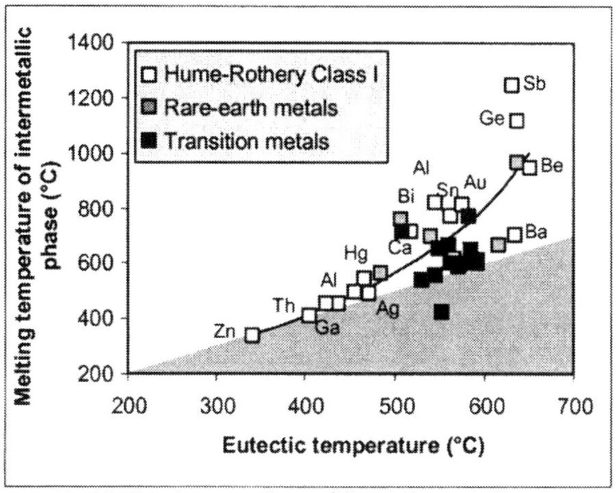

Figure 5 –Melting temperature of first intermetallic phase vs. eutectic temperatures for various elements alloyed to magnesium [4-5].

3.4. The Commercial Magnesium Alloys

The effect of alloying elements is generally proportional to the amount added, up to the limit of solubility. While alloying additions improve properties, residual elements like iron or nickel are substantially detrimental, especially when alloying is ineffective to mitigate their contribution. Table 1 [32] summarizes the effects of alloying elements in magnesium, starting with aluminum, which is at the origin of the AZ (Al-Zn), AM (Al-Mn) and AS (Al-Si) commercial magnesium alloys. In these alloys, aluminum directly stabilizes the β-$Mg_{17}(Al,Zn)_{12}$ phase without appearance of intermediate precipitates such as Guinier-Preston (GP) zones. Elements like zirconium and manganese are also frequently added for grain refinement and corrosion resistance respectively. Zirconium, however, does not reduce grain size in the presence of aluminum because both elements interact to form compounds as suggested by their electronegativity values and Equation 13. Zirconium is mainly used in the ZK alloys (Mg-Zn); a group of alloys with greater room-temperature strength, but low weldability because zinc enhances solidification cracking susceptibility, as shown in Part 3.8.

Zinc, most of the rare-earth metals, thorium and silver form GP zones in magnesium alloys after quenching from a high-temperature single-phase field and subsequent ageing. The precipitation of such fine metastable intermetallic phases often starts with small coherent particles similar to the GP zones seen in aged aluminum alloys, which upon heat treatment coarsen and loose coherency. Considerable strengthening is obtained in precipitation-strengthened alloys, which are mainly reserved for aerospace applications. In analogy with precipitation-strengthened aluminum alloys, rapid laser welding (including hybrid-laser welding) or solid-state joining processes like friction stir welding may be preferred to minimize the metallurgical changes caused by prolonged exposure to elevated temperatures.

Table 1 – Roles of alloying elements and impurities [32].

Element	Impact	Limit (wt.%)
Aluminum	Improves strength and castability	max: 10
Beryllium	Decreases tendency to oxidization	max: 0.001
Calcium	Reduces oxidation; improves sheet rollability	<0.3
Copper	Improves high temperature strength	<0.05
Iron	Greatly reduce corrosion resistance	<0.005
Lithium	Lowers density, improves ductility (formability)	< 11
Manganese	Increases strength slightly and interact with Fe to improve corrosion resistance	<1.5
Nickel	Greatly reduce corrosion resistance	< 0.002
Rare-Earth Metals	Increases strength and reduces weld cracking and porosity in castings.	
Silicon	Increases fluidity, decreases corrosion resistance	
Silver	Increases response to age hardening.	
Strontium	Increases high-temperature strength of die cast alloys	
Thorium	Increases creep strength at temperatures up to 370°C.	typically 3
Tin	Increases ductility in small amounts of aluminum.	
Yttrium	Added with other rare-earth elements to improve creep resistance at temperatures up to 300°C.	
Zinc	Improve room-temperature strength and corrosion resistance.	
Zirconium	Refines grains except in alloys with aluminum	typically 4 to 5

3.5. Alloy Physical Properties and Effects on Weld Morphology

It is now evident from Figures 2 to 4 that large additions of alloying elements will reduce melting temperature due to eutectic decompositions. Furthermore, intermetallic phases coexisting with a magnesium-rich phase will also be stabilized, because amounts of alloying elements are typically greater than room-temperature solubility. Because of the formation of substitutional solid solutions and intermetallic phases in various morphologies, physical properties such as thermal conductivity and electrical conductivity are both reduced by the alloying elements. The thermal conductivity of AZ alloys can be approximated irrespective of microstructural variations [1,5-6]:

$$k_{AZ} = 118 - 7.5(\text{wt.pct. Al}) \quad (14)$$

Equation 14 shows that thermal conductivity decreases from about 105 $W \cdot cm^{-1} \cdot K^{-1}$ for the AZ21 alloy to 72 $W \cdot cm^{-1} \cdot K^{-1}$ for the AZ91 alloys. Since density and specific heat are less affected by composition than thermal conductivity, thermal diffusion (perhaps a more suitable measure of heat transport) decreases in a similar fashion as aluminum content increases. Both the decrease in thermal conductivity (and diffusivity) and the lowering of the alloy melting temperature explain Figure 6 [4,5], which reveals that the cross-sectional area of GTA weld fusion zones decreases as alloy aluminum content increases.

Such compositional effects extend to keyhole welds, where an ionized vapor column (or plasma cavity) must be produced within the materials. Figure 7 [8] shows that morphologies of AZ91 and AM50 keyhole welds can be quite different. Compared to the AM50, the AZ91 exhibits broader fusion zones, as could be expected since this alloy remains liquid at lower temperatures (598-425°C for the AZ91 vs. 620-435°C for the AM50) and possesses a smaller thermal conductivity (55 $W \cdot m^{-1} \cdot K^{-1}$ for AZ91 vs. 74 $W \cdot m^{-1} \cdot K^{-1}$ for AM50 at 100°C). However, these physical properties do not explain why keyhole welds in the AM50 were more irregular and why pores were observed.

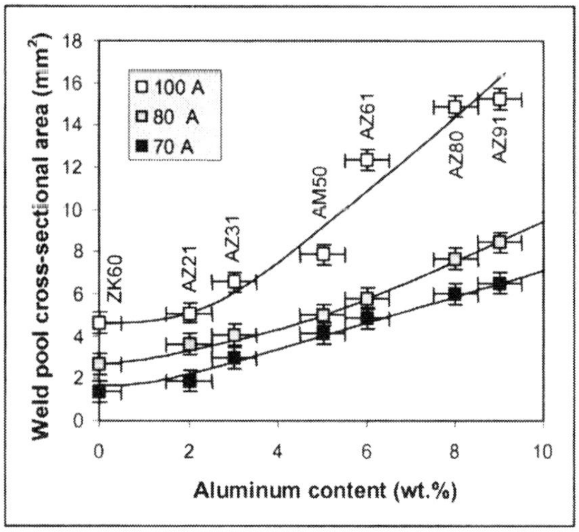

Figure 6 – Effect of aluminum on the GTA weld cross-sectional area in several commercial alloys [4,5].

In analyzing keyhole welds, basic background in plasma physics is worth providing to fully clarify the results seen in Figure 7 [8], because the absorption of the laser beam inside the keyhole is largely an electronic process, known as inverse bremsstrahlung absorption. The absorption of a plasma first increases as the population of low-energy electrons increases, then decreases as their energy approaches that of the photons, as quantified by Equation 15 [8]:

$$a_w = \frac{n_e n_i e^6 \left[1 - \exp\left(-\frac{h\omega}{2\pi k T_e}\right)\right]}{6\sqrt{3}\mu\varepsilon_0^3 c \left(\frac{h}{2\pi}\right)\omega^3 m_e^2} \left(\frac{m_e}{2\pi k T_e}\right)^{1/2} \bar{g} \qquad (15)$$

In Equation 15, n_e and n_i are the ion and the electron densities, e is the electron charge, h is the Planck's constant, ω is the angular frequency ($1.7 \cdot 10^{14}$ Hz for a CO_2 laser), k is the Boltzmann's constant, T_e is the electron temperature, and g is the average Gaunt factor, which for a CO_2 laser is well approximated by $(T_e/2116)^{0.227}$. Assuming an equilibrium state (i.e. all plasma particles are at the same temperature), electron and ion densities can be computed as a function of temperature using Saha's equation [9,17-18]. Like a chemical equation with ideal gases, population densities can be related to a temperature dependent quantity, conventionally referred as equilibrium constant. If partition functions are neglected and pressure taken as one atmosphere, Saha's equation can be simplified as:

$$\frac{n_e n_{Z+1}}{n_Z} = 2\left(\frac{2\pi m_e k T_e}{h^2}\right)^{3/2} \exp\left(\frac{-E_i}{kT_e}\right) \qquad (16)$$

where n_Z is the density of ions with charge +Z and E_i is the ionization potential (activation energy for ionization).

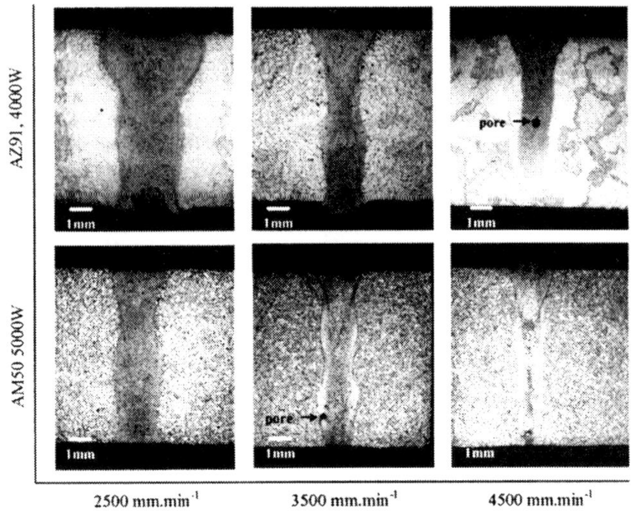

Figure 7 – Optical macrographs of laser beam keyhole welds produced in the AM50 and AZ91 alloys [8].

Aluminum's smaller first ionization potential (5.986 eV vs. 7.646 eV for Mg) improves absorption by the keyhole plasma and thus, like thermal conductivity and alloy melting temperatures, can explain the fact that keyhole welds were wider in the AZ91 alloy, as seen in Figure 7 [8]. In contrast with aluminum, the manganese in the AM50 resists ionization (6.766 eV). Manganese therefore reduces beam absorptivity and allows unabsorbed fraction of the laser beam energy to reach greater depths. These features can explain why deeper and narrower welds were seen in the AM50 alloy. Also, as weld penetration increases, so does the keyhole, which becomes unstable as simplistically explained from a force balance concept. After a critical depth, surface tension at the keyhole interface is incapable of supporting the lateral liquid column from collapsing. As shown by Matsunawa et al. [33], this instability results in significant weld pool oscillations and the entrapment of pores. Deep and narrower welds are therefore also associated with irregular morphologies and entrapped pores, as is seen in the AM50 welds of Figure 7 [8].

Compared to laser welding, arc welding is first and foremost a plasma process, where the initial air gap between the welding electrode and the workpiece is ionized to dissipate the supplied electrical energy into intense heat. For GTA arcs, Krenov [34] empirically linked the temperature of the arc central region, or arc column, to an effective ionization potential, E_{Ef} defined as:

$$T(°K) = 800 E_{Ef} \qquad (17)$$

where the effective ionization represents the contribution from a mixture of n species, and was approximated as [34]:

$$E_{Ef} = -\frac{T}{5800}\ln\sum_{i=1}^{n}\left(\frac{X_{i+}}{X_i}\right)^{1/2}\exp\left(-\frac{5800E_i}{T}\right) \quad (18)$$

Based upon Equations 17 and 18, the arc column temperature surprisingly does not depend on the process parameters. Equations 17 and 18 also show that the species with low first ionization potentials determine the temperature of the arc column. Emission spectroscopy during the GTA welding of AZ21, AZ31, AZ61 and AZ91 alloys confirmed the role of the element with low ionization potential by showing that the addition of aluminum to magnesium reduced arc temperatures, a feature largely counterbalanced by the decrease in thermal conductivity. Other research revealed that diatomic hydrogen, diatomic nitrogen, and halide fluxes pre-deposited on magnesium force the self-regulated constant current welders to raise the voltages, thus the heat inputs, and the arc temperatures. Weld fusion zones made in the presence of such gases and fluxes are therefore larger than those created with only argon, as shown in Figure 9 [23].

3.6. Alloying Effects on Hydrogen Solubility

Figure 8 [35] shows that hydrogen solubility is greatly reduced as the amount of aluminum increases. Also, the solubility gap between the solid and the liquid phases appears to be increased by the alloying elements [35], a feature which could explain why porosity has often been seen in AZ91 welds [23,25]. Because porosity is time-dependent, as seen in Part 2.3, it can be minimized using fast thermal cycles, as with a laser or an electron beam. Although process parameters may minimize porosity, weldability can always be improved whenever the gas content of the base material is reduced (die cast materials are less prone to porosity than high-pressure die cast materials, due to initial pore content).

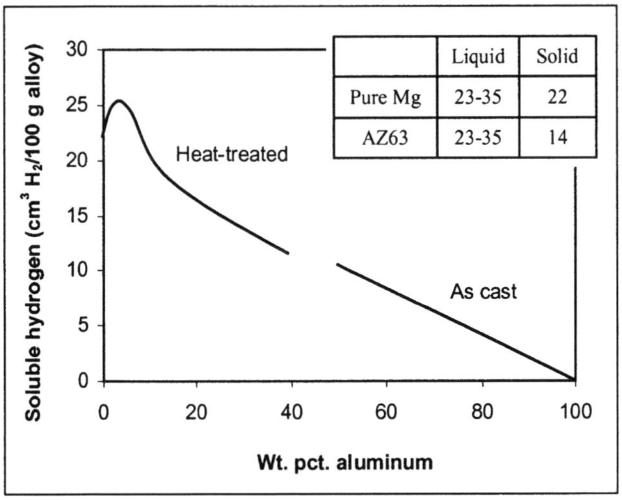

Figure 8 – Solubility of hydrogen in binary Mg-Al alloys between 675 and 700°C [35] and solubility values between solid and liquid phases of magnesium (in Table).

To clarify the role of gases, particularly hydrogen, Figure 9 [23] presents optical macrographs of GTA welds fabricated with three currents and three different gases (Ar+1%H_2, Ar+6%H_2 and N_2+1%H_2). Clearly, the presence of hydrogen in the shielding gas was a source of porosity at both 1 and 6 percent concentrations. The substitution of hydrogen by nitrogen reduced porosity, but stabilized a dual microstructure of magnesium nitride and α-magnesium. Due to the absence of significant pores in the nitrogen-made weld, hydrogen solubility appeared to be dramatically decreased by the presence of nitrogen, as it is by the presence of aluminum (Figure 8).

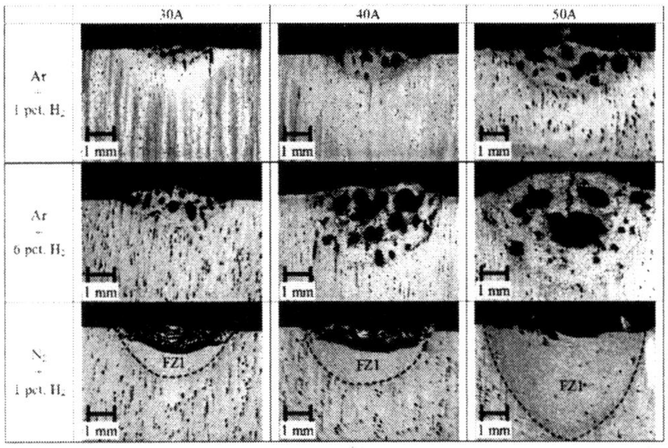

Figure 9 – Optical micrographs of bead-on-plate welds made in an AZ21 alloy using different currents and gases [23].

3.7. Vaporization Losses

As pointed out in Part 2.1, magnesium's high vapor pressure is orders of magnitude greater than that of its alloying elements. Magnesium and its alloying elements will therefore evaporate to dissimilar extents, depending upon weld surface temperatures. The relationship between rate of evaporation, G ($Kg.m^{-2}.s^{-1}$) and vapor pressure, P_v (Pa), given in the kinetic theory of gases, can help to understand evaporation losses during welding, particularly high-power density welding [36]:

$$G = 4.37\cdot10^{-5}\cdot P_v\sqrt{\frac{M}{T}} \quad (19)$$

where M is the molecular mass ($g.mol^{-1}$). Although Equation 19 applies to high vacuums, Equation 19 reveals that the element with the highest vapor pressure will evaporate the most. In the AZ alloys, magnesium and zinc will evaporate the most; aluminum will thus evaporate the least. Figure 10 [4], which shows compositions of keyhole welds in an AZ91 alloy indicates that aluminum concentrations in the welds are slightly increased [5,36]. While some beneficial effects on corrosion can be anticipated (Part 3.10), mechanical properties will practically not be altered by these slight evaporation losses.

3.8. Dendritic solidification and Alloy Cracking Susceptibility

In welding and casting, the development of the solidification microstructure is controlled by the stability of the solid-liquid interface. Growth rate, R, and thermal gradients in the liquid, G_L, are primary to determine the scale and the morphology of the solidification microstructure. At low growth rates, diffusion is usually localized to a scale comparable to the size of the microstructure, although compositions at the solidification front remains at the equilibrium compositions. At large growth rate (>10^{-2} $m \cdot s^{-1}$), local equilibrium at the solid-liquid interface can be lost, and solute trapping then sets in [37]. In rapid solidification conditions, solubility limits are in between the equilibrium solidus line and the T_0 lines, also shown in Figure 2 [28] for the Mg-Al system. This T_0 line indicates where the free energy of both the liquid phase and the α-Mg solid phase are identical, which corresponds to an idealized instantaneous freezing with no amorphous phase formation.

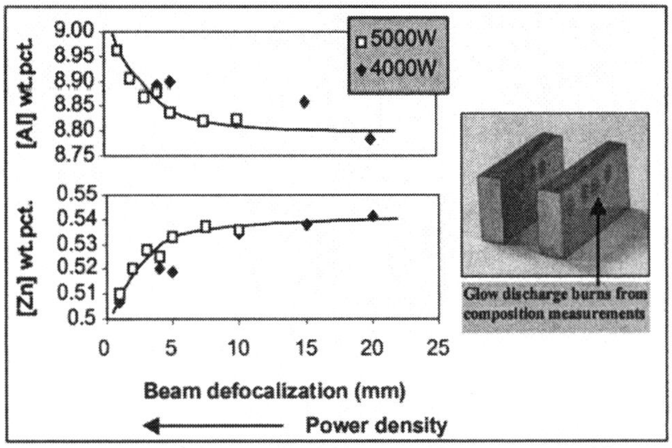

Figure 10 – CO_2 laser weld compositions measured by discharged spectroscopy throughout AZ91 weld cross-sections [4-5,36].

Practically, typical weld solidification fronts satisfy the conditions of equilibrium depicted by the phase diagram, and constitutional undercooling in its immediate vicinity determines whether the interface growth will be planar, columnar/cellular, or dendritic. In those conditions, thermal, curvature and kinetic undercoolings are usually secondary [9-10,22]. Planar front instability occurs due to constitutional undercooling when:

$$\frac{G_L}{R} \le -\frac{m_L}{D_L}\left(\frac{1-k_0}{k_0}\right)C_0 \tag{20}$$

In Equation 20, C_O is the equilibrium composition, m_L is the slope of an idealized straight liquidus line, and D_L is the solute diffusion coefficient in liquid. Due to compositional undercooling ($C_{O\ in\ AZ21} < C_{O\ in\ AZ91}$), the microstructure of AZ21 welds is generally characterized by long and coarse columnar dendrites, quite different from the "equiaxed" dendrites seen in the AZ91 weld [4-5].

When the solid-liquid interface exhibits equilibrium compositions, the temperature gradient and growth rate are important in the combined forms GR (cooling rate) and G/R, which respectfully influence scale and morphology of the solidification structure. For dendritic microstructures, the primary, λ_1, and secondary, λ_2, dendrite arm spacings are given as [22]:

$$\lambda_1 = \frac{c_1}{(G_L)^{1/2}(R)^{1/4}} \tag{21}$$

$$\lambda_2 = c_2 M^{1/3}\left(\frac{\Delta T}{G_L R}\right)^{1/3} \tag{22}$$

where c_1 are c_2 are kinetic constants and M is a mobility term, defined as follows:

$$M \approx \frac{\gamma}{\Delta S_f} \frac{D_L \ln\left(C_{Eut}/C_0\right)}{m_L(1-k_0)(C_0 - C_{Eut})} \tag{23}$$

In Equation 23, γ is the solid/liquid interfacial energy, and ΔS_f is the entropy of fusion. In Equation 22, the term $\Delta T/G_L R$, which represents the solidification time, shows that short solidification times, as produced by large growth rates (travel speeds) and large thermal gradients (focused heat source), reduce secondary dendrite arm spacing. Figure 11 [5,26], which shows dendrite arm spacing measurements in AZ91 welds, agrees with Equations 21 and 22. The cooling rates in laser welding, typically 2 or 3 orders of magnitude greater than in arc welding, produce finer and thus stronger microstructures. Equation 23 also shows that the richer compositions will also lead to finer microstructures. Equation 23 predicts that the decrease in secondary dendrite arm spacing will be as large as 40 percent when changing from an AZ21 to an AZ91 alloy.

Since the solidification of AZ alloys is well approximated using the Mg-Al binary phase diagram (Figure 2) [28], actual fractions of the α-Mg phase and the liquid phase in AZ welds will be in between the values given by the lever rule, as follows:

$$f_s = \left(\frac{1}{1-k_0}\right)\left(\frac{T_L - T}{T_F - T}\right) \tag{24}$$

and the Scheil solidification model [22], which assumes that diffusion in the solid phase is neglected whereas the interface is at the equilibrium compositions:

$$f_s = 1 - \left(\frac{T_L - T}{T_F - T}\right)^{\frac{1}{k_0 - 1}} \tag{25}$$

When back-diffusion between dendrites is considered, phase fractions in between that given by Equations 24 and 25 are obtained using the Brody-Fleming model [22]:

$$f_s = (1+\alpha k_0)\left[1 - \left(\frac{C_s}{k_0 C_0}\right)^{\frac{1}{k-1}}\right] \tag{26}$$

where α is a dimensionless diffusion parameter, which for back diffusion is well approximated as:

$$\alpha = \frac{4D_s}{\lambda_2^2}\left(\frac{\Delta T}{G_L R}\right) \tag{27}$$

While microsegregation always occurs in fusion welds, macro-segregation has been sporadically detected in some rapidly solidified laser welds [26]. As expected from Equations 25 and 26, macrosegregation gradually enriched the liquid phase in solute elements, pushing the eutectic liquid toward the hottest regions, usually located at weld centerlines and at weld surfaces. Macrosegration is sometimes well revealed by etching, as illustrated in the AM50 welds of Figure 7 [8].

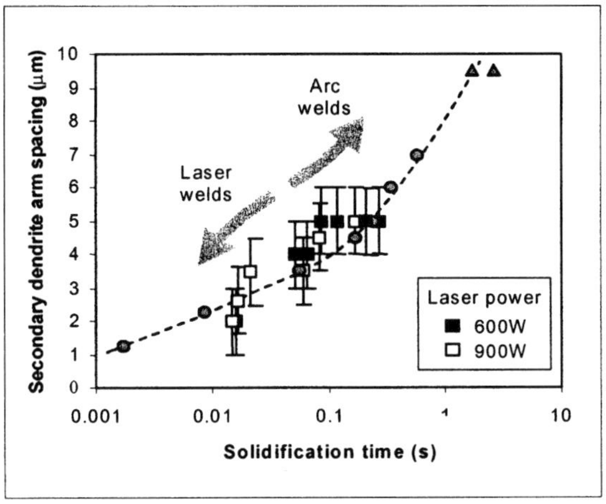

Figure 11 – Secondary dendrite arm spacing vs. solidification time (data from this study include error bars) [5,26].

When a eutectic transformation occurs at a low temperature and the partitioning coefficient is small, the freezing range is increased. From simple algebra, the freezing range in a binary alloy, ΔT, before reaching maximum solid solubility, C_{ss}, can be expressed as:

$$\Delta T = \left(\frac{1-k_0}{k_0}\right)\left(\frac{T_m^{Mg} - T_{Eut}}{C_{Eut}}\right)C_0 \qquad (28)$$

where T_m^{Mg} is the melting temperature of pure magnesium, T_{Eut} is the eutectic temperature, and C_{Eut} is the eutectic composition. Typically, freezing ranges in excess of 150°C are found in cracking susceptible alloys. In such alloys, large thermal stresses coexist within the weld and self-healing by liquid back filling is insufficient. Figure 12, which shows the freezing range at maximum solubility for several elements, reveals that elements such as calcium and zinc (located on the gray band of Figure 1) increase the freezing range radically even at very low amounts. Based upon Figure 12, it can be understood why the zinc-rich ZK alloys are not weldable.

From Equation 28, large freezing ranges are associated with small partitioning coefficients, k, and rich compositions, C_0. Recall that elements with limited solid solubility have generally small partitioning coefficients (Figure 3), which infers that large compositional gradients between the solid and the liquid phases will exist. When such gradients are established, interfacial tension and stress intensity factor, K_c, are reduced, thereby increasing susceptibility to cracking, as reproduced by Equation 30 [38]:

$$K_C = 2\left\{E\left[\gamma_o - 2N\int_{X_o}^{X} G\left(\frac{dc}{dx}\right)^2 dx\right]\right\}^{1/2} \qquad (30)$$

In Equation 30, dc/dx designates the compositional gradient, E is the Young's modulus, and G and N are material constants. Although large compositional gradients are detrimental, the same gradients are favorable to produce fine dendritic microstructures, as is shown by Equations 20 and 23. From a theoretical standpoint, it therefore appears that an optimal compositional gradient exists to compromise between a high resistance to cracking and a fine dendritic microstructure.

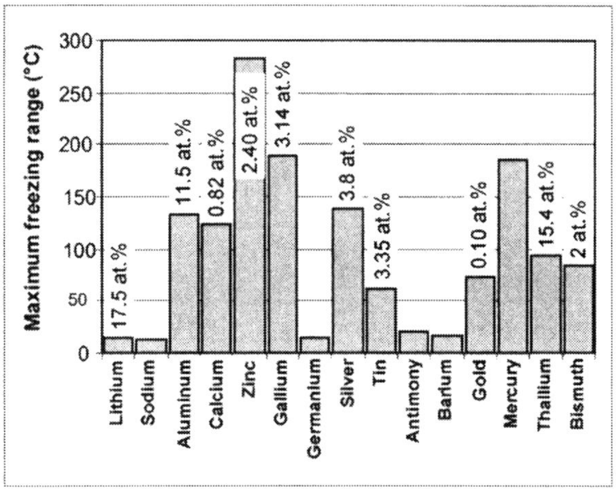

Figure 12 – Maximum freezing range for various alloying elements in magnesium. Solubility at the eutectic temperature, where freezing range is maximized is also indicated.

3.9. Hardness and Strength

Hardness truly measures resistance to indentation, but it can be related to material's yield strength, as shown by the following relationships [39]:

$$\sigma_y = \left(\frac{Hv}{3}\right)0.1^n \qquad (29)$$

where Hv is the material's Vickers hardness, σ_y its yield strength, and n is its strain-hardening exponent. In the base materials, yield strength and hardness both monotonically increase with the alloy aluminum content [4]. Figure 13 confirms that hardness increases linearly with the aluminum content as the percentage of the $Mg_{17}(Al,Zn)_{12}$ intermetallic phase increases. In weld fusion zones, strength can be therefore expected to increase with the alloy aluminum content, providing that defects such as porosity and cracks are prevented. Also, based upon Figure 13 [4-5], strength of weld fusion zones must be greater than that of the original materials, and the amounts of strengthening with respect to the initial materials must depend upon the welding parameters.

When comparing arc welds with laser welds, laser welds are usually harder because the greater power density of the laser beams makes possible the use of high travel speeds (i.e. solidification rates) to produce fine dendritic microstructures. Figure 14 [4-5] agrees with the hardness data (Figure 13), showing that tensile strength of laser welds is greater than that of comparable arc welds [4-5]. Also, note that initial strength of die cast materials can be less than that of extruded materials although second-phase strengthening is greater. Figure 14 [4-5] shows that internal defects can impact the properties of base material significantly, and welding can improve those properties.

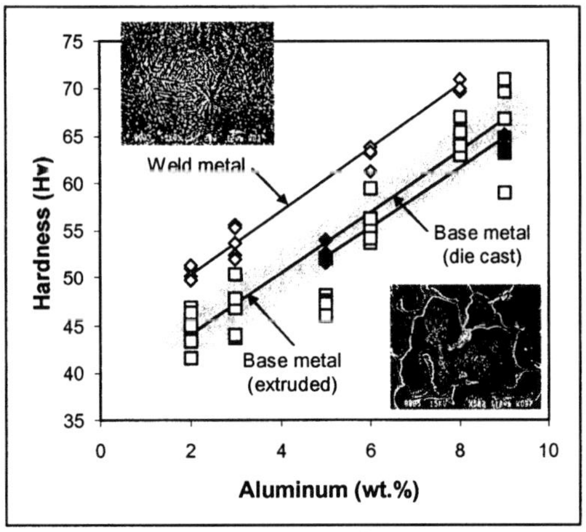

Figure 13 – Comparison between GTA weld hardness and base metal hardness [4-5].

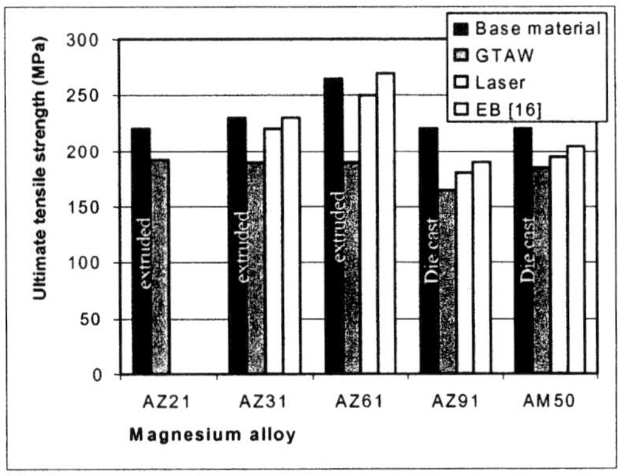

Figure 14 – Ultimate tensile stress of magnesium alloy weld joints [4-5].

3.10. Corrosion of Magnesium Alloy Welds

With a standard potential of -2.37 V with respect to a normal hydrogen electrode, magnesium is anodic with most metals. Although transition metals usually act as cathodic impurities, other metals improve corrosion resistance by forming passive layers and harmless intermetallic phases.

In the AZ alloys, the role of the $Mg_{17}(Al,Zn)_{12}$ phase may be two-fold, depending upon the alloy aluminum content [40]: (1) it acts as a barrier because it is inert to corrosion, and (2) it plays the role of a cathode for hydrogen evolution, as simply illustrated in Figure 15a neglecting the formation of passive films. As shown in Figure 15b [40], the films protecting surfaces of AZ magnesium alloys are strongly influenced by alloy chemical composition. In the near-single-phase AZ21 alloy and the dual-phase AZ91 magnesium alloys, the semi-metallic layers of alumina, Al_2O_3, and magnesia, MgO, coat the anodic magnesium phase (Figure 15b) and protect it from further corrosion. These layers are more stable than the pure phase immersed in water, and are less electrically conductive. Transport through these protective layers by diffusion must occur for the corrosion of the metal underneath to continue. The solubility of these passive films in the corrosive solution is also an important parameter regarding corrosion resistance.

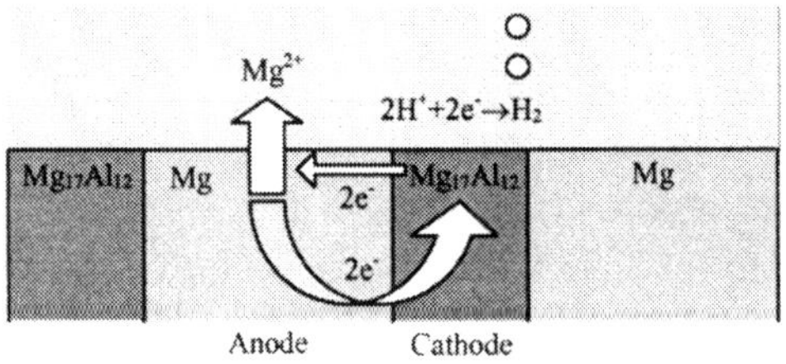

Figure 15a

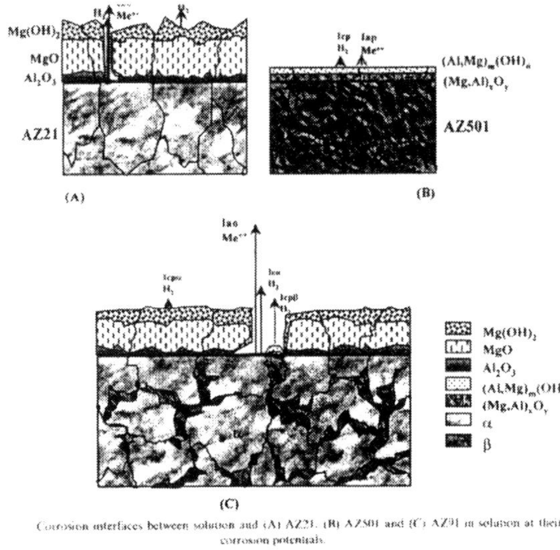

Figure 15b

Figure 15 – (a) Galvanic cell mechanism for the corrosion of AZ alloys in the absence of oxygen; (b) The corrosion of AZ alloys in solution at their corrosion potential [40].

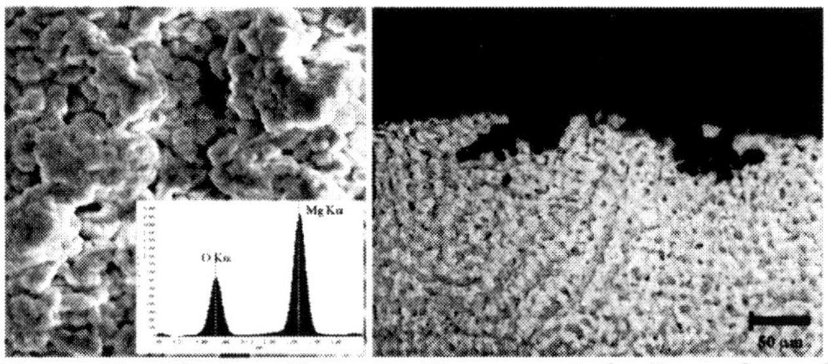

Figure 16 – Electron dispersive X-ray counts for the surface of a corroded as-extruded AZ31 weld surface and optical micrographs of the microstructure.

Like the alloy composition, the distribution and morphology of the intermetallic phases greatly affects corrosion, and that makes welds significantly different from their base materials. Figure 16, which shows the surface of a corroded weld, reveals that corrosion occurs within the fusion zone through the α-Mg dendrites. As indicated by the energy spectrum of Figure 16, magnesium is gradually converted into magnesium hydroxide, a phase which dissolves readily in most solutions, particularly those with a low hydrogen activity. Although susceptible to corrosion, the microstructures of welds are significantly more resistant. In AZ91 laser welds, the fine microstructures can be significantly stronger than that of base materials. Other than the microstructural refinement, the enhanced corrosion resistance of welds has been attributed to the retention of alloying elements in the α-Mg phase and to the new morphology and distribution of the β-$Mg_{17}(Al,Zn)_{12}$ phase, which is possibly more effective in retaining the magnesium hydroxide film attached to the bulk material.

4. CONCLUSIONS

The fusion welding of magnesium and its alloys has been explained using theory, semi-empirical correlations, and experimental results. It is shown that the alloying of magnesium is by nature restricted to a few elements (Figure 1), generally eutectic-formers that leave strengthening intermetallic phases after solidification (Figures 3 to 5). Although these second-phases enhance mechanical (Figure 14) as well as corrosion properties (Figure 15), they also affect welding considerably. Melting efficiency by either an arc plasma or a laser (Figure 6), dendritic microstructure (Figure 11), fusion zone hardening (Figure 13), solidification cracking (Figure 12) and heat-affected zone liquation (Figure 5) are caused by the necessity of alloying magnesium. For a given alloy, the extent of each also depends strongly on the welding process and the thermal history. Weld fusion zone irregularities (Figure 7), weld evaporation loss (Figure 10), alloying element segregation, and porosity (Figures 7 and 9) are all influenced by the process parameters. The rapid thermal cycles offered by high power density beams are preferred over arc welding thermal cycles. Although hydrogen pores are less likely in magnesium than in aluminum, potential pores can be eliminated through rapid cooling. Fusion welding of commercial magnesium alloys does not present surprises and is well predictable. The models introduced in this article are important to understand the welding of magnesium alloys which are currently developed.

ACKNOWLEDGEMENTS

The main author would like to acknowledge General Motors R&D for granting him the time to complete this article. Elements of this article were borrowed from research funded by Visteon Automotive Systems and the United States Army, both acknowledged for supporting magnesium welding efforts at the Colorado School of Mines.

REFERENCES

1. www.matweb.com
2. Polmear, I.J., in Light Alloys – Metallurgy of the Light Metals, Third Edition, Halsted Press-Wiley (1996).
3. Atkins, K.G., in Magnesium Technology 2003, Edited by H.I. Kaplan, TMS, 153-156 (2003).
4. Marya, M., Edwards, G.R., Marya, S.K., Olson, D.L., in Proceedings of Seventh International Symposium of the Japan Welding Society (7'WS) on "Today and Tomorrow in the Science and Technology of Welding and Joining", Nov. 20-22 2001, Kobe, JAPAN, Vol.1, Paper LM4-3, 597-602 (2001).
5. Marya, M., Ph.D. Thesis, Colorado School of Mines (2002).
6. http://www.intlmag.org
7. www.webelements.com
8. Marya, M., Edwards, G.R., Journal of Materials Engineering and Performance, 10(4):435-443 (2001).
9. Lancaster, J.F., in Metallurgy of welding, Allen and Unwin, London (1987).
10. ASM Handbook, Vol.6 – Welding, Brazing, and Soldering, edited by ASM International, 772-782 (1993).
11. Turkdogan, E.T., Canadian Metallurgical Quarterly, 41(2): 151-162 (2002).
12. Kittel, K., in Introduction to Solid State Physics, Second Edition, John Wiley & Sons, New York (1996).
13. Bakhtiyarov, S.I. Overflet, R.A., Teodorescu, S.G., in Magnesium Technology 2003, edited by H.I. Kaplan, TMS (2003).
14. Munitz, A., Kohn, G., Cotler, C., in Magnesium Technology 2002, edited by H.I. Kaplan, TMS, 303-307 (2002).
15. Steen, W.M., in Chapter 2: Laser Material Processing, Springer Verlag (1991).
16. Materials Science and Technology - A Comprehensive Treatment, edited by R.W. Cahn, P. Haasen, E.J. Kramer, VCH Verlagsgessellschaft Vol.1 – Structure of Solids, 184-216 (1992).
17. Marya, M., International Welding Institute Henri Granjon Prize (2002), Welding in the World (The Journal of the International Institute of Welding), 46(7/8):7-21 (2002).
18. Marya, M., Edwards, G.R., Welding Journal, 81(12):291s-298s (2002).
19. Marya, M., Edwards, G.R., in Proceedings of the 6[Th] International Conference Trends in Welding Research (ASM International, USA), 398-403, Pine Mountain, Georgia, USA (2002).
20. Trevisan, R.E., Schwemmer, DD., Olson, D.L., in Welding: Theory and Practice (edited by D.L. Olson, R. Dixon, A.L. Liby), Elsevier Science Publishers, B.V. (1990).
21. Aarstad, K., Syvertsen, M., Engh, T.A., in Magnesium Technology 2002, edited by H.I. Kaplan, TMS, 39-42 (2002).
22. Grong, O., in Metallurgical Modeling of Welding, (Materials Modeling Series), The Institute of Materials, University Press, Cambridge, UK (1994).
23. Marya, M., Edwards, G.R., accepted for publication in the Welding Journal (2004).
24. Popovic, Z.D., Piercy, G.R., Metallurgical and Materials Transaction A, 60A:1915-1917 (1975).
25. Zhao, H., Debroy, T., Welding Journal, 80(8):204-260 (2001).
26. Marya, M., Edwards, G.R., Science and Technology of Welding and Joining, 7(5):286-293 (2002).
27. Hume-Rothery, Acta Metallurgica 14(1):17-20 (1966).
28. ASM Handbook, Vol.3 – Alloy Phase Diagrams, edited by ASM International, (1992).
29. Saunders, N., Miodownik, A.P., in CALPHAD: Calculation of Phase Diagram, A Comprehensive Guide, Pergamon Materials Series (1998).
30. Miedema, A.R. in Inorganic Chemistry, North Holland Publishing Company, The Netherlands (1975).
31. Chelikowski, J.R., Phillips, J.C., Physical Review B, 17(6):2453-2477 (1978).
32. http://my.noranda.com/Noranda/magnesium/Magnesium+Overview/Why+Magnesium+Alloys/Overview.htm
33. A. Matsunawa, J.D. Kim, N.Seto, M.Mizutani, S. Katayama, Journal of Laser Applications, 10(6):247-254 (1999).
34. K.K. Krenov, Electric Welding Arcs, Mashgiz, Moscow and Kiev (1949).
35. E.F. Emley, in Principles of Magnesium Technology, Pergamon Press, 190-195 (1961).
36. Marya, M., Edwards, G.R., in Proceedings of 6[Th] International Conference Trends in Welding Research, 124-129, Pine Mountain, Georgia, USA (1998).
37. Kurz, W., Trivedi, R., Materials Science and Engineering, A179/A180:46-51 (1994).
38. Berry, Jr. G.J., Olson, D.L., and Matlock, D.K., Materials Science and Engineering. A148, 1-6 (1991).
39. Cahoon, J.R., Broughton, W.H., Kutzak, A.R., Metallurgical and Materials Transactions, 2:1975-1983 (1971).
40. Song, G., Atrens, A., Wu, X., Zhang, B., Corrosion Science, 40 (10), 1769-1791 (1998).

Proceedings from Joining of Advanced and Specialty Materials, 13–15 October 2003, Pittsburgh, Pennsylvania, ASM International, 2004

Investigation of the Fusion Zone Microstructure for Gas Tungsten Arc Welded Gamma Titanium Aluminide

Mario F. Arenas and Viola L. Acoff
The University of Alabama, Department of Metallurgical & Materials Engineering, Tuscaloosa, AL 35487

Abstract

Although various studies have been performed on the weldability of gamma titanium aluminides, little is known about the microstructure that evolves during the gas tungsten arc welding process. This technique has been proposed for gamma repair operations due to its flexibility and relative cost-effectiveness. In this study cast, hot-isostatic pressed (HIPed) gamma titanium aluminides alloys of composition Ti-48Al-2Cr-2Nb were welded using GTAW without preheating. Microstructural examination revealed a nearly lamellar microstructure in the base metal which was transformed to a dendritic structure with evidence of interdendritic gamma phase and metastable structures in the fusion zone. The solid state transformations that occurred were heavily dependent on the cooling rate. It was found that the selected welding parameters should ensure that the heat affected zone cooling rate will be less than 500 K/sec. Such cooling rates prevent the development of deleterious alpha-2 structures, which promote cracking. To further optimize the fusion zone microstructure, postweld heat treatment (PWHT) was employed. It was found that PWHT promotes the formation of the gamma phase and reduces the fusion zone microhardness mainly due to the reduction of the hard, brittle alpha-2 phase.

Introduction

The low density, superior strength and high stiffness at elevated temperatures of TiAl intermetallic alloys have attracted a great deal of attention from the aerospace and automobile industries (Refs. 1-5). However, it is apparent that the ability to weld these materials to themselves and to other materials is the key to make them more attractive. It is of particular interest to study the feasibility of using the gas tungsten arc welding (GTAW) technique, which promises to be an economical way to weld gamma titanium aluminides. Despite its importance, only a few papers have been published on the joining of gamma titanium aluminides using gas tungsten arc welding. For instance, L.C. Mallory et al. (Refs. 6,7) investigated the evolution of the weld fusion zone and HAZ microstructure in preheated GTA welds of a forged gamma Ti-48Al-2Cr-2Nb alloy. When preheating to 800°C was used, no cracking was detected. However, weld beads produced under identical welding conditions but without preheating all contained cracks. A microstructural analysis of the fusion zone showed cored dendritic solidification structures with single colonies of fine lamellar alpha-2/gamma (α_2/γ) constituent extending across the dendrites. The occasional presence of interdendritic gamma grains was also observed. Interestingly, the microhardness of the fusion zone and HAZ were significantly increased relative to the base metal.

T.J. Kelly (Ref. 8) studied the GTA weldability of a cast Ti-48Al-2Cr-2Nb gamma alloy. Prior to welding, all samples were HIPed and some were heat treated after HIPing at 1300°C for 20 hours. As-deposited weld metal exhibited a highly segregated solidification structure that consisted of the gamma phase, colonies of gamma plus alpha-2 ($\gamma + \alpha_2$) laminates, and the Laves phase. Solid-state cracking was observed in all welds of as-HIPed samples even with preheats as high as 600°C. However, samples heat treated at 1300°C after HIPing did not crack even without preheating because heat treating at 1300°C increased the ductility of the specimens.

In another study, Bharani and Acoff (Ref. 9) used autogenous GTA spot welding to produce crack-free welds in cast alloy Ti-48Al-2Cr-2Nb and extruded alloy Ti-46Al-2Cr-2Nb-0.9Mo without the use of preheating. They showed that it is possible to achieve crack-free welds by only applying a post-stress relief treatment at 615°C. The crack length in the fusion zone decreased as a function of increasing current level, with no cracks observed for the highest current levels. Dendritic solidification structures were produced in the fusion zone of both alloys. Microhardness profiles also indicated an increase in hardness from the base metal to the fusion zone.

Although some success has been achieved, a comprehensive understanding of the weldability of these alloys is still lacking. The objectives of the present study are to analyze GTA welding of γ-TiAl by examining the fusion zone structure for different welding currents and to optimize the fusion zone structure and properties using postweld heat treatment.

Experimental Procedure

Materials and Sample Preparation. An alloy of nominal composition Ti-48Al-2Nb-2Cr (at %) was used in this study. The alloy was received in the form of 256 mm long × 110 mm wide × 15 mm thick cast plates from GE Aircraft Engines that were HIPed at 1185°C at 107 MPa for 3 hours followed by a stabilization heat treatment at 1205°C for 2 hours, which was all performed at GE. The cast plates were sectioned into 20 mm × 15 mm × 2 mm specimens for spot welding (i.e. welding with the torch stationary at a single location for a fixed time) and 100 mm × 15 mm × 2 mm specimens for butt welds using electrical-discharge machining (EDM). The specimens were mechanically ground with 120-grit SiC paper then pickled in a solution composed of 400 ml H_2O + 40 g KOH + 40 ml H_2O_2 in order to remove surface contaminants. Before welding, a pre-stress relief treatment was performed at 615°C for 2 hours to relieve any stress produced during grinding of the re-cast EDM layer.

Experimental Apparatus. Autogenous GTA welds were produced using straight polarity, direct current. A welding time of 2 seconds was employed for the spot welds and a welding speed of 3 mms^{-1} for the butt welds. The welding equipment had a spot timer device that allowed precise control of the welding time. The welding currents used ranged from 50-150 amps and the corresponding welding voltage was 14-18 volts. Immediately after welding, the samples were visually checked for the presence of cracks. All welding was performed inside a welding chamber that was purged with argon for 10 minutes prior to welding. The welded samples were submitted to postweld heat treatment (PWHT) at two different temperatures: 1000°C and 1200°C. The purpose of choosing these temperatures was to transform the as-welded microstructures above and below the eutectoid temperature. The samples were exposed to these two PWHT temperatures for three different times: 1, 10 and 48 hours.

Materials Characterization. The welded, as-received, and postweld heat treated samples were sectioned into metallographic specimens. These metallographic samples were prepared using conventional metallographic procedures for titanium aluminides (Ref. 10). Microstructural details were revealed by a Kroll's modified etching solution comprised of 100 ml H_2O + 35 ml H_2O_2 + 5 ml HNO_3 + 10 ml HF. Microstructural characterization was performed using light microscopy, scanning electron microscopy (SEM) equipped with energy dispersive x-ray spectrometry (EDS), and transmission electron microscopy (TEM). X-Ray diffraction (XRD) was used for primary phase identification of the fusion zone and base metal. The welded specimens were prepared for XRD by carefully removing the fusion zone from the specimen using a precision saw. Measurement parameters for XRD were as follows: Cu Kα radiation, accelerating voltage 40 kV and tube current 35 mA. Diffraction patterns were obtained in the step, 2θ mode in the range of 20-90°. Microhardness data was obtained using a Knoop microhardness tester with a load of 500 grams. The Knoop hardness number (KHN) was plotted as a function of distance across the welded specimen.

Results and Discussion

Microstructural Characterization of the Base Metal. The as-received condition of the alloy consisted primarily of a lamellar microstructure composed of alternating platelets of γ and α_2, and some equiaxed gamma grains as shown in Figure 1. This microstructure is the result of the treatment for cast alloys developed by Austin et al. (Ref. 11) which involves HIPing of the as-cast alloy followed by heat treatment in the alpha + gamma field. The XRD analysis of the base metal is shown in Figure 2. In this XRD scan, peaks due to γ and α_2 phases are identified. The prominent peak at about 39° corresponds to the gamma phase (Ref. 12) and it accounts for both the γ lamellae and the equiaxed γ grains observed. After solidification, cast gamma titanium aluminides are transformed into a lamellar structure with a preferred orientation of the lamellae (see Fig. 1). The lamellae are aligned parallel to the outer surface of the ingot, i.e. perpendicular to the heat flow during solidification. This texture is reflected in the XRD measurement showing a high intensity peak in the (111) plane. The small peak at about 41° corresponds to the α_2 phase (Ref. 13), which is present in the base metal as α_2 lamellae. The weak intensity of the α_2 peak is probably due to the very thin α_2 lamellae, which is below the spatial resolution of the XRD technique, and the minimal presence of individual α_2 grains as shown in Figure 1.

Figure 1. Light micrograph of the as-received microstructure.

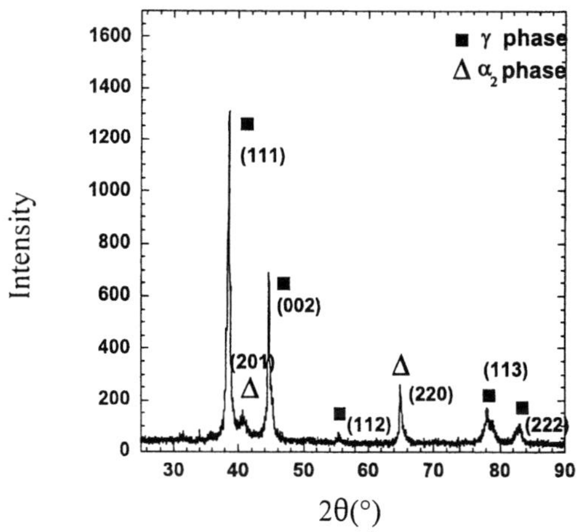

Figure 2. XRD pattern of the base metal.

Fusion Zone Microstructures. Figure 3 shows a light micrograph of a weld obtained for welding with an arc current of 50 A, where the evolution of microstructure from the base metal (left) to the fusion zone (right) is evident. This figure shows that the fusion zone microstructure was composed of columnar grains oriented perpendicular to the direction of the radial fusion boundary. This was expected since during solidification, grains tend to grow in the direction perpendicular to the solid/liquid interface, which is the direction of the maximum temperature gradient and thus the maximum driving force for solidification. Higher magnification micrographs of the columnar structure show lamellar grains and some amount of small blocky grains identified as the gamma phase (Figure 4).

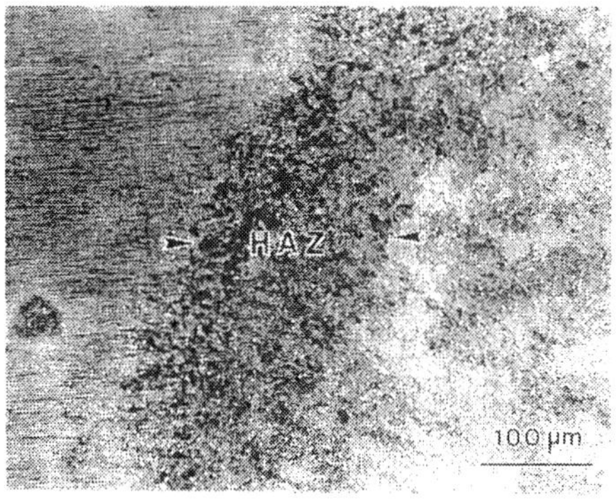

Figure 3. Light micrograph for the 50 A, spot welded specimen. HAZ denotes the heat-affected zone.

Figure 4. Higher magnification light micrograph of the columnar structure obtained for spot welding using a current of 50 amps.

The center of the fusion zone showed a markedly different microstructure. This region, which experienced the highest temperatures, exhibited a dendritic structure (Figure 5). The difference in microstructure between the weld centerline and the fusion boundary can be attributed to different temperature gradients and solidification rates between the two regions. As the welding current was increased, the volume fraction of lamellar structure increased while that for the equiaxed gamma grains decreased.

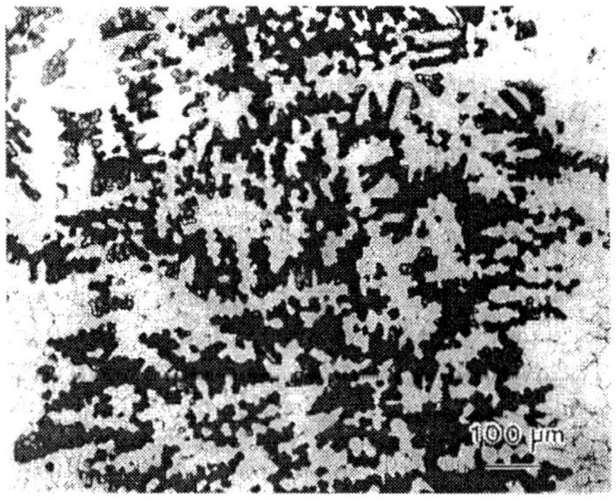

Figure 5. Light micrograph showing the center of the fusion zone for the 50 amps specimen.

Some amount of alpha-two phase, resolvable only by TEM analysis, was identified in the 55 A butt welded specimen (Figure 6). Figure 6a is a bright-field TEM micrograph of the fusion zone showing an α_2 grain with a neighboring lamellar grain. The identification of the phase was confirmed with a selected area diffraction pattern in the [0001] orientation as shown in Figure 6b. The diffraction pattern confirmed the hexagonal close-packed α_2 structure.

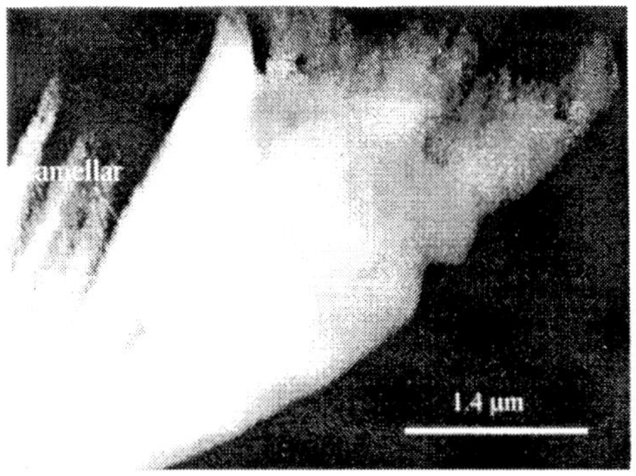

(a)

(b)

Figure 6. TEM micrograph of the fusion zone of the 55A butt welded specimen showing the α_2 phase. (a) Bright-field image where the electron beam was parallel to the (0001) α_2 grain. (b) Selected area diffraction pattern of the region shown in (a).

A high cracking susceptibility was observed for welding at 50 A. A typical crack that occurred is shown in Figure 7. The cracks appeared either immediately after welding or several seconds after welding was performed. These cracks were identified as solidification cracks and occurred at the center of the fusion zone. The center of the weld pool was highly susceptible to crack development and occasionally the cracks propagated into the base metal as shown at the bottom of Figure 7. Interestingly, SEM/EDS analysis found that the cracked surfaces were enriched with Al ranging from 51-54 at.%, suggesting that the heavily segregated interdendritic regions were initiation sites for cracks.

Figure 7. Light micrograph showing a solidification crack that developed for spot welding at 50 amps.

Since the specimens were relatively small (2 mm thick) and there was little or no external stresses exerted to the workpiece, it is believed that the major contributing factor to the susceptibility to solidification cracking was thermal stresses. The thermal stresses developed from the difference in temperature across the fusion zone as a result of the metal at the fusion zone/base metal boundary solidifying much faster than the metal at the center of the fusion zone. The thermal stresses cannot be accommodated due to the lack of ductility of the fusion zone, which results from the increased volume fraction of the brittle α_2 phase compared to the base metal. Furthermore, severe solute segregation at the center of the fusion zone provides fracture initiation sites for the cracks. However, it is believed that decreasing the cooling rate (increasing the welding current), which in turn reduces both α_2 formation and interdendritic segregation, reduces the susceptibility for solidification cracking. Traditional hot cracking was not the operative mechanism since no evidence of grain boundary liquation or fracture features associated with the occurrence of subsolid liquid microconstituents was observed.

No cracks were observed for using welding currents equal to or greater than 75 amps. This observation supports the belief that decreasing the cooling rate by increasing the welding current reduces the susceptibility for solidification cracking. Although butt welded specimens have a much higher heat input compared to spot welded specimens since spot welding is a relatively spontaneous process, it was consistently observed that the specimens spot welded at 50 A exhibited solidification cracking whereas specimens spot welded at currents greater than or equal to 75 A did not. Patterson et al. (Ref. 14) concluded that there is a critical weld cooling rate (300 K/s) below which cracking does not occur for laser welding of γ-TiAl. They observed from light microscopy that an increase in the acicular microconstituent occurred with increasing the weld cooling rate. For the present study, a change in the fusion zone microstructure with increased welding current (decreased cooling rate) was also observed. As the welding current was increased, the volume fraction of the brittle α_2 phase decreased. This microstructural observation was supported by the XRD results that are presented in the present study.

X-Ray Diffraction Analysis of the Fusion Zone. XRD was utilized to determine the different phases that developed in the fusion zone microstructure. Figure 8 shows the results for spot welds obtained at three different heat inputs: 50, 100, and 150 amps. The XRD data in this figure indicates that there was a significant increase in the α_2 phase compared to the base metal condition (Figure 2). This was revealed by the stronger peaks of the α_2 phase, especially the main peak in the (201) plane. The 50 amps specimen had the strongest α_2 peak indicating an increase in the α_2 structure compared to the 150 amps specimen. This suggests that the cooling rate plays an important role in the microstructural control of the fusion zone. In GTA welding of gamma titanium aluminides, solid-state phase transformations are strongly affected by the cooling rates. At lower cooling rates, the lamellar structure is largely predominant. In this case, the α phase transforms into a well-defined α_2/γ lamellar structure ($\alpha \rightarrow \alpha + \gamma$). Studies have shown that this reaction occurs in the two-phase $\alpha + \gamma$ phase field of the TiAl phase diagram (Ref. 15) and exhibits characteristics of a nucleation and growth process, very likely involving diffusion (Ref. 16). At higher cooling rates however, the decomposition of the α phase is suppressed in favor of the ordering of the α phase into the α_2 phase ($\alpha \rightarrow \alpha_2$) resulting in an increased amount of alpha-two in the fusion zone. The present study suggests that welding of TiAl-based alloys should be conducted at cooling rates that will not suppress the decomposition of the α phase. In this work, the heat input or the applied welding current controlled the cooling rates. Moreover, the study of other variables that can also modify the cooling rates, such as welding speed and the use of preheating, was beyond the scope of this work.

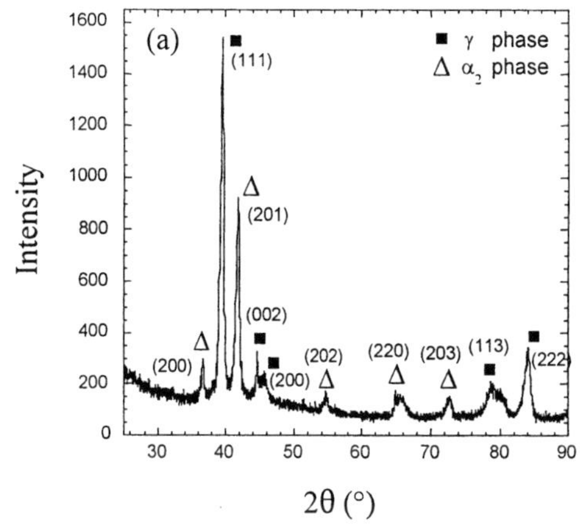

Figure 8. XRD patterns of the fusion zones obtained for different heat inputs. (a) 50 amps, (b) 100 amps, (c)150 amps.

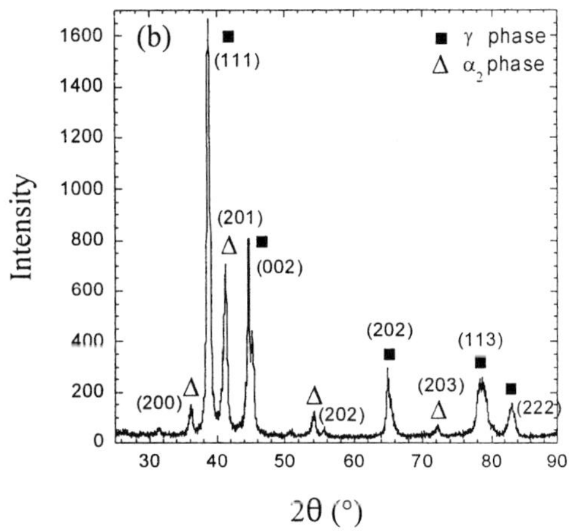

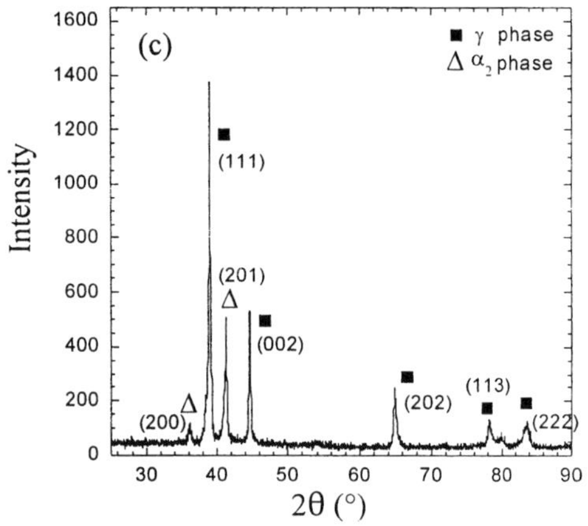

Figure 8 (continued). XRD patterns of the fusion zones obtained for different heat inputs. (a) 50 amps, (b) 100 amps, (c)150 amps.

Microstructures for the 1000°C PWHT. PWHT at 1000°C had a negligible effect on the base metal however, it significantly altered the fusion zone microstructure. A dendritic structure was still evident in the fusion zone after one hour at 1000°C. This is illustrated in Figure 9a. This figure shows that a dendritic structure is still present in the fusion zone to some extent and the region is partially transformed into equiaxed gamma grains. Higher magnification of the center of the fusion zone (Figure 9b) reveals that a fine lamellar structure, compared to the lamellar grains of the base metal, is also present within the dendrites along with some isolated recrystallized gamma grains. The dendrites appear to be in transition to forming distinct small lamellar grains. As the duration of PWHT was increased to 10 hours, coarsening and an increase in volume fraction of the equiaxed gamma structure were observed as shown in Figure 10. The grain growth of the gamma phase seems to be epitaxial as suggested by the orientation of the gamma grains which follow the direction of the prior columnar dendritic phase. Appreciable twinning is observed, which is characteristic of gamma alloys having the duplex microstructure [18].

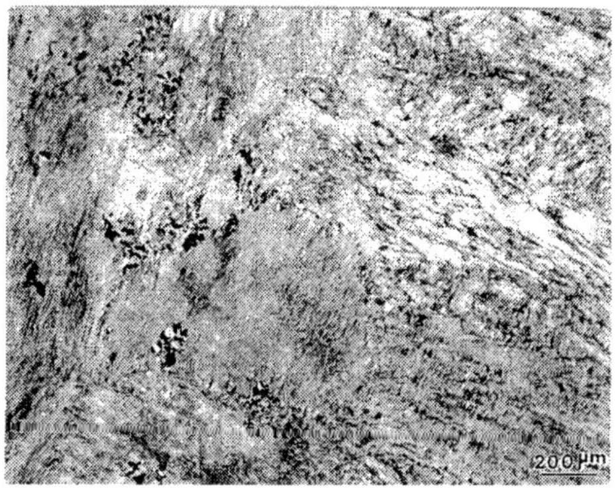

(a)

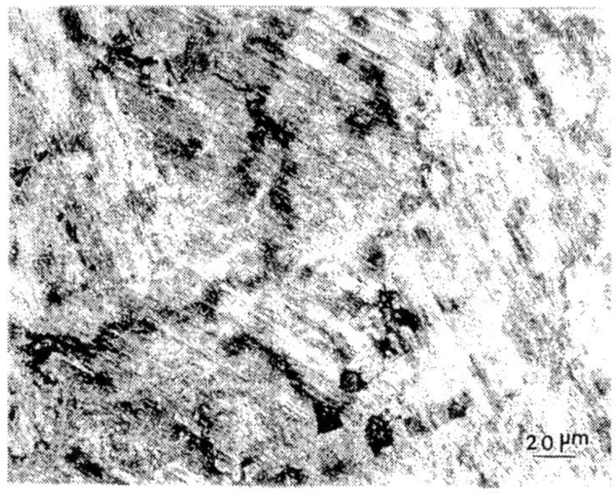

(b)

Figure 9. Light micrographs of the 1000°C PWHT (75 A specimen) for 1 hour. (a) Low magnification showing the base metal and the fusion zone microstructures. Equiaxed gamma grains are seen along with dendritic structure (b) Higher magnification reveals that a fine lamellar structure exists within the dendrites along with some isolated recrystallized gamma grains.

Figure 10. Light micrograph of the 75 A specimen after 1000°C PWHT for 10 hours.

Microstructures for 1200°C PWHT. A PWHT of 1 hour at 1200°C resulted in the dendritic microstructure of the as-welded condition being partially transformed into a fine lamellar structure with intergranular gamma grains as shown in Figure 11. Increasing the duration of heat treatment to 10 hours produced a fine, equiaxed gamma microstructure with some acicular-type morphology running through the γ grains. The dendritic structure was no longer present indicating a full microstructural transformation. The gamma grains observed were smaller compared to those obtained for 1000°C PWHT (Figure 10). In addition to the gamma microstructure, some Widmanstatten microstructure was present (Figure 12). The Widmanstatten-like structure is not single-phase gamma, but consists of both alpha-2 and gamma phases. After 48 hours at 1200°C, the entire fusion zone microstructure was transformed into a structure consisting of fine equiaxed gamma together with Widmanstatten-like structures and lamellar grains (Figure 13).

Figure 11. Light micrograph of the fusion zone after 1 hour at 1200°C (75A specimen). The dendritic structure is seen along with epitaxial gamma grains.

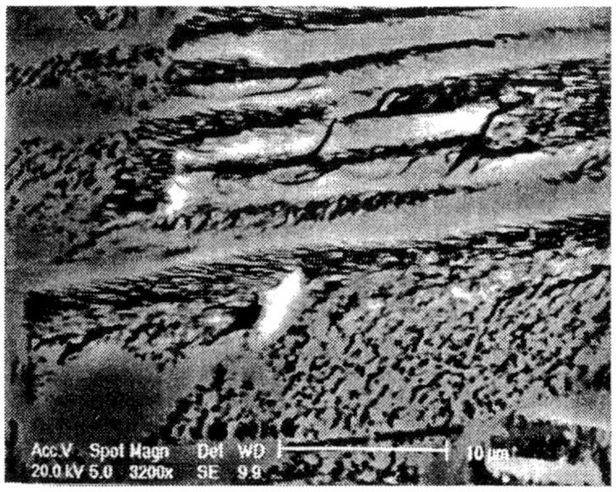

Figure 12. SEM secondary electron image showing the Widmanstatten-like structure that occurred for 10 hours at 1200°C. Both the alpha-2 and gamma phases are present.

Figure 13. Light micrograph of the 48-hour, 1200°C PWHT (115 A specimen). The fusion zone consisted entirely of gamma grains and Widmanstatten-like structures.

XRD Analysis of the PWHT Condition. Figure 14 shows XRD results for the PWHTs performed at 1000°C. Results show the strong texture of gamma titanium aluminides in the (111) plane which was not modified by the postweld heat treatment. Interestingly, the peak intensities of the α_2 phase for all PWHT conditions were lower than those obtained for the as-welded condition (Figure 2). This is indicative of a reduction in α_2 structure as a result of PWHT. Furthermore, the intensity of the α_2 peaks was sequentially reduced as the duration of the PWHT increased. Conversely, the gamma peak intensities increased when the PWHT duration was increased from 1 to 10 hours. However, they remain almost constant as the duration of PWHT was increased from 10 to 48 hours. Similar results were obtained for the 1200°C PWHT specimens.

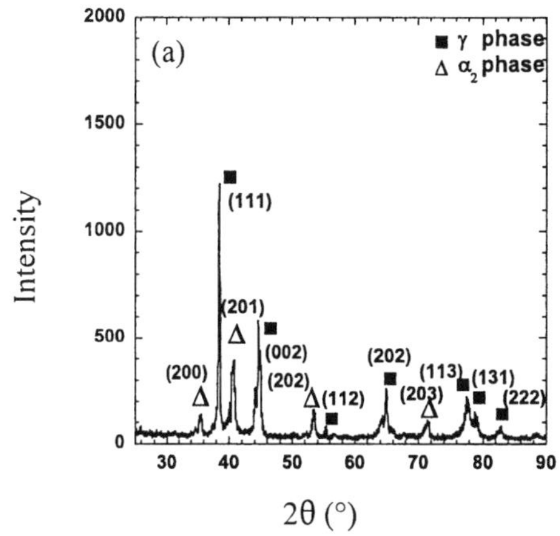

Figure 14. XRD patterns of the specimens subjected to PWHT at 1000°C for (a) 1 hour and (b) 48 hours.

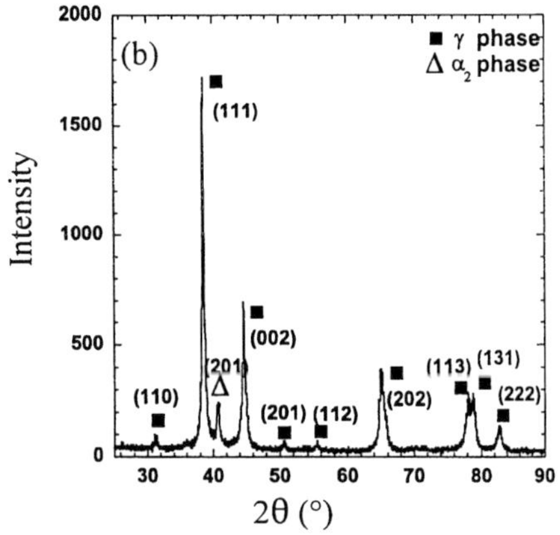

Figure 14 (continued). XRD patterns of the specimens subjected to PWHT at 1000°C for (a) 1 hour and (b) 48 hours.

Microhardness Measurements. The microhardness data were plotted as Knoop hardness number (KHN) versus distance across the fusion zone for the as-welded and each PWHT condition. Figure 15a shows the microhardness profiles for the as-welded and the 1000°C PWHT conditions. As shown, the microhardness increased from the base metal to the fusion zone. The base metal was not significantly altered by the heat treatment as denoted by the small decrease in microhardness that was observed. All postweld heat treated samples exhibited a reduction in microhardness compared to the as-welded condition. This reduction in microhardness was attributed to the presence of the equiaxed gamma microstructure that developed in the fusion zone as a result of PWHT. The scatter of the results can be explained by the presence of hard lamellar bands in the fusion zone. Heat treatment at 1200°C for 48 hours resulted in a slight increase in microhardness compared to the 10-hour specimen. This increase in microhardness can be explained by the presence of Widmanstatten-like structures that contain increased amounts of the α_2 phase. A comparison of the microhardness profiles for the 1000°C PWHT to those for the 1200°C PWHT shows that an overall lower hardness was obtained for the 1200°C PWHT.

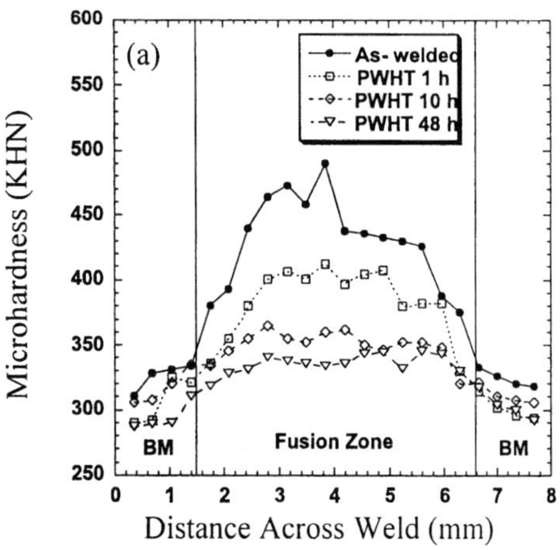

Figure 15. Microhardness profiles for (a) 1000°C and (b) 1200°C PWHT. BM denotes the base metal.

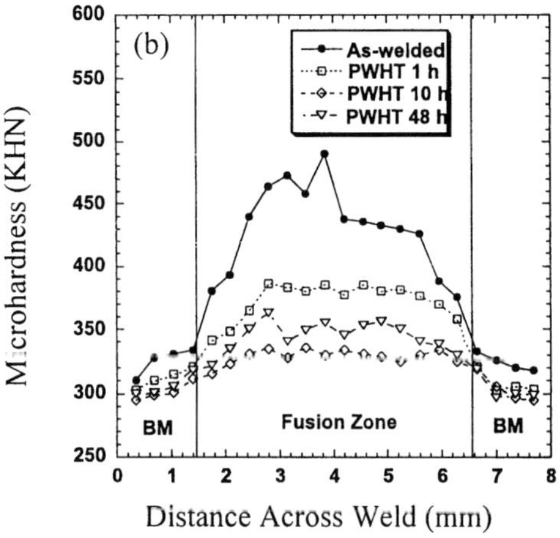

Figure 15 (continued). Microhardness profiles for (a) 1000°C and (b) 1200°C PWHT. BM denotes the base metal.

Conclusions

Autogenous GTA welding has been used to produce crack-free welds in Ti-48Al-2Cr-2Nb without the use of preheating. A procedure for obtaining sound, crack-free welds consisted of pre-stress relief treatment of 615°C for 2 hours and welding at arc currents equal to or greater than 75 A.

The γ-TiAl base metal microstructure consisted of a lamellar ($\gamma + \alpha_2$) structure with some amount of gamma grains. However, the welded microstructure exhibited columnar and equiaxed dendritic structures. The columnar grains extended from the fusion boundary towards the weld centerline, whereas the equiaxed structure developed at the center of the fusion zone. A closer examination of the dendrites showed that they were composed of a considerably fine lamellar structure compared to that of the base metal. Other microstructural constituents observed in the fusion zone were supersaturated α_2 phases. There were more of these structures present at the lower heat inputs. Less cracking was observed as the welding current was increased, which corresponded with a decrease of the α_2 phase in the fusion zone.

Postweld heat treatment was effective in altering the microstructure and decreasing the microhardness of the as-welded fusion zone. The as-welded fusion zone microstructure was transformed into a microstructure composed mainly of equiaxed gamma grains. The volume fraction and grain size of this structure increased as the duration of the PWHT increased. Both PWHTs at 1000°C and 1200°C were effective in reducing the fusion zone microhardness, however, a slightly higher reduction was observed for PWHT at 1200°C, suggesting the driving force for the formation of the gamma structure is more pronounced at higher temperatures. The fusion zone microhardness decreased to values close to that of the base metal after 48 hours of PWHT at 1000°C and after 10 hours at 1200°C. These results agree with the XRD analysis of the fusion zone which showed a reduction of the hard, brittle alpha-two phase upon PWHT.

Acknowledgments

The authors gratefully acknowledge Tom J. Kelly and GE Aircraft Engines (Cincinnati, OH) for donating the alloy used in this study. Financial support for this work was provided by the National Science Foundation under grant number DMR-9702448.

References

1. Y.-W.Kim and D.M. Dimiduk, In *Structural Intermetallics 1997*, p. 531, TMS, Warrendale, PA, (1997).
2. Y-W. Kim, In *Gamma Titanium Aluminides*, p. 637, TMS, Warrendale, PA, (1995).
3. C.M. Austin and T.J. Kelly, In *Gamma Titanium Aluminides,* p. 21, TMS, Warrendale, PA, (1995).
4. Y-W. Kim, *J. Chinese Inst. Eng.*, 22, 13-25, (1999).
5. T. Tetsui, In *Structural Intermetallics 1997*, p. 489, TMS, Warrendale, PA, (1997).
6. L. Mallory, W. Baeslack III, and D. Phillips, *Journal of Materials Science Letters*, 13, 1061-1065, (1994).
7. L. Mallory, W. Baeslack III, and D. Phillips, and T. Kelly, In *Titanium' 92, Science and Technology, vol II*, p. 1115, TMS, Warrendale, PA, (1993).
8. T. Kelly, In *Proc. 3rd International SAMPE Metals Conference*, p. 183, SAMPE, (1992).
9. D.J. Bharani and V.L. Acoff, *Metallurgical Transactions A*, 29A, 927-935, (1998).
10. W. Glatz, B. Retter, A. Leonhard, and H. Clemens, *Struers Journal of Metarialography*, 29, 3-7, (1996).
11. C.M. Austin, T.J. Kelly, K.G. McAllister, and J.C. Chesnutt, In *Structural Intermetallics 1997*, p413, TMS, Warrendale, PA, (1997).
12. P. Duwez and J.L. Taylor, *Journal of Metals*, 4, 70-71, (1956).
13. A.J. Goldak and J.G. Parr, *AIME Transactions*, 221, 639-640, (1961).
14. R.A. Patterson, P.L. Martin, B.K. Damkroger, and L. Christodoulou, *Welding Journal*, 69 (1), 39-s to 44-s, (1990).
15. M. Palm, and G. Inden, In *Structural Intermetallics 1997,* p. 73, TMS, Warrendale, PA, (1997).
16. S.A. Jones and M.J. Kaufman, *Acta Metallurgica et Materialia*, 41, 387-398, (1993).
17. F. Appel and R. Wagner, *Materials Science and Engineering R*, 22 (5), 187-268, (1998).
18. Y-W. Kim and D. Dimiduk, *JOM*, 43 (8), 40-47 (1991).

Proceedings from Joining of Advanced and Specialty Materials, 13–15 October 2003, Pittsburgh, Pennsylvania, ASM International, 2004

Bonding Mechanism in Gold Wire Crescent Bonding

Y. Zhou, X. Li, N.J. Noolu and I. Lum

Microjoining Laboratory
Department of Mechanical Engineering
University of Waterloo
Waterloo, ON, Canada N2L 3G1

1.0. Introduction

Thermosonic wire bonding is the predominant interconnecting technique amongst the wire bonding processes used for chip level packaging in the microelectronic industry [1, 2]. Thermosonic wire bonding essentially consists of the application of bonding force, ultrasonic and thermal energy simultaneously to form a ball bond (first bond of the interconnection) on the chip and a crescent bond (second bond) on the substrate. Although processes are being developed for Cu and Al wires, Au wires are most commonly used for thermosonic wire bonding. Very limited work has been published on crescent bonds relative to those on ball and wedge bonds [1].

It is widely accepted that wire bonding is a solid-state joining process [3, 4] and examination of the wire bond interface in a transmission electron microscope [5] and wire bonding at liquid nitrogen temperatures [6] substantiate the above argument. It is also proposed that the temperature rise during bonding is probably less than 100°C, at least under normal bonding conditions [6], which may be low for the bonding to be considered as a thermally activated diffusion process [7]. Based mainly on studies in Al and Au wire wedge bonding and, to a greater extent, on Au wire ball bonding, several models have been proposed to explain the mechanisms of wire bonding. Some have emphasized the importance of large interfacial deformation to remove surface contaminants by surface extension (i.e., to expose clean surfaces for bonding), regardless of the application of ultrasound, although ultrasonic softening resulted in higher deformation [6, 8-10]. This deformation theory cites that the observed bond formation at the periphery of the wire/pad contact area to discount the importance of fretting mechanism caused by interfacial sliding since it would produce bonding in the central region. According to the deformation theory, ultrasonic energy merely acts to reduce the force required to deform the wire and substrate [6, 9].

However, from a study on Au ball bonding using microsensors, Mayer et al. [11] indicated that the sliding and scrubbing caused by the relative motion at the wire/pad interface plays a significant role in the bond formation and ultrasonic softening contributes little to the development of the bond strength. These observations directly contradict the deformation theory. Studies on Au ball bonding using 100 KHz ultrasonic vibrations also showed that the micro welded regions form preferentially in the direction of the ultrasonic vibration [12]. Osterward et al. [13] also suggested the existence of relative movement between pad and wedge/wire in a footprint study of Al wedge bonds.

The classical theory of contact mechanics indicates that interfacial tangential force, even when smaller than the critical value associated with apparent sliding, will still produce an annular area with relative interfacial movement of 0.25-2.5 μm (microslip) between two bodies pressed together under a normal force [14]. This annular area at the periphery of the faying surfaces surrounds a central "stick" domain with no sliding and spreads radially inwards with increasing tangential forces [15]. When the tangential force is sufficiently high (especially when oscillating), this microslip could result in surface damage due to fretting [14]. This microslip model has been applied to account for the preferential bonding at the edge of an ultrasonic wire bond [13, 16]. In addition, Winchell and Berg [7] suggested that a wave-form material flow (initiating at the bond periphery and extending toward the centre with time) is responsible in breaking up of the surface oxides and the formation of the bond.

The strength of wire bonds is generally evaluated by peel testing and the failure is characterized mainly by two modes:

wire lift-off (fracturing through the wire/pad interface) occurs when the bond is sufficiently weak, while the wire fractures at the neck of the bond when the bond is sufficiently strong [1, 7, 8]. In the present work, microwelded regions were characterized in Au wire crescent bonding by observing the footprints left on the substrates after peeling off the wires. This facilitated an understanding of the roles of the bonding force, substrate temperature and ultrasonic vibrations on the initiation and the growth of the microwelded regions.

2.0. Experimental Methods

A Kulicke & Soffa 4524D ball bonder was used to perform the bonding experiments. The bonding operation was performed in semi-automatic mode, in which bond position was controlled manually by a multimouse while other functions were automatically performed by the bonder. The capillaries were supplied by Gaiser Tool Company (part number 1574-18-437GM). The Au wire used for crescent bonding was a 25 μm in diameter with an elongation of 1~3% and was supplied by American Fine Wire (Cat. No. AW8). The nominal composition of the wire was 99.99% gold, 4~9 ppm beryllium, and 2~5 ppm calcium. The substrate was 0.8 mm-thick Kovar sheet electroplated with a 1.25 μm thick layer of Au. A 1 μm thick Ni interlayer was deposited between the Kovar substrate and the Au plating to act as a diffusion barrier. Fig. 1 shows the typical morphology of the surface asperities of the Au plating.

The bonder was powered on for at least 30 mins prior to bonding in order to ensure thermal stabilization. After the work stage reached the preset temperature the substrate was clamped to the stage. Bonding was delayed at least 6 mins after the substrate was placed on the stage, to allow the substrate to reach the stage temperature. Crescent bonds were positioned such that the direction of ultrasonic vibrations was parallel to the axis of the bonding wire at the bonding location. After every set of 10 interconnections (ball and crescent bonds) was made with each combination of bonding parameters, the substrate with bonds was immediately removed from the stage for peel testing. A pair of tweezers was used to break the ball-bond neck before peel testing, and then the tail wire was taped to the lever of a pull tester to measure the breaking force of the crescent bond. Bonding parameters, such as ultrasonic power, bonding force, bond time, and stage temperature, were designed to produce underdeveloped crescent bonds so that wire lift-off would occur. The bond with a peel force that was close to the average value of ten samples under each combination of bonding parameters was selected for detailed study of morphological features of the footprints left on the substrates using scanning electron microscopy (SEM). The metallurgical bonding was characterized by the presence of fractured microwelds on the footprint, i.e., microparticulates of Au wire left on the footprint after peeling the crescent bond apart.

3.0. Results

Table 1 lists typical bonding parameters used for the crescent bonding, and the resultant peel force and maximum footprint width of the bonds. These results along with the morphological features of the crescent bond footprints are presented and analyzed below.

3.1. Bonding without Ultrasound

The role of bonding force on the bond initiation was studied first by varying the force values at a constant substrate temperature of 162°C, so as to alter the extent of deformation of the substrate and affect the peel strength of the crescent bonds. The ultrasonic power was disabled to eliminate the effect of ultrasonic vibrations. Fig. 2a shows the footprint of a bond made with a bonding force of 35 gf. The maximum bond width was 23 μm and the peel force was almost zero (Table 1). Fig. 2b, showing the details of the highlighted area of Fig. 2a are given in Fig. 2b. Observations of Fig. 2b show that the asperities in the center of the footprint were deformed due to normal stress. No metallurgical bonding (characterized by fractured microwelds) was observed under these bonding conditions.

Fig. 3a shows the footprint of a bond made at the same temperature (162°C) as the bond shown in Fig. 2a, but with a higher bonding force of 120 gf (Trial No. 2 in Table 1). The maximum width and peel force of the bond was about 45 μm and 1.5 gf, respectively. The increase in the bond maximum bond width in Fig. 3a, compared to Fig. 2a shows that the higher bonding force results in higher deformation of the wire. Detailed SEM observations of the deformation pattern of the asperities (Fig. 3b) indicated that the stresses experienced by the substrate varied from normal stresses at the center (judged from the undistorted asperity bases) to lateral shear stresses at the edge of the footprint (judged from laterally deformed asperities). The region that is predominantly subjected to normal stresses is termed as 'central region' and the region that is predominantly subjected to shear stresses is termed as 'outer region'. No microwelds were noticed in the footprint (Fig. 3(b)) although a low peel force (1.5 gf) was registered.

Further studies were conducted by varying substrate temperature so as to obtain higher deformation of the asperities (Table 1). The bond made at 80 gf bonding force and 262°C did not result in a significant increase in deformation and hence bond strength (Trial No. 3 in Table 1). However a bond made with a force of 120 gf at the same substrate temperature (262°C) resulted in higher maximum bond width and peel force of 50 μm and 3.7 gf, respectively (Trial No. 4). These bonding parameters promoted metallurgical bonding between the Au wire and substrate and the bonding is evident by the presence of fractured micro welds on the footprint in Fig. 4. A further increase of the substrate temperature to 362°C, at a bonding force of 55 gf, though resulted in a lower maximum bond width of 35 μm increased the peel force to 8 gf (Table 1). This increase in peel force is clearly reflected in the higher level of microwelds (Fig. 5). It can also be seen from Fig. 4a and Fig. 5a that microwelds were mostly concentrated in the outer regions of the crescent bond. Very few fractured microwelds were noted in the central region of the footprint (Fig. 5c), which was not observed when the substrate temperature was at 262°C or lower (Fig. 4).

3.2. Bonding with Ultrasound

Studies were performed on crescent bonds with the application of ultrasonic vibrations to investigate the effect of ultrasonic energy on the initiation and the growth of the metallurgical bonding (Table 1). Fig. 6a shows the footprint of a bond made at 23°C, an ultrasonic power of 1 W and a bonding force of 35 gf (Trial No. 6). Its maximum width was about 47 μm and peel force was about 5.2 gf. The ductile fracture surface as revealed by the microvoid coalescence clearly demonstrated the presence of bonded areas at the end of the long axis of the footprint (Fig. 6b). The extent of metallurgical bonding in this region was higher than that at the lateral edge of the footprint (Fig. 6c). Fractured microwelds were also observed at the central regions of the footprint (Fig. 6d), which was not seen in the bonds made at zero ultrasonic energy and low substrate temperature.

It was also observed that, with the same substrate temperature (23°C) and ultrasonic power (1 W), an increase of bonding force to 80 gf reduced the peel force to 2.7 gf even though the resultant maximum width was increased to 55 μm (Trial No. 7). In this case, the extent of bonding at the end of the long axis of the footprint was much lower, and no bonding occurred at the lateral edges or centre of the footprint. The reduction in the extent of bonding and peel force at higher bonding force suggest that the increase in bonding force impedes the effective transfer of ultrasonic vibrations in forming a metallurgical bond.

4.0 Discussion

4.1. Effect of Deformation

Solid-state bonding includes a group of joining processes that produce coalescence of the faying surfaces of solids without any melting. In theory, interatomic forces can cause spontaneous metallurgical bonding between two solid surfaces if they are both perfectly clean and atomically flat, and brought into intimate contact [3-5]. This has been confirmed by a number of experiments under ultrahigh vacuum conditions (e.g., in wafer bonding [17]). However, engineered surface features, typically have surface contaminants (oxide film, adsorbed layers and entrapped gas) and surface roughness, which prevent faying surfaces from achieving intimate contact. Therefore, overcoming these surface barriers (especially surface films) is an extremely important prerequisite for the initiation of bonding [18]. Pressure welding is one of the extensively-studied solid-state joining process, in which pressure is applied to produce plastic deformation of the metals to disrupt/displace surface contaminants and to create intimate contact between underlying fresh metals. While earlier research work focused more on the role of macro bulk deformation, it was later found that the local deformation at the faying surfaces, especially surface extension by shear strain, was more important than the macroscopic deformation [19, 20]. But it is not always possible to separate the relative effects of normal and shear stress in an experiment. In addition, relative tangential movement between the faying surfaces can greatly enhance interfacial bonding by breaking up the surface films [21]. Thermal energy is sometimes applied (as in hot pressure welding) to promote the deformation and hence metallurgical bonding because of reduced yield strengths of metals at elevated temperatures.

Wire bonding without the application of ultrasonic energy (also termed thermocompression bonding), therefore, resembles hot pressure welding for large-scale applications. Unlike pressure welding where two pieces of metal are deformed simultaneously by a pair of tools, the deformation of the wire and Au film, in this work, was promoted by the capillary on the top and constrained by the hard-to-deform Kovar core of the substrate at the bottom. The surface extension of the Au film was therefore mainly through the deformation of surface asperities directly caused by wire deformation, and this surface extension peaked at near-peripheral regions [10].

At the moment the wire first makes contact with the substrate surface, the contact area between the wire and substrate would be a single line for crescent bonding. If the bonding force and substrate temperature were high enough, the contact area would increase primarily because the deformed wire surface unfolds onto the substrate surface; this process was modelled numerically by Takahashi and Inoun [10]. However, the unfolding of the wire surface would mainly produce normal stress pressing the asperity peaks downwards, and little lateral shear stress would be experienced by the substrate asperities (Fig. 2b). Further increase in wire deformation (caused by increasing bonding force and/or substrate temperature) would produce interfacial deformation by shear stress in the existing contact area that was created earlier during the unfolding of the wire surface. As a result of progressive interface deformation, the substrate surface asperities in contact with the wire would be sheared outwards away from the contact centre (Fig. 3b). As the extension of existing contact surfaces and the formation of additional wire/substrate interfaces happened simultaneously and continuously, the degrees of lateral shear deformation of substrate surface asperities would vary from the contact center to the perimeter. The observation of microwelds mostly in the outer regions and minimal in the central regions indicates that shear stresses are required to promote metallurgical bonding.

When the substrate temperature is relatively low (162°C or lower), as observed in Figs. 2 and 3, the shear deformation of the asperities is not sufficient to initiate metallurgical bonding. The increase of substrate temperature to 262°C started to produce microwelded regions (Fig. 4) because of the increased deformation as a result of the lowered yield strength of Au. These microwelds were prevalent in the peripheral regions of the bond, where shear stresses were the highest, which is consistent with the observations in hot pressure welding that surface extension by shear stress is much more effective in breaking up surface films and hence initiating metallurgical bonding.

4.2. Effect of Temperature

Increasing temperature promoted metallurgical bonding not only by increasing deformation, but also by directly reducing surface contaminants. This is consistent with the observation that the bonds were stronger at a substrate temperature of 362°C (Fig. 5) than that at 262°C (Fig. 4). This improved bonding at higher temperature was also observed in Au wire ball bonding and has been attributed to the reduction of surface organic contamination on the Au surface as a result of thermal desorption [22]. For example, based on an AES (Auger Electron Spectroscopy) analysis, Jellison [22] reported that 12 hr heating in air at 300°C reduced the level of organic surface contamination of the as-received Au substrate by 90% while heating at 235°C did not show significant effect. However, in microelectronic packaging, the substrate temperature is usually limited to 220°C to avoid thermal damage to die-attach plastics, packaging materials, laminates and sensitive chips. This is the reason why thermocompression bonding (where only increased bonding force and substrate temperature but no ultrasound are used) is seldom used in microelectronics today, and has been replaced by thermosonic bonding [1].

4.3. Effect of Ultrasonic Energy

Comparing Fig. 2 and Fig. 3 with Fig. 6, it is evident that the application of ultrasonic energy promotes metallurgical bonding. At the outset, it could be argued that the ultrasonic energy causes higher macro deformation by which it promotes bonding. However, further observations and comparisons indicate that this may not be the case:

- When ultrasound was applied, metallurgical bonding was mainly observed at the end of the long axis of the footprint (Fig. 6), compared with the predominance of bonding at the lateral peripheral regions when no ultrasonic energy was used (Figs. 4 and 5). This indicates that ultrasonic vibration produced a different pattern of bond initiation.
- Ultrasonic vibration also promoted metallurgical bonding at the footprint center (Fig. 6d), where there was little shear deformation or metallurgical bonding observed when no ultrasound was used (Figs. 2 and 3). This may indicate the occurrence of interfacial sliding [11] since other mechanisms involving shear deformation and microslip would only result in bonding at the peripheral regions. Thermal desorption would not be a factor in this case because of very low substrate temperature.
- The observation in this work that increasing bonding force reduced metallurgical bonding at both the periphery and center was consistent with the observations by Mayer [23] that excessive bonding force would reduce interfacial sliding and hence bond strength by damping the relative movement between the wire and substrate. In other words, the correlation of bond quality and deformation may not be possible [13], which disagrees with the prediction of the deformation theory. A higher bonding force generally requires higher ultrasonic power.

The above discussion points to the conclusion that ultrasonic energy promotes bonding by mechanisms other than deformation-dominated. As suggested by Mayer et al. [23], interfacial sliding during ultrasonic bonding could play a significant role in the dispersal of the surface films and promote bonding.

5.0. Summary

A study was performed on the initiation of metallurgical bonding (characterized by the presence of microwelds between the wire and substrate) in Au wire crescent bonding. Bonding parameters, such as ultrasonic power, bonding force, bond time, and substrate temperature, were designed to produce underdeveloped crescent bonds so that wire lift-off occurred in peel testing. The morphological features of the wire footprint left on the substrate after peel testing were studied using scanning electron microscopy and compared to the peel strength obtained. The major observations are summarized as follows:

- Metallurgical bonding mainly initiated at the lateral peripheral regions (close to the perimeter) of the wire bond in the absence of any ultrasonic energy. The region of best bonding corresponds to the location of highest shear deformation between wire and substrate at the interface. This shear deformation compounded with the effects of bonding force and substrate temperature causes interfacial extension and probably disrupts surface films and exposes fresh metals for bonding.
- Metallurgical bonding also initiated at the bond central region when the substrate temperature was very high, which may be due to direct reduction of surface contaminants by thermal desorption.
- Ultrasonic vibration played a predominant role in wire bonding, producing different patterns of bond formation compared to the bonds made without ultrasound. This

appears to indicate that the application of ultrasound in wire bonding introduced other mechanism(s), such as interfacial sliding, as opposed to pure deformation because of ultrasonic softening.

6.0. Acknowledgements

The authors would like to gratefully acknowledge Natural Sciences and Engineering Research Council (NSERC) of Canada and Microbonds Inc., Markham, Ontario, Canada for their support to this work. Dr. S. Lawson and Mr. C. Carr from the University of Waterloo are appreciated for their assistance and discussions during the course of this study.

7.0. References

1. G.G. Harman, Wire Bonding in Microelectronics – Materials, Processes, Reliability and Yield (2nd Edition), McGraw-Hill, New York, NY, 1997.
2. S. Ruegg and D. Truncellito, Advanced Packaging, April, 2001, pp. 73.
3. S.-W. Chan, IEEE Trans. Comps. Hybrids Manuf. Technol. Vol. 12, 1989, pp. 39.
4. J.E. Krzanowski, IEEE Trans. Comps. Hybrids Manuf. Technol. Vol. 13, 1990, pp. 176.
5. N. Murdeshwar and J.E.Krzanowski, Metall. Mater. Trans.A, Vol.28A, 1997, pp.2663.
6. G.G. Harman and J. Albers, IEEE Trans. Parts, Hybrids Packag., Vol. PHP-13, 1977, pp. 406.
7. V.H. Winchell and H.M. Berg, IEEE Trans. Comp. Hybrids Manuf. Technol., Vol.CHMT-1, 1978, pp. 211.
8. K.C. Joshi, Weld. J., Vol. 50, 1971, pp. 840.
9. G.G. Harman and K.O. Leedy, 10th Annual Proc. Reliability Physics, Las Vegas, USA, April 5-7, 1972, pp. 49-56.
10. Y. Takahashi and M. Inoue, J. Electron. Packag., Vol. 14, 2002, pp. 27.
11. M. Mayer, O. Paul, D. Bolliger and H. Baltes, IEEE Trans. Comps. Packg. Technol., Vol. 23, 2000, pp. 393.
12. Carrass, A., and Jaecklin, V.P., "Analytical methods to characterize the interconnection quality and gold ball bonds", Proc. EuPac '96, Essen, Germany, Jan. 31-Feb. 2, 1996, pp. 135.
13. F. Osterward, K.-D. Lang and H. Reichl, ISHM '96 Proceedings, pp. 426.
14. K.L. Johnson, Contact Mechanics, Cambridge University Press, London, 1985.
15. R.D. Mindlin, Trans. ASME, Series E, J. Appl. Mechanics, Vol. 16, 1949, pp. 259.
16. G.K.C. Chen, International Microelectronic Symposium, October 30-31, 1972, Washington D.C., pp. 5-A-1-1 to 5-A-1-9.
17. U. Gösele, H. Stenzel, T. Martini, J. Steinkirchner, D. Conrad and K. Scheerschmidt, Appl. Phys. Lett., Vol. 67, 1995, pp. 3614.
18. W. Tan, Y. Zhou and H.W. Kerr, Metall. Mater. Trans. A, Vol. 33A, 2002, pp. 2667.
19. D.R. Milner and G.W. Rowe, Metallurgical Reviews, Vol. 7, 1962, pp. 433.
20. H.A. Mohamed and J. Washburn, Weld. J., Vol. 9, 1975, pp. 302-s.
21. E. Homers, British Weld. J.,Vol 6, 1959, pp. 29.
22. J.L. Jellison, IEEE Trans. Parts, Hybrids, Packag., Vol. PHP-11,1975, pp. 206.
23. M. Mayer, Microelectronic Bonding Process Monitoring by Integrated Sensors, PhD Dissertation, Swiss Federal Institute of Technology (ETH) Zurich, 2002.

Table 1 Bonding Parameters, Maximum Width and Peel Force

Trial Number	Substrate Temperature (°C)	Bonding Force (gf)	Ultrasonic Power (W)	Maximum Width (μm)	Peel Force (gf)
1	162	35	0	23	0.25
2	162	120	0	45	1.5
3	262	80	0	38	2.2
4	262	120	0	50	3.7
5	362	55	0	35	8.0
6	23	35	1	47	5.2
7	23	80	1	55	2.7

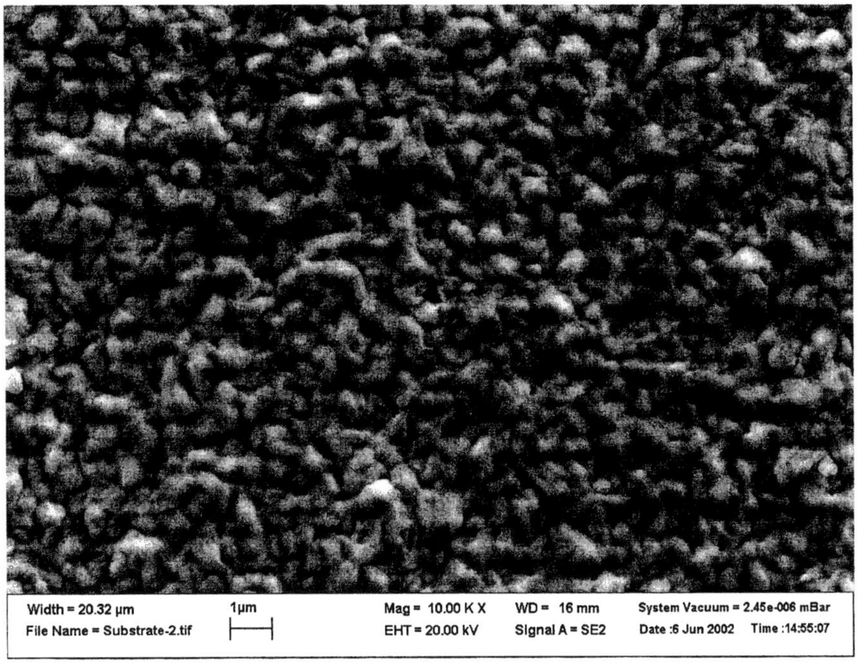

Fig. 1. Secondary electron micrograph of the surface texture of an electroplated gold surface.

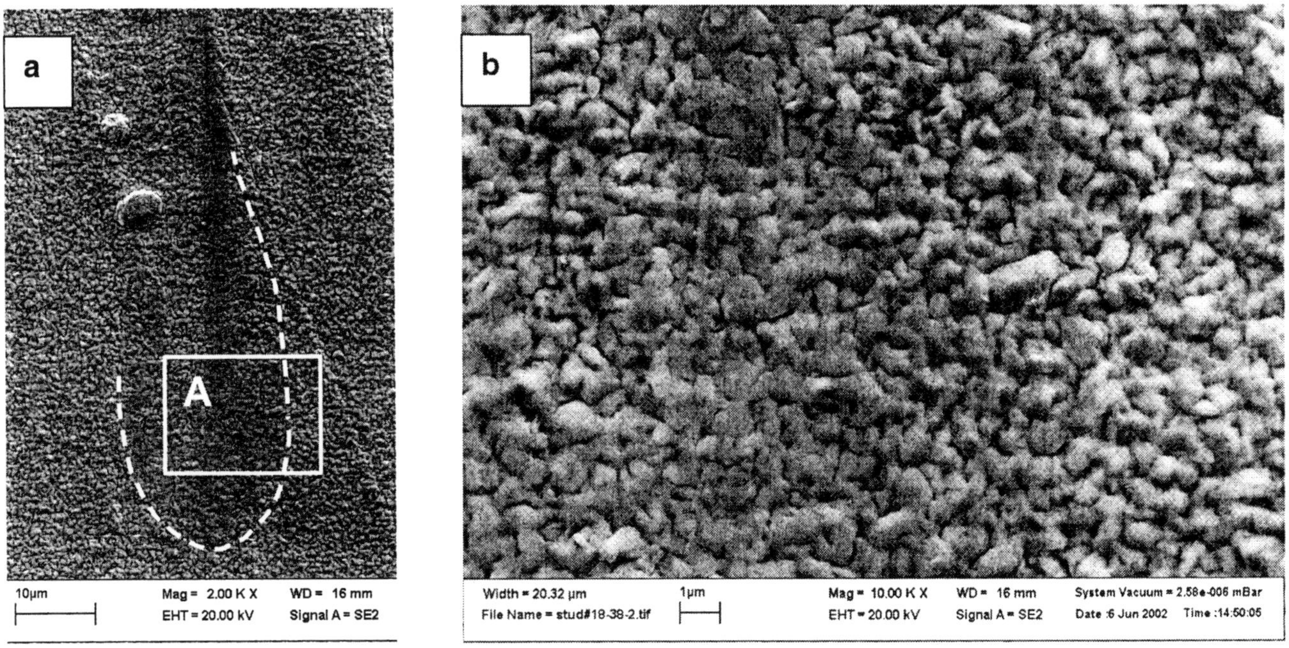

Fig. 2. (a) Footprint of a bond made at 162°C and 35 gf force without ultrasound: the maximum width was about 23 µm and the peel force was negligible. Dashed line indicates part of the footprint perimeter. (b) Details of the enclosed area A in (a).

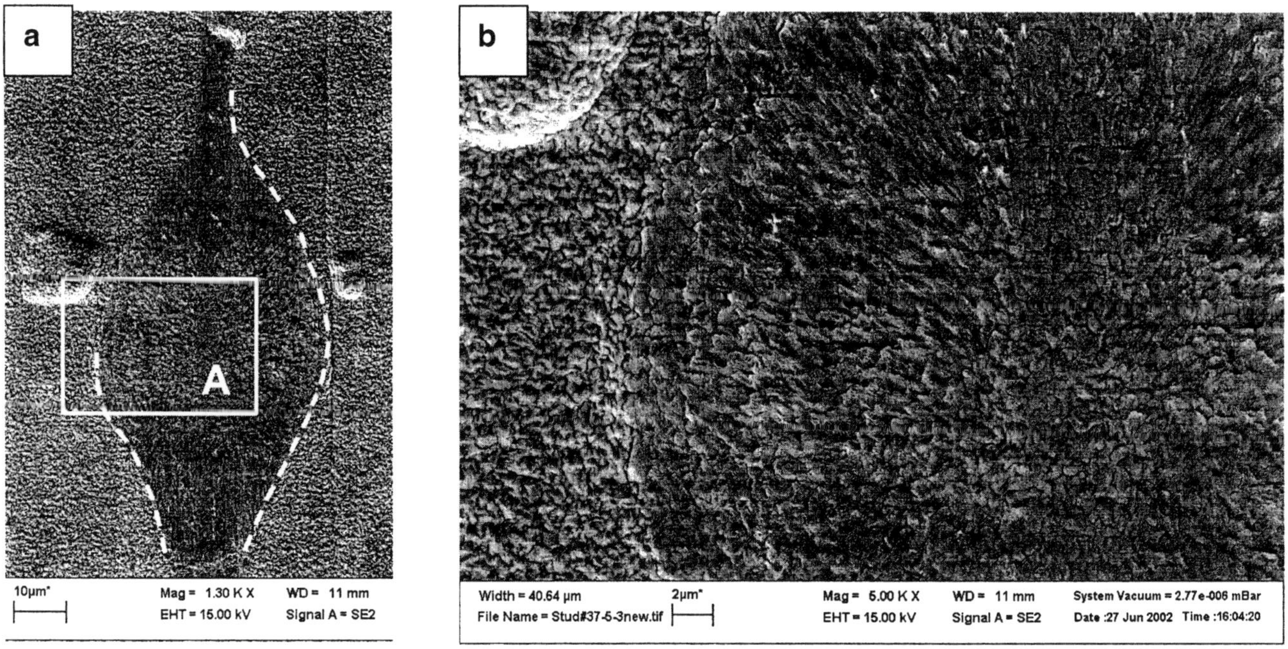

Fig. 3. **(a) Footprint of a bond made at 162°C and 120 gf force without ultrasound: the maximum width and peel force of the bond was about 45 µm and 1.5 gf, respectively. Dashed line indicates part of the footprint perimeter. (b) Details of the enclosed area A in (a).**

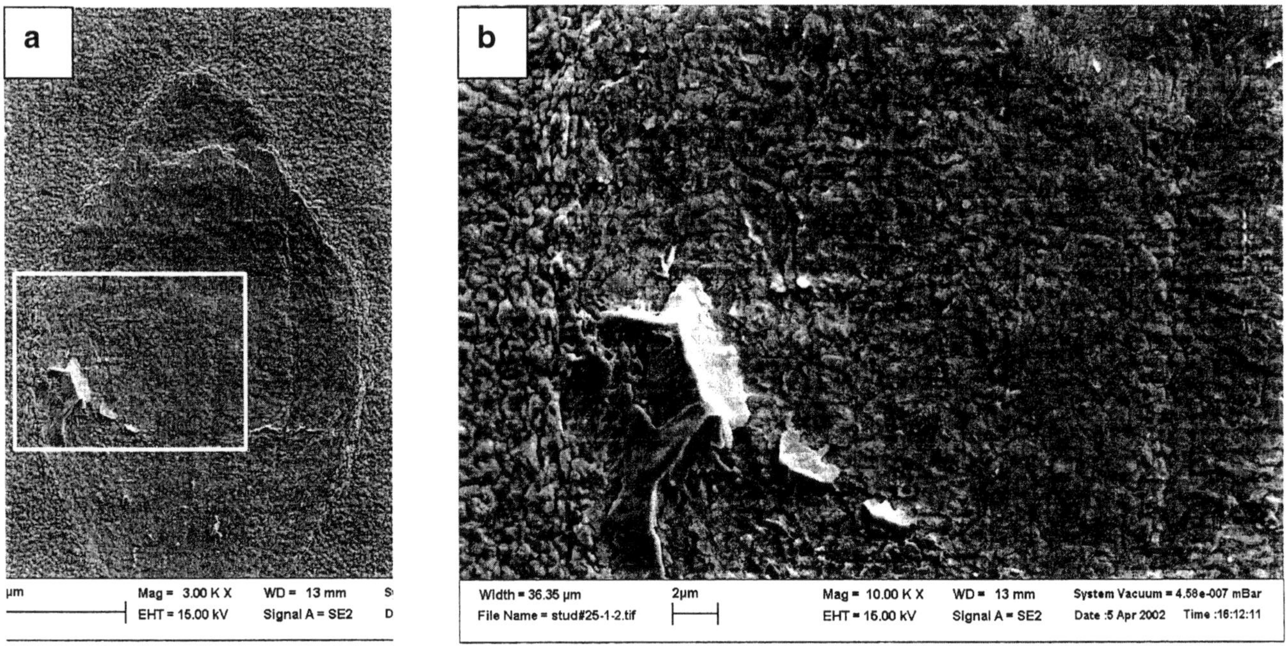

Fig. 4. **(a) Footprint of a bond made at 262°C and 120 gf force without ultrasound: the maximum width and peel force of the bond was about 50 µm and 3.7 gf, respectively. (b) Details of the enclosed area in (a) showing microwelds.**

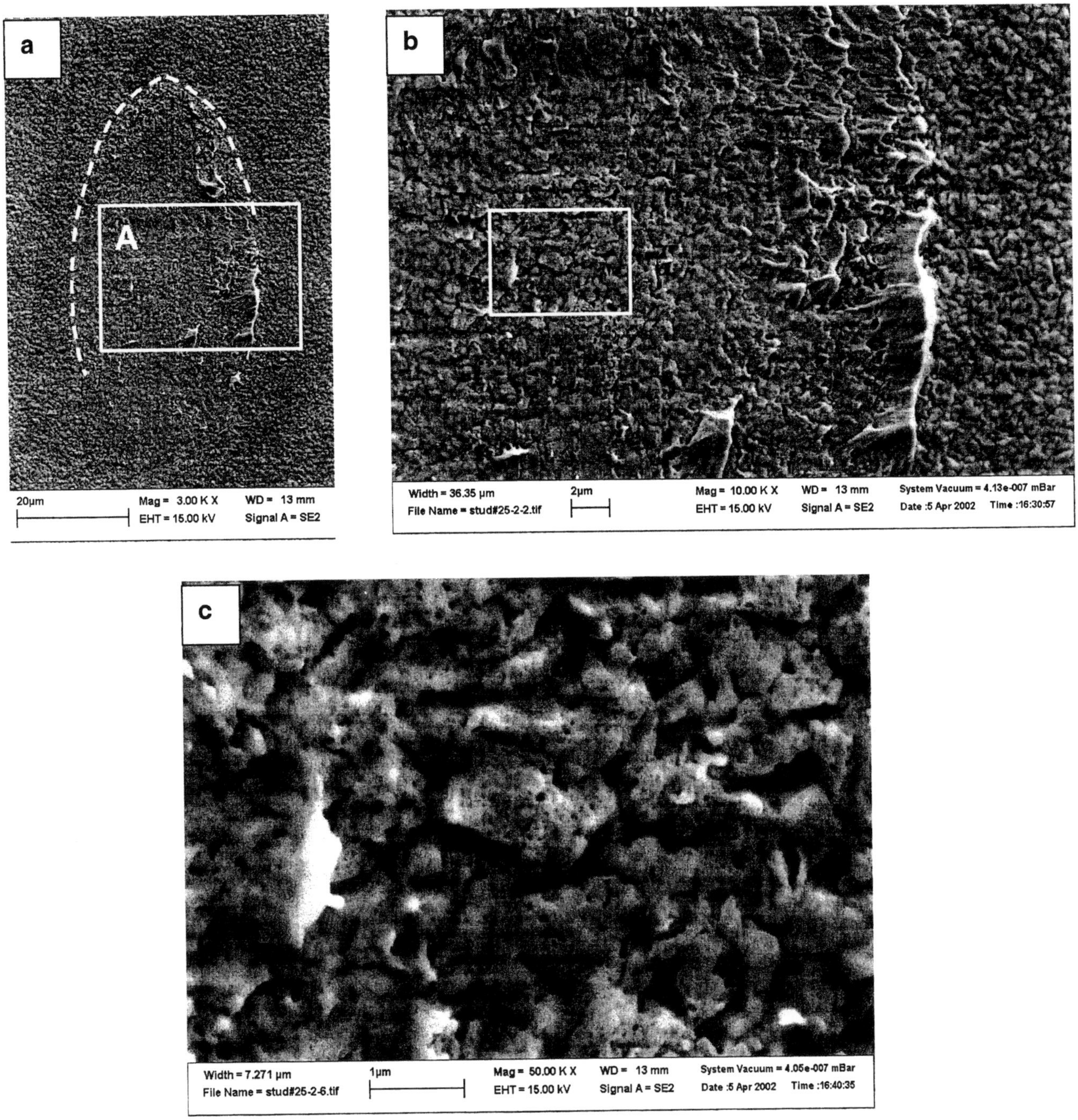

Fig. 5. (a) Footprint of a bond made at 362°C and 55 gf force wihout ultrasound: the maximum width and peel force of the bond was about 35 µm and 8.0 gf, respectively. Dashed line indicates part of the footprint perimeter. (b) Details of the highlighted area A in (a) showing microwelds at the lateral periphery of the bond. (c) Detailed of enclosed area in (b) showing microwelds in the central region of the bond.

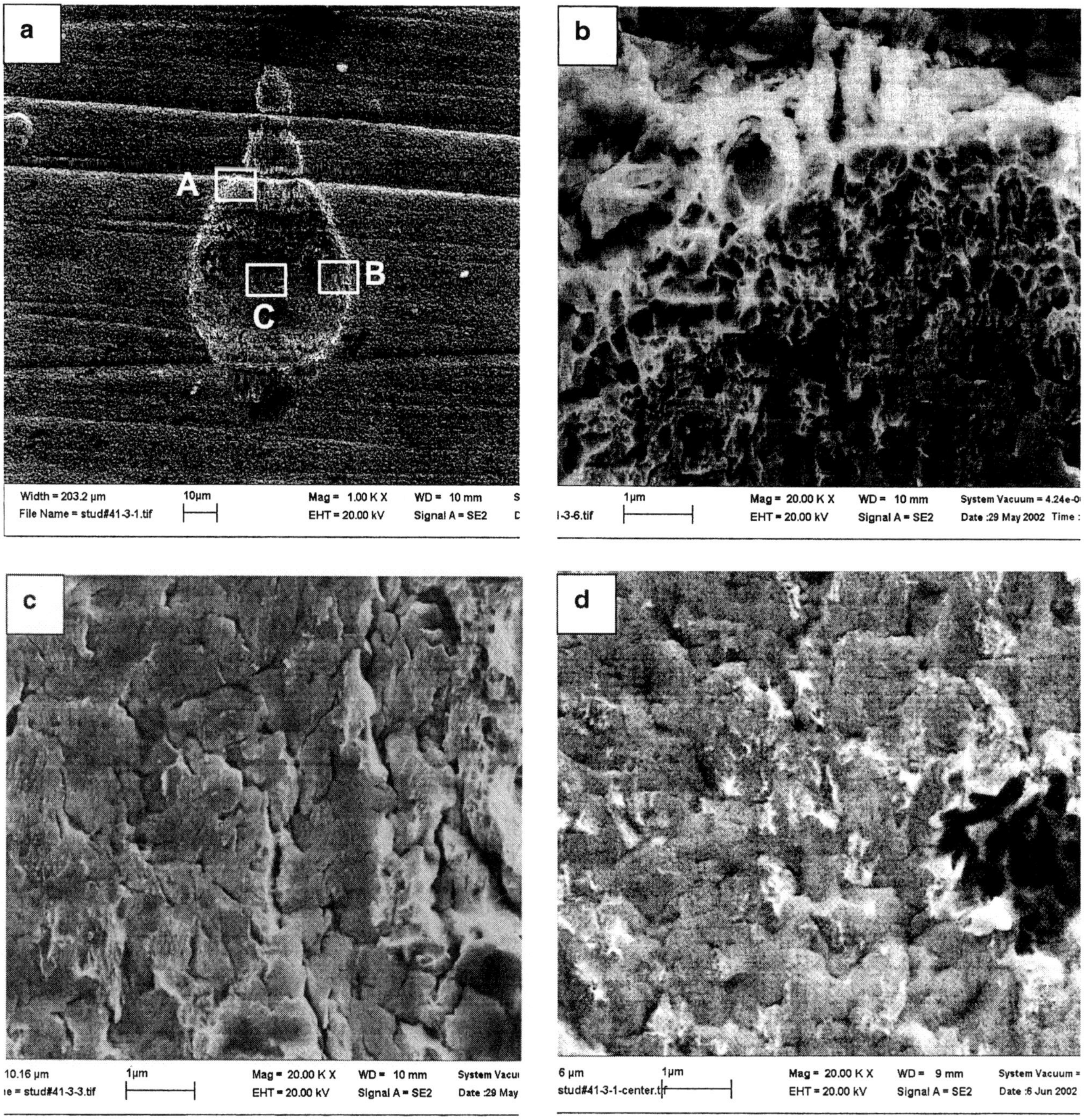

Fig. 6. (a) Footprint of a bond made at 23°C, force of 35 gf and ultrasonic power of 1 W: the maximum width and peel force of the bond was about 47 µm and 5.2 gf, respectively. (b), (c) and (d) Details of the enclosed areas A, B, and C in (a) showing microwelds.

Proceedings from Joining of Advanced and Specialty Materials, 13–15 October 2003, Pittsburgh, Pennsylvania, ASM International, 2004

Electrode Tip Life Improvement in Resistance Microwelding of Nickel-Plated Steel Thin Sheets

N. Scotchmer
Huys Industries Limited, Weston, Ontario, Canada M9L 1X9

S.J. Dong
Hubei Automotive Industries Institute, Shiyan, Hubei, China 442002

Y. Zhou
University of Waterloo, Waterloo, Ontario, Canada N2L 3G1

1.0 Introduction

Resistance microwelding is a group of microjoining processes (such as resistance spot, parallel gap, series and seam welding). These processes are commonly used for applications in electronic and medical packaging, such as lead/pad interconnections and hermetic sealing [1-4]. There are differences between resistance microwelding and "large-scale" (regular) resistance welding although the principles of the two processes are similar. For example, lower electrode force (pressure) used in resistance microwelding results in a relatively smaller contact area and higher contact resistance at the faying interfaces, which, in turn, results in lower welding current required to initiate and form a weld [5-8].

In resistance welding, a weld is formed between two metal sheets through the localized melting and coalescence of a small volume of the material(s) at the faying interface due to resistance heating generated by the passage of electric current [5]. However, the welding current will also degrade the electrode tip surfaces due to the resistance heating at the electrode/sheet interfaces. Little work has been published on electrode tip degradation mechanisms and engineering solutions in resistance microwelding. In "large-scale" resistance spot welding of Zn-coated steels for automotive applications, the primary mechanism limiting the electrode life is identified to be growth of the electrode tip face diameter [9-11]. Enlargement of contact face diameter results in reduced current density/heat generation and hence undersized welds between the sheets.

A number of damage processes that could contribute to the electrode degradation during "large-scale" resistance spot welding of Zn-coated steels have been observed and/or suggested: plastic deformation, alloying, pitting/erosion, cavitation, recrystallization, thermal shock and fatigue [9-11]. Holliday et al. [11] have investigated the relative contributions of plastic deformation, alloying and wear. The plastic flow (extrusion) of unalloyed material to the tip periphery will cause the formation of "wings" and hence increase the effective tip face diameter, which has been traditionally referred to as mushrooming. Build-up of alloyed product and/or zinc at the periphery of the electrode contact face can also result in an increase in the effective diameter. The loss of electrode material from the tip face due to the wear (pitting) process will also result in an increase in the effective diameter and a reduction in length of the electrode.

Parker et al [9] proposed that, under normal welding conditions (such as at low welding currents), the major damage process contributing to electrode degradation was electrode surface alloying and pitting, which was mainly a function of the type of coating present on the steel. Under such conditions, the use of dispersion strengthened electrode material (such as Cu-Al_2O_3) could not extend the tip life compared to the use of precipitation strengthened material (such as Cu-Cr-Zr) since the alloying and wear characteristics of both materials are similar [9]. On the other hand, the use of dispersion strengthened material could extend electrode tip life when welding with high currents or when using current stepping programs because electrode softening and hence plastic deformation becomes a more dominant damage process [9] and dispersion strengthened material provides a better high temperature strength [12].

Compared to "large-scale" resistance welding where the electrodes are internally water-cooled, the heat build-up at the electrode/sheet interfaces is worse in resistance microwelding since no water cooling is used and current density is generally higher [1-8]. Therefore, the use of dispersion strengthened electrode material (such as $Cu-Al_2O_3$) could be preferable to precipitation-strengthened material because of its better higher temperature strength. Electrode coatings represent another potential approach to life improvement, and in this regard it has been claimed that a patented TiC composite coated electrode (so called TiCap™, a trademark of Huys Industries Limited, Ontario, Canada) [13] could improve electrode life in "large-scale" resistance welding. Up to the present time, no objective experimental evidence or fundamental explanation as to the validity of this claim has been reported. In the present work, the effect of the TiC composite coating on electrode degradation was studied during series-mode micro-resistance welding of very thin sheets of nickel-plated steel to sheets of nickel. In order to compare the effectiveness of the composite coating on dispersion strengthened versus precipitation strengthened electrode material, electrodes of both types were subjected to controlled welding trials in both coated and uncoated conditions.

2. Experiments

Series-mode resistance microwelding was performed using a Unitek model HF2 power supply and model 508 weld head. The experimental setup and basic welding parameters used are described in the Reference [14]. The electrode extension (the distance from electrode holder to the electrode tip) was 1.5 mm and electrode spacing was 2.5 mm.

When direct-current power supplies are used such as the high-frequency inverter system [5, 8] employed in this work, the Peltier effect (the inverse of the thermocouple effect) can result in a higher rate of heat generation at the anode electrode than at the cathode [15]. As a result, the weld nugget at the anode can be much larger than that at the cathode. Adjusting the force at each individual electrode (e.g., by reducing the cathode electrode force or increasing the anode electrode force) can normally compensate for the Peltier effect and hence produce similar weld nugget diameters at both electrodes. In this study, a combination of 3600 g anode electrode force and 2400 g cathode electrode force was used, under which condition the cathode nuggets were just slightly smaller than the anode nuggets. It was also observed that the electrode degradation was generally more severe at the cathode; therefore; this work has focussed on the electrode degradation at the latter electrode.

Two electrode materials have been used in this work: copper alloy C18200 with a nominal composition of Cu-0.84 wt.%Cr-0.05 wt.% Zr (Cu-Cr-Zr) and C15760 with a nominal composition of Cu-1.1 wt. % Al_2O_3 ($Cu-Al_2O_3$). Both uncoated and coated electrodes were tested. The TiC composite coating is described in details in the Reference [14]. Electrodes had a diameter of 1.5 mm and a tip radius of 150 mm. The sheets to be welded were 100-μm-thick mild steel plated with pure nickel of about 7 μm thick, and 300 μm thick pure nickel.

Electrode tip life tests were performed on coupons consisting of nickel strips 9-mm wide and 30-mm long and nickel-plated steel strips 4-mm wide and 50-mm long with the nickel-plated steel in contact with the electrodes. Welding was interrupted after every 100 welds for measurement of peel force using a Quad Romulus IV universal mechanical tester and to measure nugget diameter from pullout buttons [14]. The electrode tip surfaces and the corresponding sheet areas that were in contact with the electrode tip after welding were analyzed using optical microscopy and scanning electron microscopy (SEM) and energy dispersive x-ray spectroscopy (EDX). Hardness was measured using a MHT 200 Vickers microhardness tester at 200 g load on prepared metallographic cross-sections of each electrode at locations of 0.05, 0.25, 0.5, 1.0, 2.0 and 5.0 mm away from the tip surface.

3. Results

Figs. 1 and 2 show button diameter (as an indication of nugget diameter) and peel force (as an indication of joint strength) versus the number of welds made with each type of electrode studied. Both button diameter and peel force decreased as the number of welds increased but the reduction in nugget diameter or peel force was slower for both types of coated electrode than for the respective uncoated electrodes. It also appears that the reduction in button diameter and peel force was slower for $Cu-Al_2O_3$ electrodes than those for the Cu-Cr-Zr electrodes.

It has been previously shown that joint strength is mainly determined by nugget diameter: the larger the nugget diameter, the higher the joint strength [1-3]. For an arbitrarily chosen minimum nugget diameter of 0.1 mm, the tip life was 700 and 1200 welds for uncoated and coated Cu-Cr-Zr electrodes respectively, and 800 and 1500 for uncoated and coated $Cu-Al_2O_3$, respectively (Table 1). The same trend was also found when the peel forces were examined. If an arbitrarily chosen minimum peel force of 3 kg was selected, the tip life was 700 and 1200 welds for uncoated and coated Cu-Cr-Zr electrodes, and 800 and 1300 for uncoated and coated $Cu-Al_2O_3$ electrodes, respectively (Table 1). The use of TiC composite coating increased the tip life by about 63-88% while the difference in button diameter or peel force was much less significant (about 8-25%) between Cu-Cr-Zr and $Cu-Al_2O_3$ electrodes (Table 1).

SEM examination revealed that the surface of the electrode tips was relatively smooth and clean even after a few hundreds of welds. But a flat area on the tip surface was developed during welding and the diameter of this flat face increased with increasing number of welds. The tip face also left an imprint on the corresponding sheet surface during welding (Fig. 3) and this imprint could be easily recognized. The diameter of this imprint was used as an indication of the contact areas between the electrode tips and sheets (Fig. 4) although the absolute values may be smaller than the real values because they were measured at room temperature after welding. It was obvious that this increase in electrode imprint diameter was due to the increase in tip face diameter (Fig. 4). The increase in the imprint diameter was well correlated with the decrease in button diameter (Figs. 1 and 2). It is interesting to see, from Fig. 4, that the imprint diameter was about 0.95 mm when the tip lives were reached (Table 1). This increase in diameter translates into about a 50% drop in current density (assuming the initial contact diameter was about 0.65 mm). It is believed that this reduction in current density is the direct cause of undersized nuggets. Therefore, the enlargement in tip diameter and hence reduction in current density is responsible for the reduction in nugget diameter and joint strength. This is consistent with previous observation in "large-scale" resistance spot welding of Zn-coated steels [9-11]. Fig. 4 also shows that the increase in contact area was faster for uncoated electrodes than that for coated electrodes. This is in agreement with the observations in Figs. 1 and 2 that the nugget diameter or peel force decreased at a faster rate for the uncoated electrodes. The effect of electrode material (Cu-Cr-Zr versus Cu-Al_2O_3) on the change in contact area is less significant than that of the composite coating, similar to the effect of electrode material on nugget diameter and joint strength.

4.0 Discussion

It is clear from the above results that the electrodes failed because of increased tip face diameter, which is consistent with observations in "large-scale" resistance spot welding of Zn-coated steels [9-11] in that enlarged tip face diameter results in reduced current density/heat generation and hence undersized welds between the sheets. In the following, two mechanisms (pitting and plastic deformation) that have contributed to the increase in tip face diameter in this work will be discussed in terms of interactions between the electrodes and the sheet surfaces.

4.1 Metallurgical Interactions

Detailed SEM examination has shown scattered Cu pick-up within the electrode imprints on the sheet surfaces after welding. This Cu pick-up on the sheet surfaces was clearly transferred from the tip faces since there was no Cu in the original sheets. Fig. 5 shows details of corresponding surface locations on electrode contact face and sheet surface at the 15th weld for an uncoated Cu-Cr-Zr electrode. It indicates that bonding occurred in isolated spots between electrode tip face and sheet surface during welding. The fracture of these local bonds, when electrodes were pulled away from the sheets, resulted in removal of Cu from the tip face (A, B, and D in Fig. 5c) onto sheet surface (regions A', B', and D' in Fig. 5b). The pitting (i.e., the continuous loss of electrode material) during each welding cycle would result in an enlargement of the tip face diameter. This enlarged tip (contact) face diameter would eventually result in an undersized nugget because of the decrease in current density.

Fig. 6 shows details of corresponding locations on electrode tip face and sheet surface at the 200th weld for a coated Cu-Cr-Zr electrode. Regions A and B on the tip face (Fig. 6c) contained TiC particles as shown by the elevated Ti content. Their corresponding regions A' and B' on the sheet surface (Fig. 6b) contained no Cu transfer. In this case, with coating in place, no bonding occurred at regions containing TiC particles. It could reasonably be inferred that the slower enlargement of the contact face of the coated electrodes (Fig. 4) is directly related to the low bonding tendency between TiC particles and sheet surfaces, which would reduce the rate of tip surface removal. It is believed that the TiC particles in the composite coating on the electrode surface, with a melting point of 3140°C and poor bondability with metals [16], would discourage the bonding between the tip and sheet surfaces and hence decrease the material loss from the tip face.

4.2 Mechanical Interactions

Little macroscopic plastic deformation (mushrooming) was observed in any of the electrodes [14]. Therefore, the damage process of increase in tip face diameter by mushrooming (spreading of the tip by large-scale plastic deformation) as observed in "large-scale" resistance spot welding of Zn-coated steels [9], was not a factor in this work. This may be because of the particular tip shape used: a very large tip radius could increase the resistance to bulk plastic deformation. However, the extrusion of the tip surface layer was observed at microscopic level in uncoated Cu-Cr-Zr and Cu-Al_2O_3 electrodes (Fig. 7). It is believed that the extrusion has mainly contributed to material loss, i.e., the increase in tip face diameter through break-off of these extruded layers. No obvious difference in this microscopic extrusion process was observed between the Cu-Al_2O_3 and Cu-Cr-Zr electrodes. However, the composite coating improved the resistance to microscopic extrusion since little extruded material was observed in coated Cu-Cr-Zr and Cu-Al_2O_3 electrodes.

The extrusion of the tip surface layer is obviously a result of softening because of the high temperature that the tip experienced during welding. The softened electrode tip cannot withstand the electrode force when the high temperature strength drops below the electrode pressure. A layer of recrystallized microstructure, in which the elongated extrusion structure was completely eliminated, was clear at the tip regions in both Cu-Cr-Zr and Cu-Al_2O_3 electrodes [14] although the recrystallized region in Cu-Al_2O_3 was much smaller. Fig. 8 shows the microhardness distribution in the tip surface region of sectioned electrodes that had previously made 400 welds. Both Cu-Cr-Zr and Cu-Al_2O_3 electrodes experienced large reductions in hardness (more than 50%). It appears that the hardness loss on Cu-Cr-Zr electrodes was slightly larger than that on Cu-Al_2O_3 electrodes, which agrees with the observation in recrystallized microstructure[14], but the difference in the hardness drop was quite minor. This is consistent with the tip life test results in which there was no significant difference in tip life when using different electrode materials (Cu-Cr-Zr versus Cu-Al_2O_3). New electrodes of both compositions were subjected to one-hour annealing at a range of temperatures to clarify the effect of peak temperature on softening. The results of subsequent hardness testing are shown in (Fig. 9). Therefore, the very large drop in hardness at the tip regions of used electrodes (Fig. 8) indicated that the temperature at these regions reaches well above 900°C. Therefore, the use of Cu-Al_2O_3 material could not improve the tip life because the temperature experienced during welding is even higher than the softening temperature of the Cu-Al_2O_3 electrodes.

5.0 Conclusions

- The TiC composite coating increased the tip lives of Cu-Cr-Zr and Cu-Al_2O_3 electrodes by about 70%. This may be because the TiC particles contained in the coated layer would reduce local bonding between the tip and sheet surface and hence reduce the material loss due to the fracturing of local bonds.
- Compared to the composite coating, the improvement of tip life due to the use of Cu-Al_2O_3 electrode was relatively small compared to the Cu-Cr-Zr electrode (only by about 15%). This may be due to the very high temperature experienced during welding, well above the softening temperature of the Cu-Al_2O_3 electrode.
- Pitting is believed to be the main damage process contributing to the electrode degradation, which results in material loss and hence an increase in tip face diameter. Extrusion of the tip surface layer at microscopic level may have also contributed to material loss through the fracturing of the extruded layers, in the case of coated electrodes.

6.0 References

[1] K. I. Johnson (editor), Introduction to microjoining, Abington, UK, TWI(1985)

[2] W. R. Bratschun, IEEE Trans. Comp. Hybrids Manufact. Technol., 15(6), 931-937 (1992)

[3] J. J. Fendrock and L. M. Hong, IEEE Trans. Comp. Hybrids Manufact. Technol., 13(2), 376-382 (1990)

[4] J. R. Tyler, "Seam seal resistance welding of plated microelectronic packages", in Electronic Materials and Processes (Proceedings of 1st International SAMPE Electronics Conference, Santa Clara, CA, USA, 23-25 June 1987), Ed. N.H. Kordsmeier, C.A. Harper and S.M. Lee, Corvina, CA: SAMPE, 1987, vol.1, pp. 54-69

[5] K.J. Ely and Y. Zhou, Sci. Technol. Weld. Join., 6(2), 63-72 (2001),

[6] B.H. Chang, M.V. Li and Y. Zhou, Sci. Technol. Weld. Join., 6(5), 273-280 (2001)

[7] Y. Zhou, P. Gorman, W. Tan and K.J. Ely, J. Elec. Mater., 29(9), 1090-1099 (2000)

[8] Y. Zhou, S.J. Dong and K. J. Ely, J. Elec. Mater., 30(8), 1012-102 (2001)

[9] J. D. Parker, N. T. Williams and R. J. Holliday, Sci. Technol. Weld. Join., 3(2), 65-74 (1998)

[10] P. Dong, M. Victor Li and M. Kimchi, Sci. Technol. Weld. Join.,, 3(2), 59-64 (1998)

[11] R. Holliday, J. D. Parker and N. T. Williams, Welding in World, 37(4), 186-193 (1996)

[12] A. V. Nadkarni and E. P. Weber, Weld. J., 56(11), 331s-338s (1977)

[13] US Patent No. 4480167: Welding electrode, Oct. 30, 1984

[14] S.J. Dong and Y. Zhou, Metall. Mater. Trans. A, 34A, 1501-11 (2003)

[15] Mustafa Hasir, Welding and Cutting, No.3, 116-121 (1984)

[16] L. Li, Y. S. Wong, J. Y. H. Fuh and L. Lu, Mater. Design, 22(8), 669-678 (2001)

Table 1 Electrode Tip Life in the Number of Welds

Electrodes	Cu-Cr-Zr (T_3)		Cu-Al_2O_3 (T_4)		$(T_4-T_3)/T_3$
	Button Diameter	Peel Force	*Button Diameter*	Peel Force	
Uncoated (T_1)	*700*	700	*800*	800	14%
Coated (T_2)	*1200*	1200	*1500*	1300	8-25%
$(T_2-T_1)/T_1$	*71%*	71%	*88%*	63%	

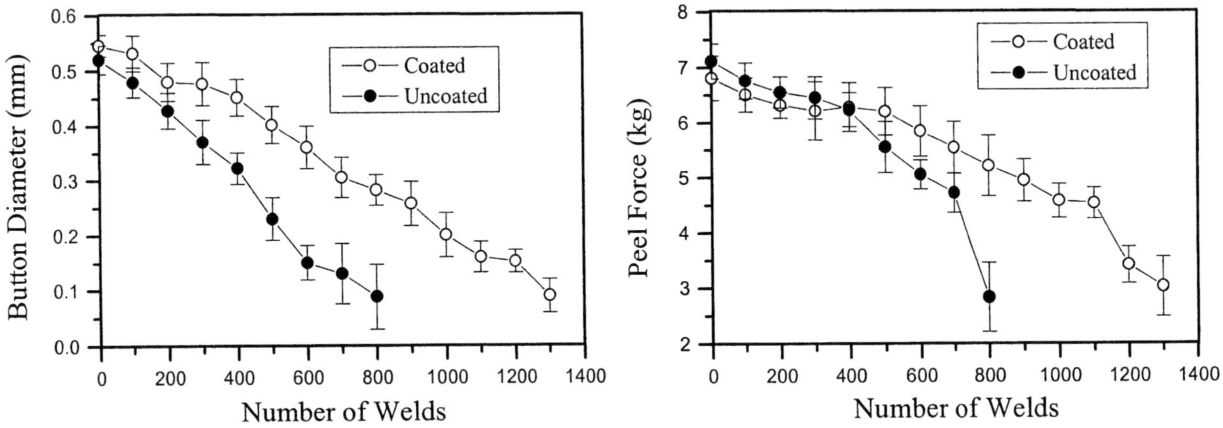

(a) (b)

Fig. 1 (a) Button diameter and (b) peel force versus number of welds using Cu-Cr-Zr electrodes

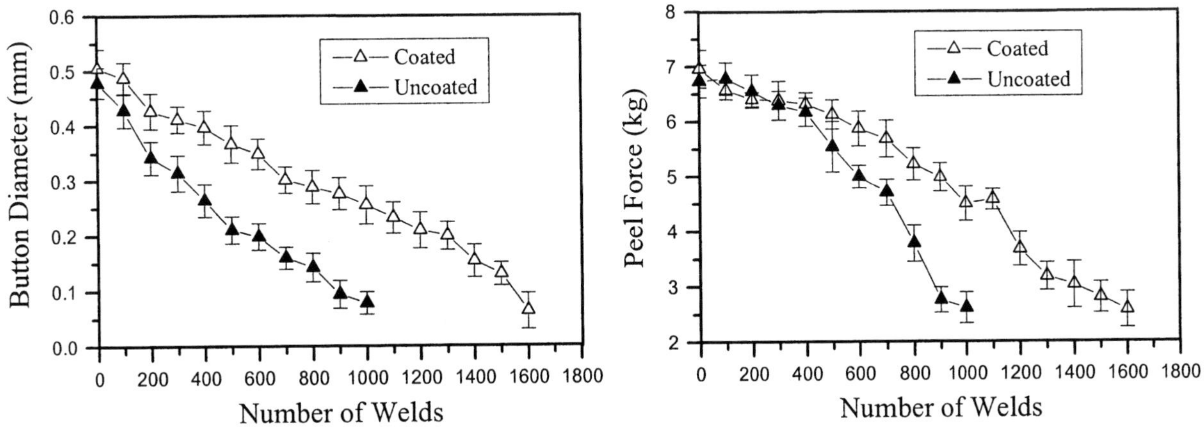

(a) (b)

Fig. 2 (a) Button diameter and (b) peel force versus number of welds using Cu-Al_2O_3 electrodes

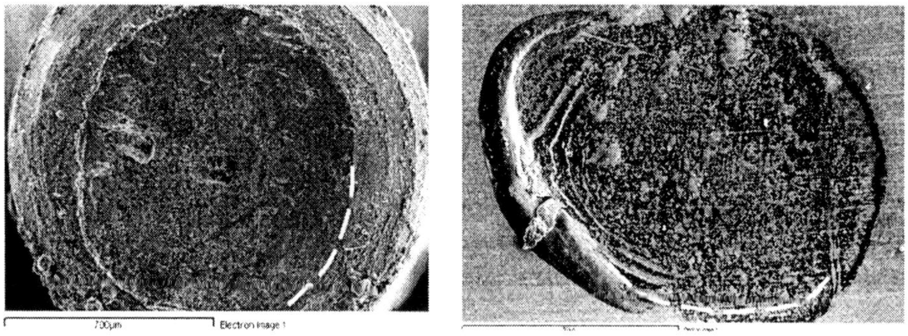

(a)

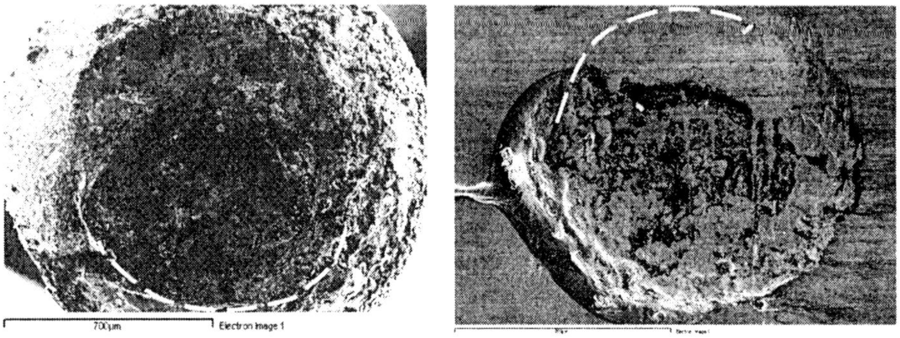

(b)

Fig. 3 Electrode tip surfaces and the corresponding sheet surfaces of (a) uncoated Cu-Cr-Zr electrodes after 800 welds and (b) coated Cu-Cr-Zr electrodes after 1300 welds. Dashed lines indicate contact areas.

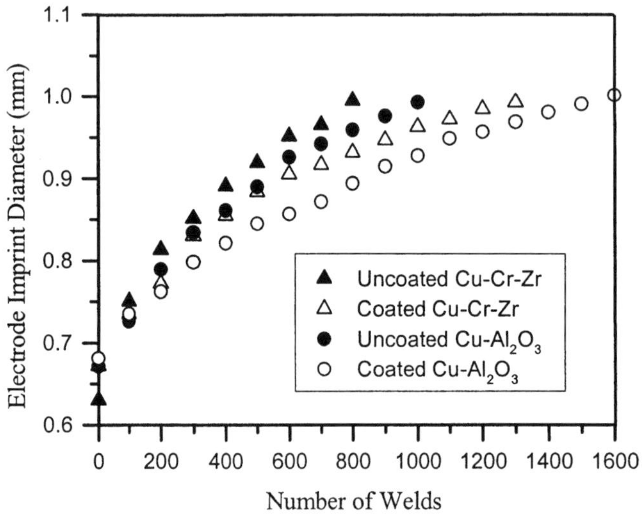

Fig. 4 Electrode imprint diameter versus number of welds

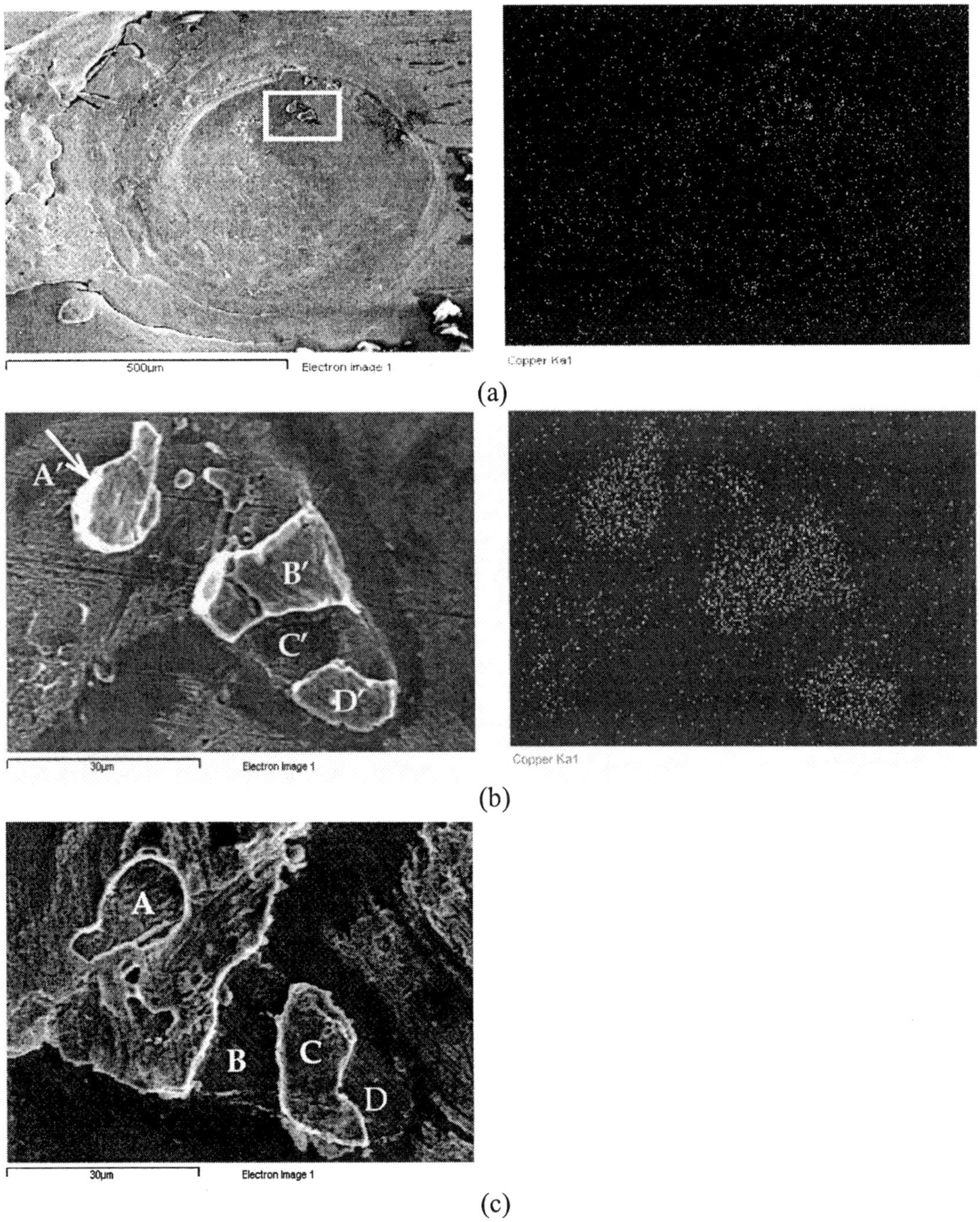

Fig. 5 (a) Sheet surface and its Cu mapping, (b) details of fractured local bonds in a highlighted box in (a) and its Cu mapping, and (c) pits formed on the corresponding tip surface of an uncoated Cu-Cr-Zr electrode

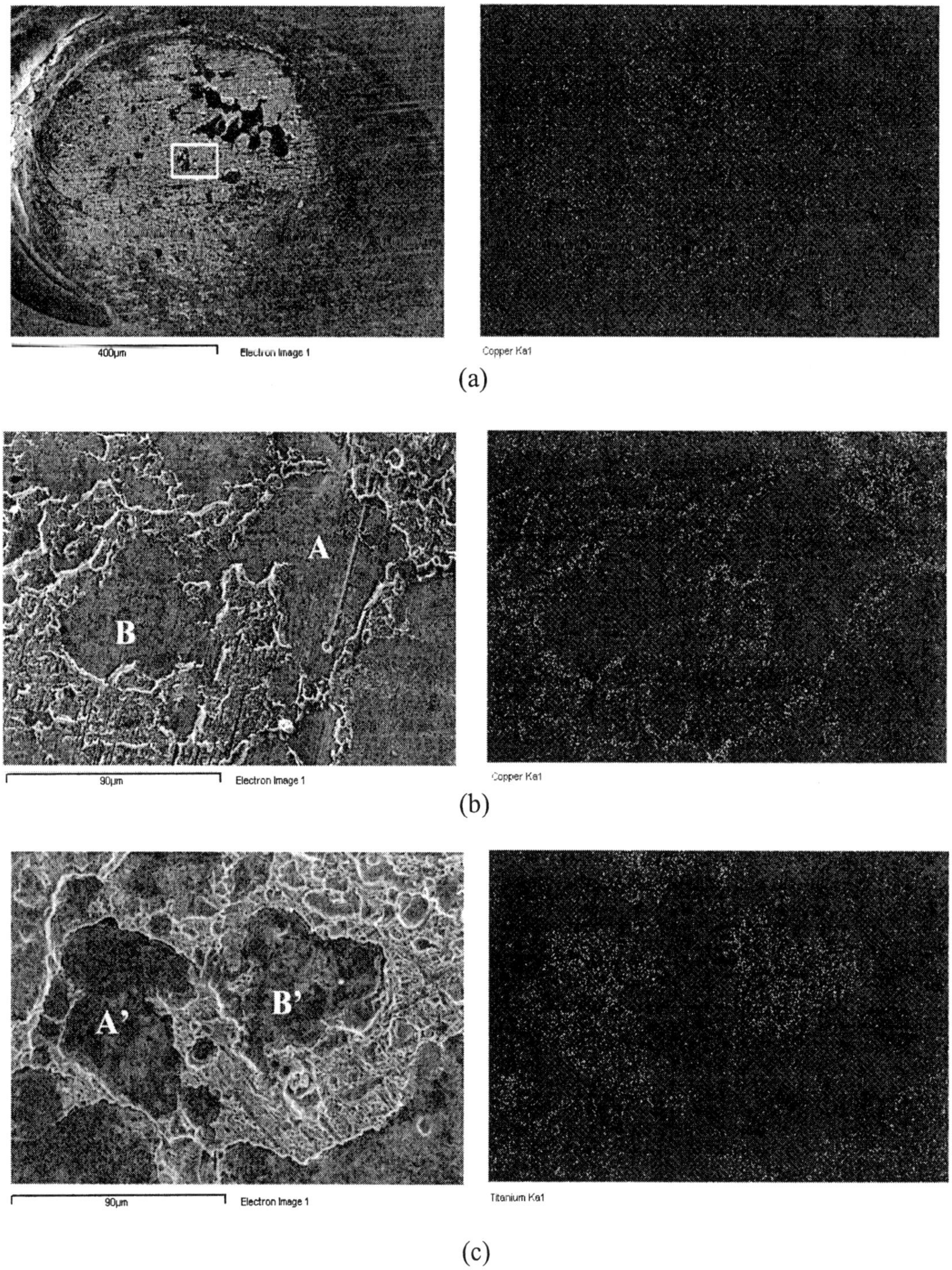

Fig. 6 (a) Sheet surface and its Cu mapping, (b) details of fractured local bonds in a highlighted box in (a) and its Cu mapping, and (c) tip surface of the corresponding coated Cu-Cr-Zr electrode

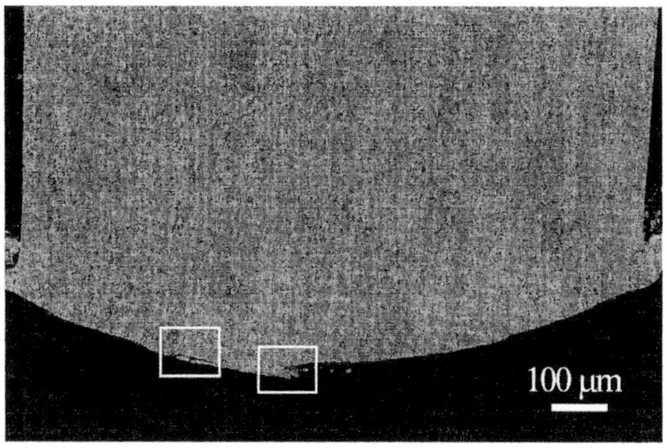

(a)

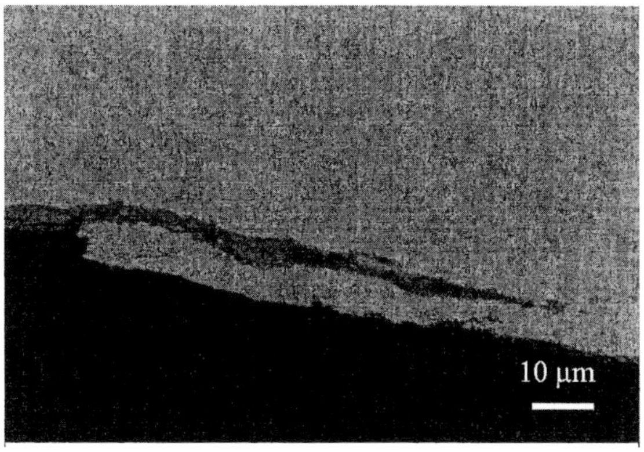

(b)

(c)

Fig. 7 (a) Extruded layers on the tip face of an uncoated Cu-Cr-Zr electrode after 400 welds, (b) and (c) details of the highlighted boxes in (a).

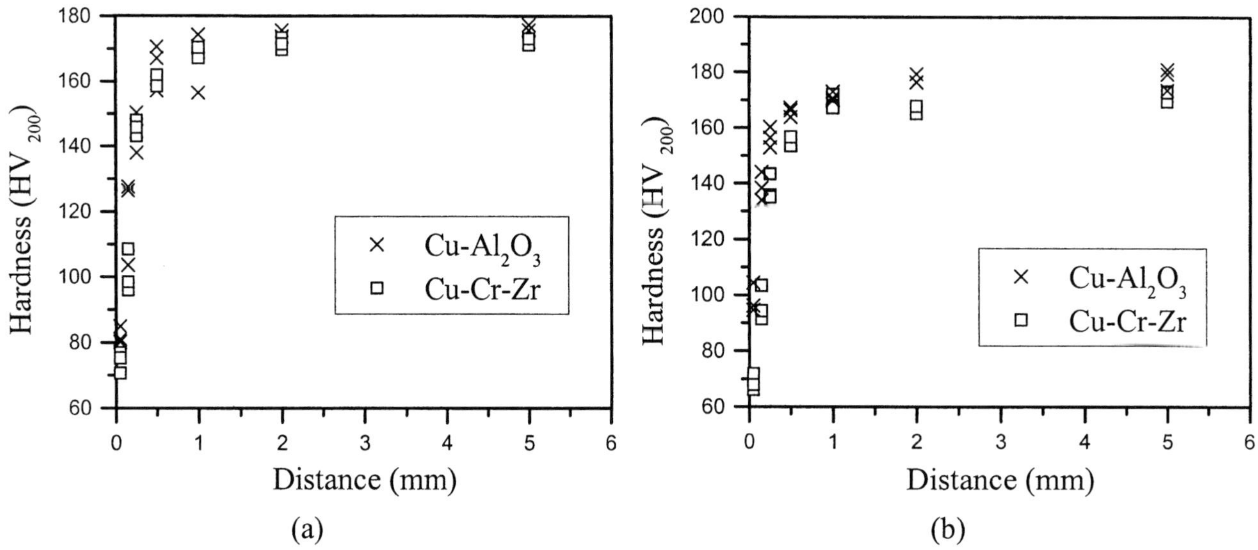

Fig. 8 Microhardness distribution at the tip regions of (a) an uncoated and (b) a coated electrode

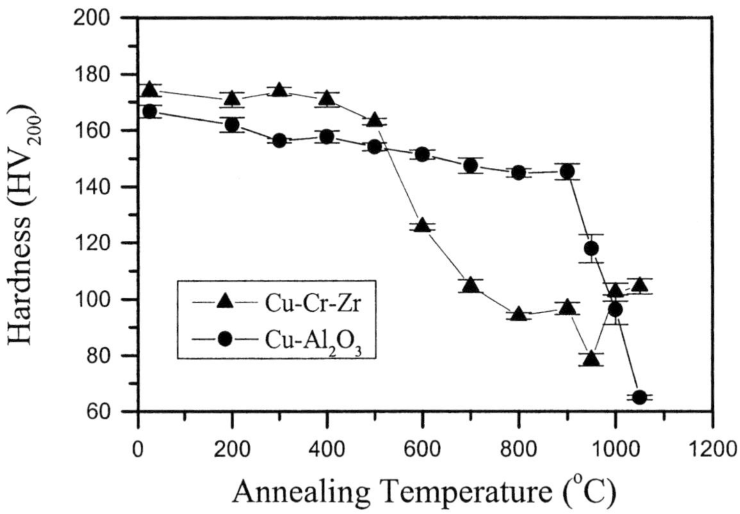

Fig. 9 The hardness of the Cu-Cr-Zr and $Cu\text{-}Al_2O_3$ electrodes after one-hour annealing